The Annotated
HODGKIN & HUXLEY

A Reader's Guide

Alan Lloyd Hodgkin and Andrew Fielding Huxley, Plymouth, c. 1948

The Annotated
HODGKIN & HUXLEY

A Reader's Guide

Indira M. Raman
and David L. Ferster

Princeton University Press
Princeton and Oxford

Published by Princeton University Press
41 William Street, Princeton, New Jersey 08540
6 Oxford Street, Woodstock, Oxfordshire OX20 1TR

press.princeton.edu

Library of Congress Cataloging-in-Publication Data

Names: Raman, Indira, author. | Ferster, David L., author. | Huxley, Andrew, 1917–2012.
    Works. Selections. | Hodgkin, A. L. (Alan Lloyd). Works. Selections.
Title: The annotated Hodgkin and Huxley : a reader's guide / Indira M. Raman and David L. Ferster.
Description: Princeton : Princeton University Press, [2021] | Includes bibliographical references
    and index.
Identifiers: LCCN 2021019815 (print) | LCCN 2021019816 (ebook) | ISBN 9780691220635 (paperback) |
    ISBN 9780691220642 (hardcover) | ISBN 9780691220659 (ebook)
Subjects: LCSH: Huxley, Andrew, 1917–2012. | Hodgkin, A. L. (Alan Lloyd) | Neurophysiology. |
    Neurophysiology—Research—History—20th century. | BISAC: SCIENCE / Life Sciences /
    Neuroscience | SCIENCE / Life Sciences / Biophysics
Classification: LCC QP355.2 .R36 2021 (print) | LCC QP355.2 (ebook) | DDC 612.8—dc23
LC record available at https://lccn.loc.gov/2021019815
LC ebook record available at https://lccn.loc.gov/2021019816

British Library Cataloging-in-Publication Data is available

Editorial: Sydney Carroll, Sophia Zengierski
Production Editorial: Terri O'Prey
Text Design: Wanda España
Jacket/Cover Design: Wanda España
Production: Jacqueline Poirier
Publicity: Matthew Taylor
Copyeditor: Danna Lockwood

This book has been composed in Minion Pro with Poetica

Printed on acid-free paper. ∞

Printed in the United States of America

10 9 8 7 6 5 4 3 2 1

In remembrance: to John W. Moore and Donata Oertel, with warm memories of their joy in neurophysiology and in life;

In appreciation: to Ann Stuart, Carl Hopkins, Larry Trussell, Craig Jahr, and Bruce Bean, with gratitude for their guidance and pleasure in their friendship;

In anticipation: to all students discovering the work of Hodgkin and Huxley, with hope that they may find beauty in these papers and inspiration to keep the flame of science alive.

Contents

Introduction

After nearly seventy years, Alan Hodgkin and Andrew Huxley's 1952 papers on the mechanisms underlying the action potential seem more and more like the Shakespeare plays of neurophysiology, works of astounding beauty that become less accessible to each successive generation of scientists. Everyone knows the basic plot (the squid dies at the beginning), but with their upside-down and backward graphs and records, unfamiliar terminology and techniques, now arcane scientific asides, and complex mathematical underpinnings, the papers become a major effort to read closely without guidance. It is our goal to provide such guidance, by translating graphs and terminology into the modern idiom, explaining the methods and underlying theory, and providing historical perspective on the events that led up to the experiments described. By doing so, we hope to bring the pleasure of reading these extraordinary papers to any physiologist inclined to read them.

But why read them at all? The mechanism of the action potential is taught in nearly all introductory classes in neuroscience and physiology and covered in every corresponding textbook. The findings from the 1952 papers are explained extensively in Bertil Hille's classic *Ion Channels of Excitable Membranes*, in which a whole chapter is devoted to Hodgkin and Huxley's experiments. Direct access to Hodgkin-Huxley type models is provided through multiple programming platforms, some specifically designed for teaching, like John Moore and Ann Stuart's *Neurons in Action*, Francisco Bezanilla's *The Nerve Impulse*, or Stephen Baylor's *Computational Cell Physiology*, and some tailored for modern computational modeling, like Michael Hines and Nicholas Carnevale's NEURON. Finally, Hodgkin himself detailed his life and his life's work, including his opus with Huxley, in his understated but precise autobiography, *Chance and Design*. Returning to the original literature is certainly not a requirement for learning and benefiting from the outcomes of the series of five papers published in 1952.

Nevertheless, the many reasons to read the original work impress themselves on us yearly when we teach them as part of a course on the historical neurophysiological literature. Directing the next generation of students through a close reading of these papers has consistently revealed that the papers—through their style, content, history, and timelessness—not only offer the substance of the experimental results, but also provide a touchstone and a perspective that can facilitate the conduct of scientific research today.

The first reason to read the papers is for the sheer pleasure of an exciting scientific saga. Much research was done through the twentieth century to try to crack the electrical code of neurons, on the premise that learning to translate biological electricity was a necessary step toward comprehending the brain itself. Hodgkin and Huxley's work was equivalent to the deciphering of the Rosetta stone of neuronal communication—a magnificent story of sleuthing and detective work, innovation and labor, setbacks and successes—which ultimately offered a gateway into understanding the language of the brain.

The second, less retrospective reason relates to the insight that can be gained from coming to know the electrical principles that govern the operation of neurons. Electrophysiological studies of neurons continue to this day; nevertheless, given the standardization and automation of the tools required, even those scientists who are engaged in electrophysiological research can be uncertain about the conceptual underpinnings of the work they do. With automation, a facility with electronics and the associated equations, which used to be *de rigueur* for anyone studying neural signaling, is falling by the wayside. Only a small subset of physiologists trained after the 1990s can build an amplifier, and even those who wish to learn the details of their art can face challenges in finding in-depth references on electricity and electronics that are relevant to neurophysiology.

Meanwhile, imaging methods, which more and more occupy the cutting edge of neuro-technology, permit neuronal 'activity' to be tracked through nonelectrical measures that translate the electrical signals of neurons into optical ones, the visual domain being the native 'language' of humans. While unquestionably powerful, the images of electrical signals are limited much as translations of poetry are: the sense is there, but the nuance is often lost. Without entering into the brain's own language, one is limited to hearing by proxy only the words that the translating tool's dictionary has the capacity to represent—calcium signals, metabolic products, and/or gene expression. Reading Hodgkin and Huxley's systematic consideration of the electrical basis of neuronal signaling expands the lexicon of the modern physiologist.

The third reason is something of a corollary to the preceding point, namely, the conscious use of language as a tool. Reading the papers allows one to see firsthand what it is like to be at a scientific frontier. The scientists must come to recognize phenomena and explore ideas for which no language yet exists; they must define terms, invent words, and develop conventions in order to create meaning that can be effectively transferred from scientist to scientist. The Hodgkin-Huxley papers beautifully illustrate the process of unveiling 'meanings' and clothing them with words: the very essence of scientific discovery and communication.

It is the logic that underlies clear communication that leads to the fourth reason to read the original work. The series of papers provide an exemplary (and arguably unparalleled) illustration of the scientific method at its best, with repeated sequences of observations, hypotheses, predictions, experimental tests, interpretations, evaluation of error, and consideration of plausible alternatives. The way in which Hodgkin and Huxley carefully amassed evidence to define biological phenomena in terms of electrical counterparts provides a template for identifying fundamental variables in all modern work. The primary data have withstood seventy years of extensive testing, and the conclusions—those presented with confidence as well as those offered with skepticism—have spawned countless subfields of neuroscience and physiology. In an era in which scientific inquiry and truth itself is under scrutiny, the value of studying a successful application of the scientific method cannot be overestimated.

The fifth, related, reason has to do with coming to understand the purpose and power of quantification and computation in science. The modern interest in scientific rigor tends to focus on statistics. The Hodgkin-Huxley papers contain not a single statistical test. Yet the research is as rigorous as any study can be. The experiments and analyses illustrate an alternative route to rigor: the collection of multiple lines of evidence, each including quantitative estimation of variables in precisely formulated hypotheses, followed by the assessment—through equations and predictive modeling—not only of the precision of the estimates but also of the validity of the variables themselves as underlying parameters. Equations in the papers are translated into logical English and vice versa, clearly demonstrating the interplay between the identification of quantifiable physical variables and the meaningful conceptualization of biological ideas. The work provides an ideal template for research.

Sixth, the papers teach us that rigorous science does not require the *elimination* of error and artifact. A remarkable aspect of the papers is the number of times Hodgkin and Huxley frankly report the shortcomings in their ability to make measurements with the desired level of precision. These deviations from the ideal—from the lack of stable series resistance compensation, to drift of the resting potential in the presence of choline, to reversal potential shifts from potassium accumulation—are called out and accounted for one by one. After the severity of the error is assessed, the scientists proceed, using their awareness of error to correct their path of experimentation and reasoning. These examples of not concealing error, yet not being paralyzed by it, also provide an invaluable guide to the modern scientist.

The seventh, though probably not final, reason to read the original papers is to develop a sense of one's place in history. The papers have been cited or otherwise referred to thousands of times, sometimes as a springboard for further work and sometimes as a point of contrast, to set the stage for an observation that defies the Hodgkin-Huxley 'dogma.' Dogma, however, is that which is accepted without evidence. The power of the Hodgkin-Huxley work is their reliance on evidence. By examining the papers directly, one can observe Hodgkin and Huxley's own awareness of the strengths and the limitations of their work, their view of certain points as conclusive, and their expectations that other points were provisional and would be supplanted. Looking at the science in its own context, in light of the 'answer key' provided by decades of hindsight, allows the modern reader to appreciate more fully the significance of discoveries in the present era—whether new results provide a higher resolution view of a previously held thought, a predictable but previously unmeasured exception to a rule, an illustration of an orthogonal phenomenon, or a genuine overthrow of earlier conclusions. They also remind us how perceptions of science can be colored by a moment in time, thereby teaching humility in the face of nature, which underlies all good science and which runs so deeply throughout these papers.

ACKNOWLEDGMENTS

This book grew out of our graduate course at Northwestern University, Great Experiments in Cellular Neurophysiology, which convinced us that a readers' guide to Hodgkin and Huxley's classic papers would be useful. We are grateful to several others who shared this conviction: our editor, Sydney Carroll, for undertaking this project with us and helping make the book become a reality; our former students and lab members for their excitement and encouragement; Professor Eric Frank, for helpful comments on an early draft of the manuscript; Yin-Peng Chen, a former student in our course, for enthusiastically entering into the spirit of this project and creating her well-researched artistic illustrations of squid; Sally Howell at the Physiological Society for permission to reprint Hodgkin and Huxley's articles; Danna Lockwood, for copyediting with attention to detail, content, and line; Professor Kenneth Miller, for discussions on differential equations; and Professor Bertil Hille, for providing original copies of the journal issues for reproduction. We are most deeply indebted to Professors Denis Baylor, Francisco Bezanilla, and Bertil Hille, for carefully reading the full text and providing invaluable comments, insights, and perspectives, from which the book benefited greatly.

We would also each like to acknowledge our parents. From Indira: Dr. Varadaraja V. Raman, for cultivating my pleasure in science, history, and language, and Dr. Marilú Raman, for teaching me the equation for a straight line and for showing me how to be a teacher. From David: Paul and Dorothy Ferster, for teaching me to use the right tool for the right job, both in the laboratory and on the page.

Indira M. Raman
David L. Ferster
Wilmette, Illinois
April 2021

How to Use This Book

The Hodgkin-Huxley papers can be read with a range of interests, from the physiological to the computational to the historical. Recognizing that different readers may have distinct backgrounds and goals, we have written annotations to offer clarification and contextualization of as much of the text as is practical. The notes explain experimental methods, results, and analyses, as well as technical and mathematical approaches, and occasionally include commentary on the scientific method or writing style. Because some readers may choose to skip some notes or to read them discontinuously, each annotation is self-explanatory, with references to other notes or sections provided as necessary. Readers of the original text may therefore refer to the annotations whenever supplementary information would be useful to their purposes. The Historical Background tells the story of the research and the ideas that set the stage for the five papers; the appendices at the end of the book offer general mathematical and scientific information for reference.

The Annotated
HODGKIN & HUXLEY
A Reader's Guide

Historical Background

The quest for a scientific understanding of electrical signaling in the nervous system began more than a century and a half before the 1952 papers of Hodgkin and Huxley. In 1791, Luigi Galvani (1737–1798) reported his discovery of 'animal electricity'—the electrical processes somehow generated by biological tissue to transmit signals from nerves to muscles. Like many scientific discoveries, the initial observation was incidental: while a metal scalpel was in contact with a nerve in the leg of a decapitated frog, the muscle contracted whenever a nearby frictional machine—a device that generated static electricity by rubbing two materials like glass and wool together—emitted a spark. Apparently, it was one of Galvani's assistants who first noticed the coincidence of the two events. As Galvani reported,

> He, wondering at the novelty of the phenomenon, immediately apprised me of the same, wrapped in thought though I was and pondering something entirely different. Hereupon I was fired with incredible zeal and desire of having the same experience, and of bringing to light whatever might be concealed in the phenomenon. Therefore I myself also applied the point of a scalpel to one or other crural nerve at a time when one or other of those who were present elicited a spark. The phenomenon always occurred in the same manner: violent contraction in individual muscles of the limbs, just as if the prepared animal had been seized with tetanus, were induced at the same moment of time in which sparks were discharged. (Galvani 1791, trans. Green 1953 p. 24)

Galvani and his assistants eventually determined that this striking phenomenon was reproducible, and, subsequently, that simply touching the metal contacting the nerve with a different kind of metal was sufficient to induce contraction. From these observations, Galvani developed a theory, building on ideas of the Ancient Greeks, of an electric fluid inherent to the nerves, whose role was to generate muscle contraction. To defend his proposal against its most vocal opponent, Alessandro Volta (1745–1827), who asserted that electricity originated exclusively from dissimilar metals in contact and never from living organisms, Galvani conducted an experiment referred to as 'contraction without metals' (*contrazione senza metallo*). He reported that contraction could be induced simply by bringing the end of the nerve into contact with a nearby (damaged) muscle, which apparently depolarized the nerve sufficiently for it to fire (Galvani 1791, trans. Green 1953; Mauro 1969; NONC p. 163–173).[1]

Investigations of a possible electrical component to nerve signaling ('activity') and muscle contraction continued throughout the nineteenth century, although the controversy over the existence of animal electricity persisted for decades. These studies drew on electrical principles being discovered at the time—notably the relation among voltage, current, and resistance, articulated by Georg Ohm (1789–1854) in 1827—as well as new technologies in the form of measurement devices. After the development of the galvanometer as a tool to detect current, Leopoldo Nobili (1784–1835) and, later, Carlo Matteucci (1811–1868) demonstrated that current flow could be detected between two electrodes placed on an injured and an intact region of muscle—essentially the conditions of Galvani's contraction without metals; these results offered the first evidence of a voltage difference between the interior and exterior of the cell, or 'resting potential.' Remarkably, Matteucci later recanted, after difficulty reproducing his results in nerves. The line of research was continued by Emil Dubois-Reymond (1818–1896), who ultimately

---

[1] In addition to original articles, this section repeatedly cites four books, which are referred to by their initials:
NONC: *Nineteenth Century Origins of Neuroscientific Concepts.* E. Clarke and L. S. Jacyna (1987)
MII: *Membranes, Ions, and Impulse.* K. S. Cole (1968)
CDD: *Chance and Design: Reminiscences of Science in Peace and War.* A. Hodgkin (1992)
HNA: *The History of Neuroscience in Autobiography. Vol. 4.* A. F. Huxley (2004)
Full citations are given in the references.

recognized the unit of electrical signaling as a transient reduction in current flow between a region of intact nerve and its cut end, which he named 'the negative variation' (*der negative Schwankung*). Stated in modern terms, because a healthy region of resting nerve membrane has a substantial transmembrane potential (of about −60 mV) and a damaged region has a potential near zero, a voltage difference exists across these two sites, making it possible to measure current flowing between them. During an action potential, the transmembrane voltage of the intact region approaches zero, reducing the voltage difference. Consequently, the measured current changes from a high to a low value during electrical activity. The negative variation was thus the signature of the action potential (Schuetze 1983; NONC pp. 189–190, 196–211).

A student of Dubois-Reymond, Julius Bernstein (1839–1917), improved the measurement technique and reconstructed the time course and conduction velocity of the negative variation in nerve bundles (Bernstein 1868). Later, drawing on the work of Walther Nernst (1864–1941), Bernstein proposed the 'membrane theory,' arguably becoming the father of membrane biophysics. The theory stated that cells consisted of electrolytes encapsulated by a membrane that was relatively impermeable to all ions except potassium, which was found to be permeant through ion substitution experiments. Bernstein deduced that an electrical potential would exist across such a membrane and further proposed that the permeability of the postulated membrane would break down during electrical activity or 'irritability' (*der Reizung*). The resulting redistribution of ions would produce 'action currents' (*Actionsströme*), accounting for the negative variation (Bernstein 1902; MII pp. 6–9).

Evidence in favor of the membrane theory came some years later from Rudolf Höber (1873–1953), who was working under the guidance of Nernst (Höber 1910, 1912). Höber succeeded in measuring resistances of preparations of red blood cells by applying alternating currents to them under different conditions. He found that intact cells indeed had high resistances, but only when a low-frequency current was applied. The frequency sensitivity of the intact cell led to the concept of the membrane as a capacitor, a circuit element that filters low- but not high-frequency currents. In contrast, after hemolysis, the resistance of the red blood cell preparation was measured to be low regardless of frequency, indicating that the cell interior was conductive. These results provided evidence that the protoplasm was indeed composed of electrolytes. Bernstein's son reported that his father was highly gratified by Höber's experimental support of his theory. (MII pp. 6–7)

The biophysical characteristics of cell membranes, including the value of the capacitance and the conditions that generated the selective permeability (which was lost in dead cells), were studied further in a variety of cells, including guinea pig muscle and liver cells (Philippson 1921), kelp (Osterhout 1922), blood cell suspensions (Fricke 1925), and algae (Blinks 1928). By applying alternating currents to cell preparations and observing the phase shifts and frequency-dependent resistances, the results from different cells converged on the conclusion that most cell membranes had a specific membrane capacitance close to $1\,\mu F/cm^2$ and a thickness of ~33 Å (Fricke 1925). (MII p. 8)

The demonstration that living phenomena, including bioelectricity, could be described by physical laws motivated the drive to identify an electrical circuit whose properties would mimic a living cell. According to Cole (1968), perhaps the first such equivalent circuit (Philippson 1921) included a parallel resistance and capacitance, representing the membrane, in series with another resistance, representing the highly conductive cell interior (MII p. 9). The stage was set for the heroic age of membrane biophysics.

According to his memoir *Chance and Design*, Alan Lloyd Hodgkin (1914–1998) became interested in the field of physiology as an undergraduate at Trinity College at Cambridge University. Hodgkin's boyhood had been shaped by ornithological, botanical, and other natural pursuits, and physiology ran neck-and-neck with zoology as his specialization of choice for the research phase of his studies; botany, too, was a subject he apparently relinquished with some regret. Hodgkin began his independent research in physiology in 1934 (at age 20) on the question of whether the membrane of excitable cells underwent an increase in conductivity in association with the electrical signaling.

His father George, who had died in 1918, had also studied Natural Sciences at Trinity, and Hodgkin cites the studies of his father's friend and classmate Keith Lucas as particularly influential in his (Alan's) choice of research direction (C&D p. 63). Between 1904 and his premature death in 1916, Lucas published 21 papers in the *Journal of Physiology* on the excitability of muscle and nerve, including the all-or-none nature of the contraction of skeletal muscle fibers, the refractory period, and a theory of excitation (Lucas 1909a, 1909b, 1910). Lucas was not only prolific but also precise about defining terms, in a style that seems to be echoed in Hodgkin's later scientific writing. In his 1910 paper, Lucas wrote:

> The word excitation has by somewhat loose usage become applicable to all or any of the successive processes which constitute the connecting links between the application of a stimulus to a nerve or muscle and the appropriate final response. The application of the stimulus is not infrequently spoken of as excitation. The immediate local effect of the stimulus is called by the same name. The disturbance which is conducted away from the seat of application of the stimulus is often called the wave of excitation. A muscle is even said to be excited when it contracts in consequence of a stimulus applied to its motor nerve. It seems therefore that we are bound to define precisely at the outset what is meant in this place by a theory of electric excitation.
>
> When an electric current is passed through part of a muscle fibre or nerve fibre there must be produced in the fibre a local physical alteration which is the immediate consequence of the current. This physical alteration provides the necessary condition for starting a disturbance which is then propagated away from the seat of application of the current. A theory of electric excitation means, as here used, a theory of the physical nature of that local alteration within the fibre which constitutes the necessary condition for starting the propagated disturbance. It is not a theory of the nature of the propagated disturbance, though no doubt it may ultimately lead to such a theory. Still less is it a theory of the more remote disturbance which constitutes contraction. (Lucas 1910)

More contemporary scientific influences on Hodgkin's choice of research topic were the early studies of cell capacitance by the American botanist Winthrop van Osterhout (1871–1964) and his student and collaborator Lawrence R. Blinks (1900–1989), who studied *Nitella*. Cells from this freshwater alga produce action potentials of a few seconds in duration, later found to depend on efflux of chloride followed by potassium (Gaffey & Mullins 1958). Hodgkin particularly recalls being influenced by the study of Blinks (1930), which stated:

> It is an interesting property of the cells of *Nitella* to be stimulated by electric current, and to transmit that stimulus as a negative variation, giving a typical diphasic action current comparable to that observed in muscle and nerve. (Blinks 1930)

Using a Wheatstone bridge (see Appendix 3.4), Blinks measured the resting resistance across the long axis of cylindrical *Nitella* cells, as well as the change in resistance upon stimulation. Blinks concluded his paper with the following description relating current (the negative variation), conductance (the resistance change), and voltage (potential difference) during activity:

> The study of these phenomena is not complete, but the outstanding effects on resistance may be indicated. The crest of the negative variation is really a depression of the P.D. [potential difference] at the contact nearly to zero. At the same time the resistance across the protoplasm is likewise greatly lowered and may fall momentarily to 0.1 megohm [from >3 megohms], about as in contact with 0.1 M KCl. This suggests that the cathodic stimulation may consist in the movement of sufficient K^+ ions in an outward current (from the sap to the external solution) to reach approximately such a concentration just outside the protoplasm. (Blinks 1930)

Hodgkin states that Blinks' study motivated him to begin the research phase of his studies by investigating conductance changes in nerve (C&D p. 63). Working with frog myelinated sciatic nerve, in which a short segment was cooled sufficiently to block firing, he found that stimulating an action potential on one side of the block could increase the excitability of the nerve beyond the block. In the first of a pair of papers, Hodgkin comments on his own study with a succinct encapsulation of the scientific method:

> The experiments described in this section do not throw much light upon the mechanism of summation, but they must precede any detailed analysis of the process. (Hodgkin 1937a)

In other words, observation is the necessary precursor to hypothesis. Hodgkin's second paper provides the detailed analysis. Complementing his observations with a series of calculations based on cable theory, he provided evidence that excitability must spread as a result of current flowing in local circuits: down the axons, out the membranes, backward along the outside of the axon, and back in across the membrane. By this electrotonic mechanism, a local change in voltage could be transmitted to more distant regions (Hodgkin 1937b).

In an amusing side note that contrasts vividly with twenty-first century scientific training, Hodgkin's memoir recalls the response of Joseph Barcroft, the head of the laboratory in which he worked, to Hodgkin's query about whether his manuscript required approval before submission to a journal:

> He was quite taken aback and explained first that we did not do anything like that in Cambridge and, second, that anything I wrote was entirely my own affair. (C&D p. 68)

Hodgkin's published papers drew the attention as well as the skepticism of the American physiologists Herbert Gasser (1888–1963) and Joseph Erlanger (1874–1965), who shared the Nobel Prize in Physiology or Medicine in 1944 for their collaborative work on conduction velocity in axons. Erlanger expressed polite scientific doubt about Hodgkin's results, based largely on the absence of detectable ephaptic stimulation of neighboring axons in his own experiments; Gasser invited Hodgkin to spend a year in his lab at the Rockefeller Institute. Accordingly, Hodgkin went to New York. He spent the following summer (1938) at the Marine Biological Laboratory at Woods Hole, on Cape Cod in Massachusetts, where he met and worked with Kenneth Cole (C&D pp. 74–78).

Like Hodgkin, Kenneth S. (Kacy) Cole (1900–1984) was influenced by Blinks and vice versa; indeed, Cole is thanked for making measurements reported in Blinks (1928). Cole completed a PhD in physics in 1926 and gained experience working on the electrical properties of cell membranes with Hugo Fricke (1892–1972) during the summer of 1923. Ultimately, Cole became interested in the relationship between 'irritability,' electronics, and oscillations. His first published work on nerve, a note to *Science* in 1934 (Cole 1934), started with a statement on the scientific power of Fourier analysis and went on to analyze the system in terms of resistance, capacitance, and inductance. His experimental observations of frog nerve responses to stimulation deviated from the expected linear behavior of these elements. He therefore proposed a modified equivalent circuit for the membrane, which included a capacitance and a hypothetical resistance that, like a capacitor, would change its impedance as a function of the frequency of applied current. In this scenario, a conductance change in response to high-frequency alternating current would lead to the damped oscillation that is the action potential. The fundamental implication, which followed from the nature of a capacitor, was that *current* was the underlying independent variable that drove the conductance change.

Measurement of the conductance changes, and the stimuli that drive them, therefore became paramount. Cole, with collaborator Howard Curtis (1906–1972), repeated Blinks' 1930 studies in *Nitella* cells with an interest in testing whether alternating current of different frequencies affected the conductance. Recognizing that Blinks' longitudinal measurements across the long, cylindrical cells might be dominated by protoplasmic rather than transmembrane impedance, Cole and Curtis connected a Wheatstone bridge in the perpendicular orientation and made recordings of the transverse resistance across the ~0.45 mm diameter of the cell. They concluded that the cell membrane had negligible conductance at rest (Curtis & Cole 1937). Upon shock-stimulation of an action potential that propagated slowly (1 cm/sec) along the length of the cell, however, the impedance of *Nitella* decreased, such that the conductance peaked on the rising phase of the extracellularly recorded action potential. Cole and Curtis emphasized the distinction between the extracellularly recorded action potential and the actual transmembrane voltage, but they suggested that such a voltage must exist that would be 'intimately related' to the transmembrane conductance. They illustrated another modified equivalent circuit for a membrane that included a capacitor in parallel with a serial combination of a battery and resistor (Cole & Curtis 1938).

Comparable transverse recordings to investigate the basis of excitability of animal cells promised to be informative, but nerve fibers on the physical scale of *Nitella* cells were unknown. In 1936, however,

John Zachary Young (1907–1997) described the squid (*Loligo forbesi*) nervous system as having giant fibers arising from giant cells and forming giant synapses (Young 1936). Cole and Curtis recognized the squid giant axon, which was about 0.6 mm diameter, as a potentially useful preparation for the study of nerve impulses (Curtis & Cole 1938). Working with these axons, they made recordings of the capacitance and conductance across the axon during rest and activity, comparable to those they had made in *Nitella*. In what is arguably the most famous figure from their work together, they illustrated the conductance changes that occurred during a propagating action potential (Cole & Curtis 1939, reproduced in the final paper of the Hodgkin-Huxley 1952 series). Despite the conductance change, the capacitance remained constant throughout the action potential, indicating that the membrane itself did not break down during activity. The mechanism of the conductance change, however, remained a mystery. In their discussion, Cole and Curtis commented, somewhat prophetically,

> In contrast to the *Nitella* results, it will be noticed that for the squid axon the recovery of the action potential is completed considerably before that of the membrane resistance, but it seems likely that when this difference can be explained the whole phenomenon of excitation and conduction will be fairly well understood. (Cole & Curtis 1939)

Interestingly, Hodgkin apparently walked in on Cole and Curtis's classic experiment. In his memoir, he recalls

> arriving in Cole and Curtis's room there and seeing the increase in membrane conductance displayed in a striking way on the cathode ray tube. (C&D p.115)

At Rockefeller, Hodgkin had been working on single axon fibers from the crab, a preparation he had started on back in Cambridge; at Woods Hole, he was introduced to the preparation of the squid giant axon by Cole. His memoir quotes a letter to his mother from mid-June 1938:

> As you know, I spend my time working with single nerve fibre from crabs, which are only about 1/1000 of an inch thick. Well, the squid has one fibre which is about 50 times larger than mine and Cole has been using this and getting results which make every one else's look silly. Their results are almost too exciting because it is a little disturbing to see the answers to experiments that you have planned to do coming out so beautifully in someone else's hands. No, I don't really mean this at all, what I do dislike is the fact that at present English laboratories can't catch squids so that I don't see any prospect of being able to do this myself. (C&D p. 119)

While at Woods Hole, Hodgkin used both squid and crab to conduct a straightforward test of the local circuit hypothesis. If the current indeed flowed in loops, he reasoned, then the speed of propagation of the action potential would be influenced by the magnitude of the resistance *extracellular* to the axon. This idea gave rise to specific testable predictions: increasing the external resistance should impede current flow in the circuit, slowing the conduction velocity; reducing the external resistance should facilitate current flow, speeding conduction. The predictions were fulfilled. The results provided strong evidence that voltage indeed spreads along the axon as a consequence of local circuit currents, carried by ions, flowing through the resistances within the axon, out across the membrane, extracellularly along the axon, and back across the membrane into the axon (Hodgkin 1939).

In Hodgkin's final weeks at Woods Hole that summer, he and Cole worked together on a fundamental problem that interested them both: measuring the resting resistance of the squid axonal membrane (Cole & Hodgkin 1939). Hodgkin's memoir quotes part of a letter to his mother, 'This is the first time that I ever collaborated with any one, and I never realized till now how much nicer it is than working alone.' (C&D p.117) Upon returning to Trinity College in Cambridge, Hodgkin began what were intended to have been three years of pure research. That autumn (1938), he supervised a laboratory practical class for physiology students, which included the 20-year-old Andrew Huxley.

Unlike Hodgkin, who confessed 'I have always been rotten at making things' (C&D p. 71), the propensities and enthusiasms of Andrew Fielding Huxley (1917–2012) were mechanical. Huxley begins his memoir with a quote by his grandfather Thomas Huxley, the famous proponent of Darwin's theory of evolution, followed by a comment on its pertinence to himself:

> T.H. Huxley wrote a short autobiography which includes the following passage:
>
> 'As I grew older, my great desire was to be a mechanical engineer, but the Fates were against this; and, while very young, I commenced the study of Medicine under a medical brother-in-law. But, though the Institute of Mechanical Engineers would certainly not own me, I am not sure that I have not, all along, been a sort of mechanical engineer *in partibus infidelium*. . . . The only part of my professional course which really and deeply interested me was Physiology, which is the mechanical engineering of living machines.'
>
> Much of the same could be said of me: my boyhood interests were mainly mechanical, and I entered Cambridge University with the intention of specializing in physics and becoming an engineer. My subsequent interest in physiology is exactly described by the phrase 'the mechanical engineering of living machines,' and a substantial part of my work has been the design and construction of instruments needed for my research. (HNA p. 284)

Huxley describes a childhood of working with Meccano, microscopes, and a lathe that he kept and used all his life. During his first two undergraduate years at Cambridge, he studied physics, chemistry, and mathematics. On the advice of a senior classmate, he chose physiology as an elective science course. Huxley writes,

> [Ben Delisle Burns] told me that physiology was a lively subject in which even in the first year newly discovered things, and things still controversial, were taught, unlike the situation in physics or chemistry. (HNA p. 290)

Huxley went on to specialize in physiology, and among the courses he took at the beginning of the research phase of his training was the laboratory practical taught by the young Hodgkin.

Meanwhile, having purchased new equipment with an unexpectedly large grant of £300 from the Rockefeller Institute, Hodgkin returned to recording from crab axons (C&D p. 124). He describes an early experiment, intended primarily to test out his DC amplifier, in which he estimated the magnitude of the action potential relative to that of the resting potential. These voltages had to be measured between the outside of the axon and a region where the membrane potential had likely been brought to zero, either by injury, as in the days of Dubois-Reymond, or by increasing external potassium ions to match the intracellular concentration; with this method, only relative rather than absolute magnitudes could be compared. Bernstein's membrane theory predicted that the action potential would bring the membrane potential from a resting negative value (now known to be near −60 mV) to a value close to zero (e.g., −10 mV). Thus, the magnitude of the action potential (in this example, 50 mV) might be close or even equal to, but never greater than, the magnitude of the resting potential (in this example, 60 mV).

Hodgkin's results, however, suggested something quite different: the action potential appeared to have a *greater* magnitude than the resting potential. With his student Huxley joining him for some of the experiments, Hodgkin repeated the experiments not only in crab but lobster (C&D pp. 130–131). During the following summer (1939), he went to the Laboratory of the Marine Biological Station at Plymouth, in Devon, England, where squid could be caught for experiments. Huxley joined him, having turned down a research offer that would have let him pursue his interest in microscopy, and began his research at Plymouth with some unpromising experiments on the viscosity of squid axoplasm (HNA pp. 291–292). Hodgkin recalls,

> Huxley said that he thought it would be fairly easy to stick a capillary down the axon and record potential differences across the surface membrane. (C&D p. 133)

Huxley's memoir, however, credits Hodgkin for the idea:

> Hodgkin suggested pushing an electrode down inside so as to record the membrane potential directly between axoplasm and external fluid. (HNA p. 292)

The method worked, and the results were dramatic. In Huxley's words,

> We immediately found that the amplitude of the action potential was much greater than the resting potential, so that the internal potential went considerably positive at the peak of the action potential. This was contrary to the then current belief, although Hodgkin already had hints of an 'overshoot' from external recordings on single fibers from crabs and lobsters, although this was not published until later. (HNA p. 292)

Hodgkin continues,

> Andrew Huxley and I were tremendously excited about the potentialities of the technique and started other tests. . . . However, within three weeks of our first successful impalement, Hitler marched into Poland and I had to leave the technique for eight years until it was possible to return to Plymouth in 1947. (C&D p. 133)

Thus, after making these extraordinary observations in August 1939, Hodgkin and Huxley were forced to suspend their research, owing to England's entry into the war in Europe. Both entered military service shortly thereafter—Hodgkin worked on radar and Huxley worked on gunnery—but they published an initial report of their findings in October (Hodgkin & Huxley 1939), in what Hodgkin termed 'a cautious note to *Nature*' (C&D p. 135). They reported that the total amplitude of the action potential was about 90 mV. The resting potential was near −50 mV; thus, the action potential overshot 0 mV, reaching about +40 mV at its peak. Hodgkin and Huxley did not emphasize this result, however. The technique was novel, and they pointed out that while the 90-mV total amplitude was likely reliable, the absolute voltage values might be skewed by liquid junction potentials. Nevertheless, both scientists were fully aware that the switch in polarity of the membrane potential was too substantial to be attributable to measurement error and that it constituted evidence against the hypothesis that the electrical signal was a complete breakdown in the selective permeability of the membrane. Instead, it suggested an alternative, active process driving the voltage to positive values, which they would not be able to explore experimentally until after World War II.

Meanwhile, in America, Cole continued his studies of the squid axon. In the summer of 1938, at the end of his collaboration with Hodgkin, Cole had observed what looked like oscillations following the action potential (C&D p.117). Oscillations are reminiscent of resonance, and resonance in electrical circuits can be generated by an inductor and capacitor in parallel (see Appendix 2.6). Cole, who had made meticulous measurements of capacitance, was still seeking the correct equivalent circuit for an excitable membrane, and the oscillations made the idea of an inductor-like element in the membrane seem plausible. In a series of papers, Cole explored this idea, adding an inductor into his evolving equivalent circuit of the membrane (Cole & Baker 1941; Cole 1941). Cole and Curtis also figured out how to insert an electrode into the squid axon to measure transmembrane potential. Unlike Hodgkin and Huxley, who had used a silver wire coated with silver chloride, Cole and Curtis used a 'needle' electrode, a glass micropipette filled with a potassium chloride solution isotonic with seawater. The capacitance of the glass introduced a lag, for which Cole and Curtis compensated electronically. Even slight overcompensation of such circuits, however, can lead to oscillations ('ringing'); in this case, the ringing overlaid the oscillation-like voltage swings of the action potential, distorting its waveform and exaggerating its magnitude. In 1942, Cole and Curtis therefore reported (incorrectly) that the absolute voltage of the action potential, measured from resting potential to peak, could be as much as 150 mV. They recognized the significance of the overshoot in refuting the original formulation of Bernstein's hypotheses, but its explanation eluded them:

> Thus during the passage of an impulse the membrane potential is momentarily reversed in sign, so that the outside may be as much as 110 millivolts negative with respect to the inside. This fact throws doubt on the simple explanation of the action potential as a passive depolarization of the membrane or abolition of the resting potential. With the further observations of wide variability in the size of the

action potential with little if any change of the resting potential, it is reasonable to suppose that a separate mechanism is responsible for the production of each. Thus the resting potential may be an electrical measure of the energy made available by metabolism and the action potential an index of the ability of the membrane to utilize this energy for propagation. (Curtis & Cole 1942)

Regarding mechanism, Cole and Curtis did experiments designed to test whether the potentials that they measured were sensitive to ions in the bathing solution, as might be expected for a conductance-based phenomenon. Indeed, the idea was already afoot that sodium ions might be responsible for the depolarizing phase of the action potential, a possibility that came to be known as 'the sodium hypothesis.' Cole and Curtis failed, however, to detect much responsiveness of the action potential to the loss of external ions, including sodium:

Removing all ions by circulating isosmotic dextrose increased the potential only slightly (3 to 5 millivolts) higher than it was raised by removal of potassium alone. Likewise, the height of the action potential was not appreciably affected by these procedures. (Curtis & Cole 1942)

Instead, they raised the possibility of an inductance-based resonance:

However, there may be an explanation of this phenomenon on the basis of a passive depolarization. A membrane inductance has been observed, (Cole and Baker, '41) in this fiber of 0.2 henries per cm.^{2} and this, in conjunction with the membrane capacity of 1 microfarad per cm.^{2} (Curtis and Cole, '38) forms a resonant circuit. It has been possible to explain several phenomena of peripheral nerve on the basis of an equivalent membrane circuit involving capacity, resistance, and inductance (Cole, '41). The explanation of the present phenomenon in terms of this equivalent circuit is not available, but it seems possible that a complete solution of the problem on the basis of the cable equations may yield an adequate explanation. (Curtis & Cole 1942)

These observations, by distinguished and reputable scientists, of action potentials with peaks far surpassing the predicted sodium equilibrium potential and waveforms insensitive to changes in sodium concentration, made it seem highly unlikely that nerve activity resulted from an increase in membrane permeability to sodium ions. The insensitivity of the action potential to external sodium ions was also propounded by Rafael Lorente de Nó (1902–1990), a member of Gasser's department at the Rockefeller Institute who conducted extensive studies on the question. The reason for these erroneous observations was that the perineurium that ensheaths axonal fibers (within the epineurium that surrounds nerves) contains a layer of epithelial cells that form a diffusion barrier to ions. The ion-exchange experiments were therefore flawed. Even after evidence for an ion impermeable membrane began to accumulate, resistance remained strong. As late as 1950, in a paper rather boldly titled, *The ineffectiveness of the connective tissue sheath of nerve as a diffusion barrier*, Lorente de Nó wrote:

The concept that the connective tissue sheath, or rather the epineurium, of frog nerve is an effective diffusion barrier was introduced by Peng and Gerard ('30). The concept was dismissed by Lorente de Nó, who, from his observations on the action of a number of substances upon frog nerve, concluded that 'it is utterly impossible to believe that the connective tissue sheath of frog or bullfrog nerve could act as a diffusion barrier that would delay for considerable periods of time the penetration of solutes into the nerve (Feng and Gerard, '30),' and that 'the connective tissue sheath is freely permeable to solutes, be they ionized or not' (Lorente de Nó, '47a, vol. 1, p. 23). Recently, however, the concept that the epineurium of frog nerve is an effective diffusion barrier has been reintroduced in the literature by several authors. (Lorente de Nó 1950)

He recognized the legitimate problem this novel concept posed for the scientific literature, to which he had been no small contributor:

It must be realized that the statements made by Feng and Liu, Hodgkin, Huxley and Rashbass and Rushton have created an exceedingly serious situation. If the epineurium of frog nerve were an effective barrier to diffusion of any solute (Feng and Liu), and in particular an effective barrier to the diffusion of ions (Hodgkin, Huxley), and if the epineurium should play an immediate role in determining the electrical characteristics of nerve (Rashbass and Rushton), then, all the work that has been done in the past with intact nerve trunks would stand in need of radical revision, because all the results heretofore obtained would have been vitiated by exceedingly important sources of error. Indeed, there would be in the literature on nerve physiology hardly a single important observation that could stand uncorrected. (Lorente de Nó 1950)

Hodgkin and Huxley both recall that the arguments against the sodium hypothesis before and during the war influenced their interpretation of their 1939 result; Huxley adds that the then-prevailing view that hydrated potassium ions were smaller than hydrated sodium ions, which intuitively accounted for the selective potassium permeability through a sievelike mechanism, further discouraged a serious consideration of the sodium hypothesis (HNA p. 296). Thus, when they—according to Huxley, mostly Hodgkin (HNA p. 292)—wrote a fuller report of these experiments toward the end of the war (Hodgkin & Huxley 1945), they did not raise the possibility that the overshooting action potential resulted from an increase in sodium permeability. Instead, they offered four alternative suggestions: (1) an increase in anion permeability, (2) a change in the dipole orientation of the membrane, (3) an effect of inductance (*à la* Cole), and (4) an emf or battery in series with the capacitor, rather than in parallel. Each hypothesis ended with a critique, however, revealing their skepticism about all the possibilities:

[on anion permeability] Such a state of affairs is theoretically possible, but does not seem at all probable, since it is hard to imagine that the concentration or mobility of lactate or any other organic ion would be sufficient to swamp the contributions of K+ and Cl- to the membrane potential.

[on a dipole switch] This is not an impossible assumption, although it is a little hard to imagine that such a change would leave the membrane capacity unaltered during activity.

[on inductance] We are reluctant to accept the idea of a genuine inductance in the membrane, since it is difficult to attach any physical significance to such a concept.

[on the series-capacity hypothesis] This hypothesis has not been developed in any detail and may not bear quantitative investigation. (Hodgkin & Huxley 1945)

Hodgkin writes that, in retrospect, both he and Huxley 'came to regret the discussion in that paper,' particularly regarding omission of the possibility of a transient selective permeability to sodium, but notes that 'things looked rather black for the sodium hypothesis both then and several years later' (C&D p. 252). In later years, Cole also referred to his own initial recordings of intracellular action potentials with Curtis, acknowledged the error in the 1942 paper, and commented on his colleagues' response to it:

Our action potentials (Curtis and Cole 1940) were quite variable and inconclusive but we soon had word from Hodgkin and Huxley that they had done much the same thing, at about the same time and probably for much the same reason, but much better. . . . These results we fully confirmed (Curtis and Cole 1942) except for the published action potential of 168 mV which I came to believe was probably the result of an overcorrection for the electrode and the amplifier input capacities. When Hodgkin and Huxley were able to publish their work in full after the war (1945), they most generously spoke of their confirmation of *our* work. (MII p. 145)

Huxley recalls that his own distrust of the sodium hypothesis was finally alleviated in October 1945 by a lecture at the Royal Society given by August Krogh (1874–1949), in which he reported on radioactive tracer studies that had indicated that membranes were not as impermeable to sodium as had previously been believed. 'From then on,' he writes, 'the sodium hypothesis was under active discussion between Hodgkin and myself.' (HNA p. 295) Huxley expounds further on the Krogh lecture and its consequences in a retrospective he wrote on Hodgkin's life:

In particular, he [Krogh] mentioned the exchange of sodium across cell membranes, contradicting the previous belief that cell membranes were completely impermeable to sodium ions. This implied the continuous activity of a 'sodium pump' extruding the sodium that entered the cell passively down its electrochemical gradient. It occurred to me that if the action of this pump were temporarily interrupted, sodium ions would continue to enter, tending to cause the interior to go electrically positive, and that this might be the origin of the overshoot. When I mentioned this idea to Hodgkin, he immediately pointed out that it was totally inadequate because, if the rate of entry of sodium ions were sufficient to cause the known rapid rise of internal potential in an action potential, the energy required to expel the sodium that would be entering continuously at rest would be far more than could be provided by the known oxygen consumption of nerve. So we began to discuss the related hypothesis that the overshoot was due to the increase in membrane permeability postulated by Bernstein being highly specific for sodium ions. (Huxley 2000)

Interestingly, as far back as 1902, Ernest Overton (1865–1933) published a paper entitled *Über die Unentbehrlichkeit von Natrium- (oder Lithium-) Ionen für den Contractionsact des Muskels (On the indispensability of sodium or lithium ions for muscle contraction).* As the title indicates, Overton reported that muscles did not twitch without extracellular sodium. This discovery, however, was somehow lost from the stream of science that influenced early neurophysiology and was not recovered until after publication of the Hodgkin-Huxley 1952 series of papers. Indeed, Huxley states that, had he and Hodgkin been aware of Overton (1902), they would likely have considered the sodium hypothesis more seriously much sooner (HNA p. 292).

After the war, in late 1945, Hodgkin and Huxley restarted experiments in Cambridge on crab axons, as squid were locally unavailable. Their thinking about the basis of excitability, and about physiology in general, had changed. Hodgkin writes in his memoir,

> I found it much harder to give tutorials in Trinity College than before the war. This was partly because I had forgotten a good deal and partly because I had ceased to believe in some of the principles that had once seemed to hold physiology together. The constancy of the internal environment remained as important as ever, but the ways in which constancy was achieved had become more complicated. It was also clear that much that I had read and taught before the war had been wildly oversimplified, if not downright wrong. . . . I suppose that after five years working as a physicist I had little use for biological generalizations and always wanted to concentrate on the physicochemical approach to physiology. This didn't go down well with most medical students. (C&D p. 262)

Among their changes in perspective, as noted, was the willingness to reconsider which ions permeated the membrane, and under what conditions. Through the following year, they studied the changes in potassium permeability of crab axons during activity. Hodgkin found that bathing a stretch of axon in higher potassium concentrations yielded a higher conductance of the axon (Hodgkin 1947). He reasoned that the converse might also be true, such that a stimulus-evoked increase in conductance might indicate an accumulation of external potassium. Hodgkin and Huxley tested this idea by evoking trains of action potential in axons bathed in tiny amounts of fluid, so that any ions extruded from the axoplasm would not diffuse away. Indeed, the conductance was increased, leading them to conclude that potassium 'leaked' from the axon when action potentials were fired. Their thinking extended beyond the flux of potassium and, for the first time, they openly revisited the sodium hypothesis in the discussion:

> The simplest way of accounting for the leakage of potassium during activity is to assume that the nerve membrane becomes temporarily permeable to sodium or to one of the internal anions. (Hodgkin & Huxley 1947)

They proceed with a rough calculation based on the idea that an increase in sodium permeability drives a potassium efflux as various equilibria are restored, but which ends with the comment:

> The preceding argument is put forward only because it is the simplest qualitative explanation of the facts. There are several reasons for believing that the true situation is more complicated. (Hodgkin & Huxley 1947)

The retrospectives provide an interesting addition to the even-toned voice of the papers. Hodgkin writes that he and Huxley spent the winter of 1946–1947 coming up with possible mechanisms for the action potentials they had recorded. The sodium theory was high on their list, although the means by which sodium would cross the membrane was a mystery. Their best guess was that a carrier system—molecules with negative charges or dipoles that would attract sodium ions—might shuttle the ions across the membrane. During his war service, Huxley had cultivated a great facility with numerical methods, computed on mechanical Brunsviga calculators (see Appendix 5), and when research slowed and then stopped owing to an unusually severe British winter, exacerbated by shortages of both food and coal, he used the time away from the lab literally to crank out, with mittened hands, theoretical waveforms of propagating action potentials (C&D p. 269–271). Hodgkin recalls,

> In these theoretical action potentials the reversed potential difference at the crest of the spike depended on a selective increase in sodium permeability and a low internal concentration of sodium ions. Huxley felt all along that this was a likely mechanism, but I was more doubtful, partly because there seemed to be quantitative discrepancies, and partly because I hankered after a mechanism which would give a transient reversal, so accounting for repolarization, oscillations, and the transient nature of the action potential. We tried various mechanisms that I thought might operate in this way, but Huxley shot them all down, leaving a rise in permeability to sodium ions, or perhaps to an internal anion, as the most likely cause of the reversed potential. (C&D p. 269–270)

In other words, Huxley saw that sodium permeability was necessary, but Hodgkin knew that the hypothesis, as it stood, was insufficient.

It is also notable that Hodgkin, unlike Huxley, hesitated to discredit the report of his former collaborators, Curtis and Cole, regarding the insensitivity of the action potential to the removal of external sodium. Even though he himself had recorded data to the contrary, Hodgkin apparently did not dismiss the Americans' work until he heard of the results obtained by his pre-war friend and scientific colleague, Bernard Katz.

Hodgkin first met Bernard Katz (1911–2003) early in his research career. Their work had converged, as both had found evidence for nonpropagated, subthreshold responses in crab axons, an observation that seemed to conflict with the all-or-none theory of excitability (Katz 1937; Hodgkin 1938) and which—like the local circuit theory—was met with doubt by Erlanger and colleagues in America. Although Katz worked in the group of A.V. Hill (1886–1977) at the University College London (UCL), rather than in Cambridge near Hodgkin, the two young scientists continued to discuss their work regularly and interacted directly at Plymouth before the war. In a letter to his mother in July 1939, Hodgkin wrote,

> Katz, a refugee who works on nerve, has been down here for a few days, and I have seen a good deal of him. He is going to Australia in a fortnight to work with Eccles in Sydney. He is a very good person to talk science with. (C&D p. 132)

Katz, who was born of a Russian-Jewish family in Germany and who became stateless upon fleeing to Britain, indeed went to Australia. After the war broke out, he obtained British citizenship and served in the Royal Australian Air Force (as a radar officer). After the war, he returned to UCL, where he remained for the rest of his life. Hodgkin writes,

> Towards the end of 1946, Bernard Katz sent me a manuscript in which he showed, among other things, that crab axons became inexcitable in salt-free sugar solutions (Katz 1947). As this agreed with my own experience I began to think that Curtis and Cole's (1942) result must have been wrong and that there was hope for the sodium theory. (C&D p. 270)

When Hodgkin finally was able to return to experiments on the squid axon at Plymouth in the summer of 1947, Huxley was absent for the simple reason that he was getting married. Ironically, therefore,

Hodgkin's direct tests of the sodium hypothesis were ultimately done without Huxley. Hodgkin initially worked alone, but Katz joined him in the autumn. They recorded action potentials from squid axons with the intracellular recording technique that Hodgkin and Huxley had worked out before the war and measured the effects of changing the extracellular concentration of sodium ions. The introduction to the paper—which appeared more than a year later, in March 1949, owing to publication delays—begins with an assurance that the action potential overshoot is real, and includes an appropriate nod to Huxley:

> [T]here is now little doubt that the membrane potential of certain types of nerve fibre does undergo an apparent reversal which cannot be reconciled with the classical form of the membrane theory. Several attempts have been made to provide a theoretical basis for this result (Curtis & Cole, 1942; Hodgkin & Huxley, 1945; Höber, 1946; Grundfest, 1947), but the explanations so far advanced are speculative and suffer from the disadvantage that they are not easily subject to experimental test. A simpler type of hypothesis has recently been worked out, in collaboration with Mr. Huxley, and forms the theoretical background of this paper. (Hodgkin & Katz 1949)

The simple, testable hypothesis of sodium permeability is then explained in a straightforward fashion, but acknowledges that the mechanism of such permeability remains unknown:

> According to the membrane theory excitation leads to a loss of the normal selectively permeable character of the membrane, with the result that the resting potential falls towards zero during activity. This aspect of the theory is at variance with modern observations and must be rejected. However, a large reversal of membrane potential can be obtained if it is assumed that the active membrane does not lose its selective permeability, but reverses the resting conditions by becoming highly and specifically permeable to sodium. The reversed potential difference which could be obtained by a mechanism of this kind might be as great as 60 mV. in a nerve with an internal sodium concentration equal to one-tenth of that outside. The essential point in the hypothesis is that the permeability to sodium must rise to a value which is much higher than that to potassium and chloride. Unless this occurs the potential difference which should arise from the sodium concentration difference would be abolished by the contributions of potassium and chloride ions to the membrane potential. The hypothesis therefore presupposes the existence of a special mechanism which allows sodium ions to traverse the active membrane at a much higher rate than either potassium or chloride ions.
>
> A simple consequence of the hypothesis is that the magnitude of the action potential should be greatly influenced by the concentration of sodium in the external fluid. Thus the active membrane should no longer be capable of giving a reversed e.m.f. if the external sodium concentration were made equal to the internal concentration. On the other hand, an increase of membrane reversal would occur if the external sodium concentration could be raised without damaging the axon by osmotic effects. (Hodgkin & Katz 1949)

The predictions were fulfilled, and the data provided convincing, if indirect, evidence that the upstroke of the action potential depends on sodium. Hodgkin and Katz therefore concluded that the membrane indeed must become permeable to sodium at the time of the action potential, which would account for the change in transmembrane voltage. To express this idea quantitatively, Hodgkin and Katz built on the work of Goldman (1943), which assumed that the electric field was constant through the membrane (hence the name 'constant-field theory') and that ions could—somehow—diffuse through a lipid membrane the way they diffused in aqueous solution. They derived an equation that defined how the transmembrane voltage depended on the relative permeabilities of sodium, potassium, and chloride. With this equation, now known as the Goldman-Hodgkin-Katz voltage equation, they estimated that the resting membrane was twenty-five times more permeable to potassium than to sodium; during the action potential, the permeability ratio switched almost completely, so that sodium was twenty-fold more permeant. Permeability, however, which had the simple units of a rate—cm/sec—remained a physically enigmatic quantity.

Further analysis of the results gave extra information: the rate of rise of the action potential was proportional to sodium concentration. Drawing on simple relations between the capacitance and the rate of voltage change, Hodgkin and Katz were able to calculate, for the first time, the total inward ionic current carried by sodium during the upstroke of the action potential. Nevertheless, this transmembrane sodium flux was still only inferred, and a number of phrases in the paper, including the 'special mechanism' invoked in the introduction, still carry the skepticism initially voiced by Hodgkin:

> [I]t is much more difficult to accept the assumption that the active membrane can become selectively permeable to sodium. We therefore suggest that sodium does not cross the membrane in ionic form, but enters into combination with a lipoid soluble carrier in the membrane which is only free to move when the membrane is depolarized. Potassium ions cannot cross the membrane by this route, because their affinity for the carrier is assumed to be small. An assumption of this kind is speculative but not unreasonable. . . .

> . . . The experiments described in this paper are clearly consistent with the view that the active membrane becomes selectively permeable to sodium, and thereby allows a reversed membrane e.m.f. to be established. The evidence is indirect, and the sodium hypothesis cannot be pressed until more is known about the ionic exchanges associated with nervous transmission. (Hodgkin & Katz 1949)

Clarity about these ionic exchanges could only be achieved by a direct measure of the purported sodium current. Obtaining such data, however, promised to be difficult. First, the accumulating evidence that sodium permeability—however mysteriously it might be initiated and accomplished—would be associated with inward sodium current on action potential upstroke imposed a complicating biological twist onto Ohm's law. In a simple ohmic situation, current, I, is equal to voltage, V, divided by resistance, R. Plotting current as a function of voltage therefore would give a straight line with a slope of $1/R$; since resistances cannot be negative, the slope would be positive. A sodium current that was tiny at the resting potential but large as the voltage approached zero millivolts would create a region of negative slope—an unstable situation associated with positive feedback—which was not easily measurable. Second, the variables were not clearly limited to current, voltage, and resistance. Hodgkin and Cole, together and separately, had spent years describing the basic electrical properties of the axon, Cole pursuing an equivalent circuit including capacitance and inductance, and Hodgkin analyzing local circuit currents along the length of the axon. Their and others' work illustrated the complexity of 'cable properties' and the associated equations that quantified the relationships among factors controlling the temporal change and spatial spread of voltage. Given that action potentials propagate, any measurement of current would have to contend not only with the instability of current as a function of voltage, but with the variation of current as a function of distance.

The first problem, of unmeasurable variables under unstable conditions, might be overcome by some sort of negative feedback to counteract the positive feedback; the second, of spatial variations in voltage, might be resolved by preventing propagation by making the entire axon generate an action potential at once. In summer 1947, working alone at Plymouth on the effects of sodium on squid axon action potentials, Hodgkin wrote to Cole proposing some experiments to achieve such spatial control, which they might undertake jointly during Hodgkin's upcoming trip to America:

> I am also interested in the possibility of stimulating an axon with a diffuse electrode in such a way that the axon is excited uniformly over a length of one or two centimetres. This might give useful information about the nature of the active process uncomplicated by propagation and local circuits. What are your plans and views? (C&D p. 281)

Cole replied,

> I am sure that you will be excited to hear that we spent the whole summer with an internal electrode 15 millimetres long and about 100 microns in diameter . . . The two principal ideas are first the use of the central outside region with a guard region on each side, and second the use of a feedback circuit to control either the current flow in the central region or the potential difference in that region to the desired value. (C&D pp. 281–282).

Cole describes this setup more picturesquely in his retrospective:

> As a phenomenological description, it could be said that the axon had been robbed of its ancient right to propagate an impulse by eliminating the local circuit currents, $\partial^2 V/\partial x^2$, by which an active region normally reached ahead to move itself along the axon. (MII p. 244)

Regarding the question of feedback control—the essence of the voltage-clamp technique—Cole further explains that such methods had become pervasive during the technological advances associated with the war:

> The control concept had been highly developed during World War II, principally with feedback electronics. It was widely applied afterward. . . . In general the difference between the actual and the desired position of a system is used to control the power to reduce this error. (MII p. 246)

Hodgkin, too, was considering feedback methods soon after the war, as Huxley explains in his autobiography:

> Both Hodgkin and Cole suspected that the all-or-none character of the nerve action potential was due to a current-voltage relation in the membrane that was continuous but included a region of negative slope which caused positive feedback and therefore instability. Such a feature would make it difficult to measure the current-voltage relation. I remember a discussion with Hodgkin, probably in 1945, in which he pointed out that it would be necessary to use electronic feedback to an internal electrode so as to control the internal potential ('voltage clamp') and to make it undergo stepwise changes. I replied that it would be just as good to feed current from a low-impedance source, but Hodgkin had realized that this would be an imperfect arrangement since the electrode would become polarized by the high current density that would be needed. (HNA p. 297)

As it turned out, Hodgkin and Cole did not have the chance to collaborate during the former's visit in spring 1948, but they did meet and discuss their science. Hodgkin told Cole of his results with Katz on sodium entry into the axon on the upstroke of the action potential, and Cole and his then-collaborator George Marmont showed Hodgkin their initial work with feedback control. Marmont, who remained focused on the idea that conductance changes were dependent on *current*, succeeded in supplying current feedback through a single electrode and recorded what would now be called escaping spikes. These current responses resembled delayed, inverted action potentials, resulting from an initially stable experimental 'clamp' that was gradually overridden by large ionic currents (Marmont 1949). Cole worked separately on a modification of the same setup and attempted to control, or 'clamp,' the membrane voltage. The traces he later published showed that depolarization of the membrane was associated with an inward followed by an outward current, but the waveforms were too distorted to withstand quantitative analysis (Cole 1949). Hodgkin provides a recollection that is fairly generous:

> [Cole and Marmont] showed me the results they had obtained the previous summer at Woods Hole with the membrane current or potential of a giant nerve fiber under the control of electronic feedback. I gathered that Marmont was more enthusiastic about current control and that, perhaps for this reason, they had not done many experiments with voltage control. However, the results which Cole showed me clearly illustrated the essential features of records obtained with the 'voltage-clamp' technique. (C&D p. 282)

Huxley, however, recounts that Cole had, in fact, run into the problems anticipated by Hodgkin:

> Cole, together with Marmont, was the first to make experiments of this type [feedback control] on the squid giant fiber in the summer of 1947 (Cole, 1949). However, their experiments were limited: Marmont had originally devised the apparatus with the intention of controlling the membrane current and Cole had made an addition which made it possible to use it to control the internal potential. Using it in this voltage-control mode, they did show that the current-voltage relation is continuous with a region of negative slope (Cole, 1949), but they did not analyze the current into components carried by different ions; further, their apparatus was not a true voltage clamp since they controlled the current by feedback

from the same internal electrode by which current was injected. This effectively provided a low-impedance source from which potential changes were applied to the internal electrode and the results were therefore distorted by electrode polarization, as Hodgkin had foreseen: the long-lasting outward current during what should have been a constant raised internal potential declined because the potential of the axoplasm did not follow perfectly the potential applied to the wire. (HNA p. 298)

Cole, too, later commented on the shortcomings not only of the technique but of his conceptual framework at the time, which limited his ability to interpret the results:

> The early inward current flowing against the resting potential had come to be expected, but again I was greatly disappointed not to find a steady state negative resistance. Even though the extent of my ignorance and confusion was more clearly revealed, I was very pleased by the direct records of the amplitude and form of the currents. They gave good basis for at least a qualitative explanation of the initiation, rise, and recovery of the action potential and its propagation (Cole 1949). (MII p. 259)

Indeed, although the flow of technical information from Cole to Hodgkin is recalled in most of the retrospectives and biographies, the flow of conceptual information in the opposite direction is less strongly emphasized. It is clear that Cole showed Hodgkin his proto-voltage clamp setup and experiments early 1948, but at the time Cole presumably was still convinced of the results from his 1942 paper with Curtis, which appeared to rule out a primary role of sodium ions in generating the action potential. Cole apparently did not read Hodgkin and Katz's results as offering an explanation of his own recordings and instead remained focused on current as the independent variable. He writes,

> [Hodgkin] could not convince me that my data had any other interpretation than an inward current arising from a linear small outward current. (MII p. 268.)

Cole then explicitly recounts that he argued strongly against a carrier model when Hodgkin showed him his own yet-unpublished results with Katz from the previous summer (MII p. 269). The record is silent, however, regarding how (or whether) Cole or Marmont might have tried to account mechanistically for the inward currents they had recorded, before or even after Hodgkin provided them with evidence that the action potential upstroke was associated with a membrane permeability to sodium ions.

It was the conceptual framework built up by Hodgkin and Huxley that made the next phase of their experiments—the one that provides the basis for the five classic papers of 1952—proceed rapidly and smoothly. Huxley recounts that Hodgkin had begun to consider the novel idea that permeability was sensitive to *voltage* itself, based on his early, contentious work on subthreshold responses in crab axons:

> The prewar experiments [Hodgkin 1937c] in which Hodgkin had seen the local responses of crustacean nerve fibres when stimulated by a shock just too weak to start a full-sized impulse had led him to believe that the increase in permeability during the action potential was not itself an instantaneous change but was graded with the change in internal potential. As the internal potential was raised the increase in permeability (even if it allowed all species of ions to enter or leave, as supposed by Bernstein) would tend to raise the internal potential so that a point could come at which the situation was unstable: any small rise in potential would cause a permeability increase that would cause an additional potential rise, and so on in an explosive manner until a new equilibrium was reached. This instability would be the cause of the all-or-none character of the action potential. Hodgkin [in 1946] conceived an experiment in which current was passed between a long wire in the inside of a nerve fibre and the external solution; the potential of the interior would be monitored and a feedback amplifier would control the current so that the potential underwent a predetermined time course. This arrangement is referred to as a 'voltage clamp' because it is usually used to bring the internal potential from its resting level to another level and to 'clamp' it at this level for a substantial period. (Huxley 2000)

In the summer of 1948, Hodgkin returned to Plymouth to resume experiments on squid axons and to pursue a method of voltage control that overcame the polarization problems of single electrode-mediated feedback. Hodgkin and Katz developed the double spiral electrode; Huxley, who arrived six weeks later, began to build the circuit Hodgkin had previously sketched out. They obtained some initial recordings, which they presented at a conference in Paris the following April, some months before Cole's initial voltage-clamp study was published in October. In summer 1949, Hodgkin and Huxley returned to Plymouth to continue their research:

> At first squid were in poor supply and we took a few weeks to get going. But by mid-July 1949 Katz had joined us, there was a good supply of living squid and in the next month we obtained virtually all the voltage-clamp records that we used to illustrate the papers published in 1952. I believe we were able to do this quickly and without leaving too many gaps because we had spent so long thinking about the kind of system which might produce an action potential similar to that in nerve. We also knew what we had to measure in order to reconstruct an action potential. (C&D p. 289–290)

Katz was involved in the early experiments but soon turned his attention to muscles, apparently for aesthetic reasons. In a twenty-first century interview, his collaborator, Paul Fatt (1924–2014), recalled the summer of 1948:

> I went to Plymouth and there they were, Katz and Hodgkin and Huxley, the three of them, working with squid axons. . . . And somehow I got attached to Katz because he wasn't happy with all of this, he liked to see action potentials and here they are suppressing them; they're actually stumping them, they won't have action potentials. He liked action potentials. And he wasn't going to be on this analysis. . . . [T]hey were all worried about not being able to get squid. Getting squid and dissecting it. Oh, Katz I think, was dissecting them; that's the only thing he was doing because he didn't like this no action potentials. (Fatt 2013)

After rapidly gathering data that summer, Hodgkin and Huxley took two years to analyze and write the five resulting papers; four were submitted in October 1951, appearing in print in April of the following year, and the fifth, which included the computational analysis, was submitted in March 1952 and published in August. Although the completed opus did not provide the molecular explanations that Hodgkin and Huxley had hoped for—which Hodgkin describes as 'initially a disappointment' (C&D p. 291)—the results offered solid, quantitative evidence for a series of revolutionary insights.

Revolutions, however, are rarely accepted without opposition. In June 1952, Hodgkin attended a symposium at the Cold Spring Harbor Laboratories where he presented the work he had recently completed with Huxley and Katz. Hodgkin's brief report of the meeting in his memoir savors of understatement:

> At all events I had to work hard for the privilege of being there. After my talk, someone, possibly Ralph Gerard, organized an evening session with Cole and perhaps a dozen nerve people there and cross-questioned me step by step on the details of our five papers; this took several hours. (C&D p. 324)

Regardless of who accepted or even fully grasped the ideas discussed that evening, Hodgkin and Huxley had largely solved the more-than-a-century-old puzzle of the basis of bioelectricity. They had demonstrated that sodium and potassium ions flow across the membrane, not shuttled by carriers, but diffusing through voltage-sensitive, time-dependent, ion-selective, conductance-resembling pathways, which would later be molecularly identified as ion channel proteins. These extraordinary results, their precise quantification, and their exquisitely multifaceted interpretation—acknowledged by a Nobel Prize to Hodgkin and Huxley in 1963—accounted for virtually all the key observations of the field, including the action potential itself. The discipline of neurophysiology entered a new era.

Paper 1

Hodgkin, A. L., Huxley, A. F., and Katz, B. (1952) Measurement of current-voltage relations in the membrane of the giant axon of *Loligo. Journal of Physiology* 116:424–448.

In the full series of papers, the primary question under investigation was what are the factors that control the movement of ions across excitable membranes and thereby give rise to the action potential. If the underlying variables and their effects on ionic flux could be identified and described quantitatively, it would become possible not only to attain a conceptual understanding of electrical signaling but also to predict how an excitable membrane would behave under any set of novel conditions. To achieve this end, it was necessary to determine whether sodium and potassium were indeed the ions whose flux generated the voltage changes that constituted the action potential. In this first paper, Hodgkin, Huxley, and Katz (abbreviated HH&K) begin with an outline of their general conceptualization of the membrane as an electrical circuit. They then present the technical advances that will allow them to control the membrane potential of a short stretch of the axon—that is, to voltage-clamp the membrane. These innovations include the experimental chamber containing the squid giant axon as well as the electrodes and amplifier that generate feedback. After ascertaining that the axonal preparation can generate an action potential in the unusual situation in which propagation is prevented, HH&K make the first recordings of ionic currents at a series of experimentally determined, fixed membrane potentials. These permit the first quantitative measurements of voltage- and time-dependent ionic currents in excitable membranes.

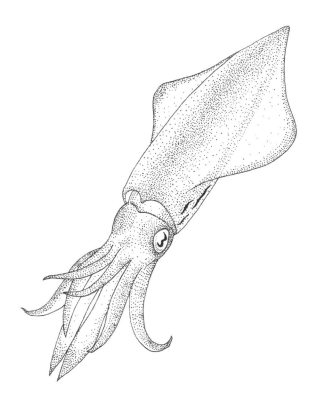

J. Physiol. (1952) 116, 424–448

MEASUREMENT OF CURRENT-VOLTAGE RELATIONS IN THE MEMBRANE OF THE GIANT AXON OF *LOLIGO*

By A. L. HODGKIN, A. F. HUXLEY and B. KATZ

*From the Laboratory of the Marine Biological Association, Plymouth,
and the Physiological Laboratory, University of Cambridge*

(*Received 24 October* 1951)

The importance of ionic movements in excitable tissues has been emphasized by a number of recent experiments. On the one hand, there is the finding that the nervous impulse is associated with an inflow of sodium and an outflow of potassium (e.g. Rothenberg, 1950; Keynes & Lewis, 1951). On the other, there are experiments which show that the rate of rise and amplitude of the action potential are determined by the concentration of sodium in the external medium (e.g. Hodgkin & Katz, 1949*a*; Huxley & Stämpfli, 1951). Both groups of experiments are consistent with the theory that nervous conduction depends on a specific increase in permeability which allows sodium ions to move from the more concentrated solution outside a nerve fibre to the more dilute solution inside it. This movement of charge makes the inside of the fibre positive and provides a satisfactory explanation for the rising phase of the spike. Repolarization during the falling phase probably depends on an outflow of potassium ions and may be accelerated by a process which increases the potassium permeability after the action potential has reached its crest (Hodgkin, Huxley & Katz, 1949).

Outline of experiments

The general aim of this series of papers is to determine the laws which govern movements of ions during electrical activity. The experimental method was based on that of Cole (1949) and Marmont (1949), and consisted in measuring the flow of current through a definite area of the membrane of a giant axon from *Loligo*, when the membrane potential was kept uniform over this area and was changed in a stepwise manner by a feed-back amplifier. Two internal electrodes consisting of fine silver wires were thrust down the axis of the fibre for a distance of about 30 mm. One of these electrodes recorded the membrane potential, and the feed-back amplifier regulated the current entering the other electrode in such a way as to change the membrane potential suddenly and

1. *'the nervous impulse is associated with an inflow of sodium and outflow of potassium'* HH&K begin with a brief summary of evidence for 'ionic movements' and the sodium hypothesis, but ionic-specific currents had not yet been measured and their existence was still questioned (see Historical Background). In studies of *Loligo* (squid; Rothenberg 1950), *Carcinus* (crab; Keynes & Lewis 1951a) and *Sepia* (cuttlefish; Keynes 1951; Keynes & Lewis 1951b), radioactive sodium or potassium had been applied either outside or inside axons, and the accumulation of radioactivity on the opposite side of the membrane had provided direct evidence that the 'nervous impulse' was associated with the inward transfer of sodium ions and the outward transfer of potassium ions. While Rothenberg had reported that potassium was not freely exchanged, Keynes and Lewis disagreed. Thus, the debate about which ions cross the membrane during electrical excitation was still ongoing.

2. *'the rate of rise and amplitude of the action potential are determined by the concentration of sodium'* The experiments by Hodgkin and Katz (1949) made use of the relatively new technique of intracellular recording in the squid giant axon, which Hodgkin and Huxley had used to make the first direct measurements of absolute transmembrane voltage during an action potential (Hodgkin & Huxley 1939, 1945). Hodgkin and Katz demonstrated that decreasing the sodium concentration in the solution bathing the axon reduced the peak amplitude of the action potential, slowed its rise time, and prolonged its duration (see Historical Background). Huxley and Stämpfli (1951) found similar results in myelinated axons from the frog. These observations provided support for the still-contentious idea that an influx of sodium ions was responsible for the upstroke of the action potential.

3. *'probably depends on an outflow of potassium ions and may be accelerated by a process which increases the potassium permeability'* While sodium substitution experiments strongly suggested that the action potential rising phase depended on sodium influx, evidence that potassium efflux formed the basis for repolarization was inferred indirectly from the 'leakage' of potassium during activity (Hodgkin & Huxley 1947). Here, HH&K state the possible mechanisms for potassium-dependent repolarization: a simple return to resting permeability following a transiently heightened sodium permeability, or an active elevation of potassium permeability above that at rest.

4. *'Outline of experiments.'* The series of papers is so extensive and unusual that it deviates in several places from the traditional structure of scientific papers, starting with this general overview of the five papers.

5. *'laws which govern movements of ions during electrical activity.'* The word 'laws' is reminiscent of physics, and HH&K indeed sought to identify biophysical laws, in the form of equations defining the precise relationships among measurable quantities that describe and ultimately predict the behavior of the system under study.

6. *'The experimental method was based on that of Cole (1949) and Marmont (1949)'* The idea of using 'feedback' to maintain zero net current flow across a membrane during an action potential was under discussion and active investigation by several physiologists of the era. In the United States, Cole and Marmont both worked on developing such a method to study the squid giant axon (see Historical Background). In their papers, both Cole and Marmont acknowledged the technical shortcomings of their new methods, from space clamp errors to series resistance errors, but did not address them. In the present work, HH&K face identical problems but respond differently. In a few crucial instances, they improve on the technique substantially to address the initial shortcomings. In other cases, errors are noted, described, and quantified, and their effects on measurements and conclusions are evaluated in detail. They are neither dismissed, nor are they considered to be problematic enough to terminate the line of work. To the modern reader, what unfolds is an extraordinary demonstration of how much can be learned from careful treatment of the available data when the quality and limitations of those data are well examined and understood.

7. *'a definite area'* a defined or restricted area.

8. *'a giant axon from Loligo'* The squid giant axon had been identified as a useful preparation for the study of nerve impulses by Young (1936). Hodgkin and Huxley began using this preparation soon thereafter, making the first intracellular recording of an overshooting action potential (Hodgkin & Huxley 1939; see Historical Background).

hold it at the new level. Under these conditions it was found that the membrane   9
current consisted of a nearly instantaneous surge of capacity current,
associated with the sudden change of potential, and an ionic current during
the period of maintained potential. The ionic current could be resolved into
a transient component associated with movement of sodium ions, and
a prolonged phase of 'potassium current'. Both currents varied with the
permeability of the membrane to sodium or potassium and with the electrical
and osmotic driving force. They could be distinguished by studying the effect   10
of changing the concentration of sodium in the external medium.

The first paper of this series deals with the experimental method and with
the behaviour of the membrane in a normal ionic environment. The second
(Hodgkin & Huxley, 1952a) is concerned with the effect of changes in sodium
concentration and with a resolution of the ionic current into sodium and
potassium currents. Permeability to these ions may conveniently be expressed
in units of ionic conductance. The third paper (Hodgkin & Huxley, 1952b)
describes the effect of sudden changes in potential on the time course of the
ionic conductances, while the fourth (Hodgkin & Huxley, 1952c) deals with
the inactivation process which reduces sodium permeability during the falling
phase of the spike. The fifth paper (Hodgkin & Huxley, 1952d) concludes the
series and shows that the form and velocity of the action potential may be
calculated from the results described previously.   11

A report of preliminary experiments of the type described here was given at
the symposium on electrophysiology in Paris (Hodgkin et al. 1949).

Nomenclature

In this series of papers we shall regard the resting potential as a positive
quantity and the action potential as a negative variation. V is used to denote
displacements of the membrane potential from its resting value. Thus

$$V = E - E_r,$$

where E is the absolute value of the membrane potential and E_r is the absolute
value of the resting potential, with signs taken in the sense outside potential
minus inside potential. With this choice of signs it is logical to take $+I$ for
inward current density through the membrane. These definitions make
membrane current positive under an external anode and agree with the accepted
use of the terms negative and positive after-potential. They conflict with the
common practice of showing action potentials as an upward deflexion and are
inconvenient in experiments in which an internal electrode measures potentials
with respect to an external earth. Lower-case symbols (v_n) are employed when
it is necessary to give potentials with respect to earth, but no confusion should
arise since this usage is confined to the sections dealing with the experimental   12
method.

9. *'hold it at the new level.'* The preceding sentences give a brief definition of the two-electrode voltage clamp: one electrode is used to measure the intracellular voltage; a second electrode is used to inject current into the axon to change the voltage and to keep, or 'hold,' it at the desired level. The use of the word 'hold' gives rise to the phrase 'holding potential,' which generally refers to the voltage at which the membrane potential is maintained before step voltage changes (or other manipulations) are applied. The use of two electrodes was a key innovation of Hodgkin's that minimized technical problems inherent to the single electrode voltage clamp configuration used by Cole and Marmont (see Historical Background and Appendices 3.2 and 3.3).

10. *'Both currents varied with the permeability of the membrane to sodium or potassium and with the electrical and osmotic driving force.'* This is a verbal statement of the equation that the papers move toward, namely Ohm's law as an expression for current: $I_{ion} = g_{ion}(V_m - E_{ion})$, or current is conductance times driving force, where the driving force depends on both the membrane potential and the concentration gradient of the permeant ion. The concentration term figures in the equation eventually known as the Goldman-Hodgkin-Katz current equation (see Paper 2, note 80).

11. *'calculated from the results described previously.'* The five papers can be thought of in shorthand as (1) the voltage-clamp paper, (2) the independence paper, (3) the tail-current paper, (4) the inactivation paper, and (5) the modeling paper.

12. *'no confusion should arise'* When measuring the transmembrane potential, which HH&K call E, it is arbitrary whether one reports the voltage inside the axon relative to the outside (which makes the resting potential negative) or the outside relative to the inside (which makes the resting potential positive). HH&K chose the latter. Unfortunately for the modern reader, conventions have changed, and consequently the polarities of all voltages throughout the papers are the reverse of what we use today. Thus, for HH&K, the resting potential, E_r, is approximately +60 mV. HH&K also report the voltage, V, as a *displacement from the resting potential*. Thus, their V is reversed *and* shifted relative to the modern V, such that $V = 0$ mV, or no displacement from the resting potential, translates to about −60 mV in modern conventions; their $V = +10$ mV means a ten-millivolt *hyperpolarization* relative to rest, corresponding to the modern −70 mV. HH&K's 'v' corresponds to the modern meaning of transmembrane potential, defined as the result when 'an internal electrode measures potentials with respect to an external earth,' or the grounded external solution. (Note that 'earth' and 'ground' are synonyms, used in different parts of the world.) Because of the reversed conventions for voltage, the polarity of currents recorded under voltage clamp is also inverted relative to modern conventions: For HH&K, the inward flux of cations is represented as positive. In modern conventions, inward cationic currents are represented as negative.

HH&K point out that their convention is consistent with extracellular recordings of the action potential, in which a depolarization of the membrane (resulting from an influx of positive ions) appears as a negativity at the electrode, corresponding to the original measurements of action potentials as a 'negative variation' (see Historical Background). They recognize, however, that their choice conflicts with the usual way of depicting the action potential as a positive deflection (as was done in Hodgkin & Huxley 1939). Therefore, despite their choice of expressing an action potential as a negative excursion of V, they still *plot* their action potentials with the deflection going upward.

Throughout the annotations, the figures have been redrawn according to modern conventions: membrane potentials are given as absolute voltage with respect to an external ground (v), assuming a resting potential of −60 mV (an approximation, given that resting potentials varied across axons); current traces are inverted; and current-voltage curves are flipped horizontally and vertically. In the notes, membrane potentials that are translated into modern conventions are denoted by '*abs.*' for absolute. Also, for consistency of terminology in the notes, the word 'depolarization' is used for any positive change in absolute membrane potential (even those that bring the membrane potential above 0) and 'hyperpolarization' for any negative change in absolute membrane potential. Formally, however, 'depolarization' refers only to a change toward a membrane potential of 0 mV and 'hyperpolarization' to an increase in the magnitude of membrane potential, regardless of sign.

Theory

Although the results described in this paper do not depend on any particular assumption about the electrical properties of the surface membrane, it may be helpful to begin by stating the theoretical assumption which determined the design and analysis of the experiments. This is that the membrane current may be divided into a capacity current which involves a change in ion density at the outer and inner surfaces of the membrane, and an ionic current which 13 depends on the movement of charged particles through the membrane. Equation 1 applies to such a system, provided that the behaviour of the membrane capacity is reasonably close to that of a perfect condenser:

$$I = C_M \frac{\partial V}{\partial t} + I_i, \tag{1}$$

where I is the total current density through the membrane, I_i is the ionic current density, C_M is the membrane capacity per unit area, and t is time. In most of our experiments $\partial V/\partial t = 0$, so that the ionic current can be obtained directly from the experimental records. This is the most obvious reason for using electronic feed-back to keep the membrane potential constant. Other 14 advantages will appear as the experimental results are described.

EXPERIMENTAL METHOD

The essential features of the electrode system are illustrated by Fig. 1. Two long silver wires, each $20\,\mu$. in diameter, were thrust down the axis of a giant axon for a distance of 20–30 mm. The greater part of these wires was insulated but the terminal portions were exposed in the manner shown in Fig. 1. The axon was surrounded by a 'guard ring' system which contained the external electrodes. Current was applied between the current wire (a) and an earth (e), while the potential difference across the membrane could be recorded from the voltage wire (b) and an external electrode (c). The advantage of using two wires inside the nerve is that polarization of the current wire does not affect the potential recorded by the voltage wire. The current wire was exposed for a length which corresponded to the total height of the guard-system, while the voltage wire was exposed only for the height of the central channel. The guard system ensured that the current crossing the membrane between the partitions A_2 and A_3 flowed down the channel C. This component of the current was determined by recording the potential difference between the external electrodes c and d.

Internal electrode assembly

In practice it would be difficult to introduce two silver wires into an axon without using some form of support. Another requirement is that the electrode must be compact, since previous experience showed that axons do not survive well unless the width of an internal electrode is less than $150\,\mu$. (Hodgkin & Huxley, 1945). After numerous trials the design shown in Fig. 4 was adopted. The first operation in making such an electrode was to push a length of the voltage wire through a $70\,\mu$. glass capillary and twist it round the capillary in a spiral which started at the tip and proceeded toward the shank of the capillary. The spiral was wound by rotating the shank of the capillary in a small chuck attached to a long screw. During this process the free end of the wire was pulled taut by a weight while the capillary was supported, against the pull of the wire, by a fine glass hook. A second hook controlled the angle at which the wire left the capillary. When sufficient wire had been wound it was attached to the capillary by application of shellac solution,

13. **'the membrane current may be divided into a capacity current . . . and an ionic current'** Here, HH&K state their crucial assumption that the capacitance of the membrane and the conductance of the membrane to ions together form a parallel *RC* circuit. This assumption, in turn, dictates the relationship between the basic quantities of charge, current, capacitance, resistance, and voltage as well as the flow of ions through each element of the circuit changes the membrane voltage.

Recall that a voltage, *V* (units of volts, or energy per unit charge), builds up when positive and negative charges, *Q* (units of coulombs), are separated; because of their mutual attraction, the separation sets up a potential for unlike charges to reunite. The cell membrane maintains the separation, or 'stores' the charge, and therefore acts as a capacitor, *C* (units of farads, equivalent to coulombs/volt); a larger area of membrane provides a proportionately larger capacitance. The greater the amount of separated charge over a given area of membrane, the higher the voltage. Thus, $V = Q/C$. In the context of excitable membranes, the charge takes the form of ions. HH&K will demonstrate that ions flow across the membrane through a yet-to-be-defined resistance *R* (units of ohms, Ω), the inverse of which is conductance, *G* (units of mhos, or in modern terminology, siemens). Current is the movement of charge, *I*, which can also be expressed as a change in charge over time, d*Q*/d*t* (units of coulombs/sec, equivalent to amperes). These principles are reviewed in Appendix 2.1.

HH&K assume a circuit diagram of the membrane (see Historical Background) consisting of a parallel arrangement of resistance and capacitance, meaning that current can flow across the membrane in two ways. First, Ohm's law, $V = IR$, dictates that at any voltage, an ionic current, I_i, of magnitude V_m/R_m flows through the membrane resistance. Second, the nameless law, $Q = CV$, and its derivative d*Q*/d*t* = *C* d*V*/d*t* (assuming constant capacitance), dictates that current flows across the membrane capacitance as the transmembrane voltage *changes* (when d*V*/d*t* is nonzero). This flow of ions, which HH&K call the capacity current, I_c, is also termed a *displacement* current (see Appendix 2.1).

I_i, as will be demonstrated, is a function of voltage and time. I_c is proportional to $\partial V/\partial t$ (the derivative of voltage with respect to time). The partial derivative symbols, ∂, indicate that voltage is normally not only a function of time but also of distance along the axon. However, a key feature of the experimental configuration (described in *Figure 1*; see note 15) is that the segment of the axon under study is made to be isopotential (of *spatially* uniform voltage). When adjacent portions of the membrane have the same potential, by definition, the derivative of voltage as a function of distance becomes zero, and $\partial V/\partial t$ reverts to a simple derivative with respect to time, d*V*/d*t*. With the voltage at every point on the axon being identical, no current flows longitudinally along the axon (see notes 28 and 44). In this circuit model, the total membrane current, I_{total}, becomes the sum of the two currents, I_c and I_i, as stated in Equation 1.

14. **'the most obvious reason . . . to keep the membrane potential constant.'** Since $I_c = C(dV/dt)$, when the voltage is constant, d*V*/d*t* = 0 and no capacity current flows. Consequently, all the current is ionic current because $I_{total} = I_c + I_i$. When voltage steps are used, capacity current only flows during the brief (< 0.1 msec) period when the voltage is changing and subsides to zero once the voltage becomes constant. This situation is ideal for measuring the ionic current in isolation, which is what HH&K set out to do.

cut close to the capillary and insulated with shellac in the appropriate regions (Fig. 4). The next operation was to wind on the current wire, starting from the shank and proceeding to the tip. Correct spacing of current and voltage wires was maintained by making small adjustments in the position of the second glass hook. When the current wire had been wound to the tip it was attached to the capillary, cut short and insulated as before. The whole operation was carried out under a binocular microscope. Shellac was applied as an alcoholic solution and was dried and hardened

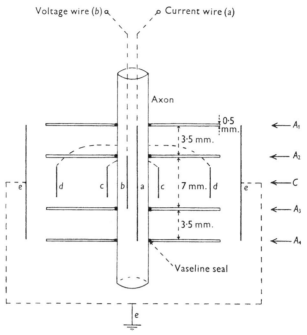

Fig. 1. Diagram illustrating arrangement of internal and external electrodes. A_1, A_2, A_3 and A_4 15 are Perspex partitions. a, b, c, d and e are electrodes. Insulated wires are shown by dotted lines. For sections through A and C, see Figs. 2 and 3.

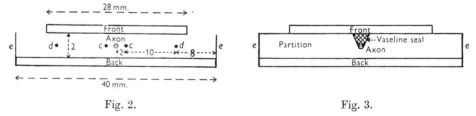

Fig. 2.   Fig. 3.

Fig. 2. Central channel of guard system. Section through C, Fig. 1. c and d are silver wires, e is a silver sheet. All dimensions are in mm.

Fig. 3. Partition of guard system. Section through A_1, A_2, A_3 or A_4, Fig. 1.

by baking for several hours under a lamp. Insulation between wires and across the shellac was tested so as to ensure that the film of shellac was complete and that the wires did not touch at any point. The exposed portion of the wires was then coated electrolytically with chloride. The electrode was first made an anode in order to deposit chloride and was then made a cathode in order to reduce some of the chloride and obtain a large surface of silver. This process was repeated a number of times ending with an application of current in the direction to deposit chloride. In 16 this way an electrode of low polarization resistance was obtained.

15. *Figure 1.* This figure, along with *Figures 2* and *3*, shows the 'guard system' into which the axon is placed. The guard system is a chamber with separate compartments and several electrodes. The purpose of the compartments is to isolate a short stretch of axonal membrane, (the central region), across which the voltage can be effectively controlled, and through which the current can be precisely measured. HH&K use five electrodes, *a, b, c, d,* and *e,* and can measure the voltage difference between any two. The solid lines are the exposed (uninsulated) portions of the electrodes; the dotted lines connect back to the amplifier (or between electrodes) but are insulated with shellac and therefore are electrically inert. Electrode *a* is the current-passing electrode; electrode *b* is the voltage-recording electrode. These are twisted around a 70 μm glass capillary so that they can both be in the axon without touching each other (see diagram in *Figure 4* and photos in *Figure 5*).

The goal is to 'clamp' the voltage across a section of membrane: stated formally, to maintain a region of membrane at the *command potential.* In Cole's earlier version of the voltage clamp (see Historical Background), a single electrode had been used both to record the membrane voltage and to inject current. While such a configuration can hold the *electrode* voltage at the given command voltage, the complication is that the *membrane* voltage will be nonuniform owing to 'edge' effects at the end of the electrode.

With a single electrode (at *left* in *Figure A*), the low (near zero) resistance of the electrode wire forces the surrounding axoplasm and the membrane to be nearly isopotential. When current is passed through the electrode, however, a *difference* in potential develops between the membrane surrounding the wire and the membrane just beyond the ends of the wire. That difference causes a current to flow longitudinally down the axon instead of flowing through the membrane near the ends of the wire (curved arrows). As a result, the membrane potential near and beyond the ends of the wire is no longer isopotential, and the voltage measured through the wire is a combination of the voltages in these different regions of the axon.

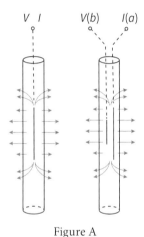

Figure A

The two-electrode configuration (at *right* in *Figure A*) addresses this problem. The shortened voltage-sensing electrode (*b*) detects only the voltage within the well-controlled (isopotential or 'space-clamped') midsection of the electrode pair, away from the ends of the longer current electrode (*a*). Even though the poorly clamped membrane at the ends of electrode *a* undergo indeterminate changes in voltage and membrane current, the amplifier does whatever is required to keep the *central*, isopotential portion of the membrane (around electrode *b*) at the command voltage.

Electrode *a* provides current not only to the central, isopotential region, but also to the poorly clamped regions of membrane at the extremes of the electrode. HH&K wish to measure only the membrane current through the well-clamped portion of the axon (see note 18). To do so, they use the partitions A_2 and A_3, and electrodes *c* and *d*. Because the partitions electrically isolate the well-clamped axonal segment around electrode *b*, current flowing out of this region necessarily flows between electrodes *c* and *d*. The voltage between these electrodes, $v_c - v_d$, is therefore proportional to the current flowing from the well-clamped region, i_c. By measuring the resistance of the seawater between electrodes, r_{cd}, HH&K can calculate the current as $i_c = (v_c - v_d)/r_{cd}$. As noted, r_{cd} was 74 Ω in seawater, but it changed and therefore had to be remeasured for each different external solution and temperature.

The two-electrode configuration overcomes additional problems inherent to the single-electrode technique. A single silver-silver chloride wire cannot apply the *milliamperes* of current required to clamp a squid axon without undergoing significant polarization, which is a change in the electrochemical junction potential (voltage) at the interface between the metal surface and the axoplasm.

(continued on page 27)

In order to test the performance of the electrode it was immersed in salt solution and the current wire polarized by application of an electric current. In theory this should have caused no change in the potential difference between the voltage wire and the solution in which it was immersed. In practice we observed a very small change in potential which will be called 'mutual polarization'. Leakage between wires was a possible cause of this effect, but other explanations 17 cannot be excluded.

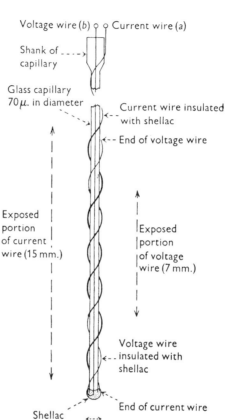

Fig. 4. Diagram of internal electrode (not to scale). The pitch of each spiral was 0·5 mm. The exposed portions of the wires are shown by heavy lines.

The general appearance of the electrode inside a giant axon is illustrated by Fig. 5. These photographs were obtained at an early stage of the investigation, and the axon was cleaned less carefully than in later experiments. The internal electrode differed from those finally employed in that both wires were wound from the shank of the electrode and that the pitch was somewhat greater.

Guard system

The general form of the guard system is shown in Figs. 1–3. It consisted of a flat box made out of Perspex which was divided into three compartments by two partitions A_2 and A_3 and closed with walls A_1 and A_4. The front of the box was removable and was made from a thin sheet of Perspex which could be sealed into position with vaseline. V-shaped notches were made in the two end walls and in the partitions. The partitions were greased and the notches filled with an oil-vaseline mixture in order to prevent leakage between compartments (Figs. 1 and 3). The guard ring assembly was mounted on a micromanipulator so that it could be manoeuvred into position after the electrode had been inserted. The outer electrode (*e*) was made from silver sheet while the

(continued from page 25)

The nonzero resistance of the electrode produces an additional voltage drop when current is applied. The resulting errors in the voltage measurement introduce an equivalent error in the control of the membrane potential. The two-electrode configuration circumvents these problems: no substantial current flows through the voltage sensing electrode, so it experiences negligible polarization (see Appendix 3.3).

16. *'to deposit chloride.'* Each electrode is a silver (Ag) wire that has been previously coated with silver chloride (AgCl) by applying a voltage to it (relative to another reference wire) while dipped in a saline solution containing chloride ions (Cl^-). When the Ag is made electrically positive, Cl^- reacts with it to precipitate AgCl onto the wire. Once in the axon, whenever the electrode is made negative relative to the ground electrode, e, Cl^- ions are liberated from the wire and into the solution. The resulting current flows through the solution, and negative ions are ultimately deposited on the membrane capacitor, making the inside of the axon more negative, or hyperpolarizing it. When the electrode is made positive, it attracts Cl^- and current flows in the opposite direction. In HH&K's terminology, the anode attracts anions and the cathode collects cations. Since the Cl^- is easily transferred on and off the wire, the electrode is said to have 'low polarization resistance.' The movement of Cl^- ions from the wire and into solution requires a small amount of energy, which accounts for the electrode junction potential. The magnitude of the junction potential is altered by any changes in the concentration or identity of ions in solution.

17. *'mutual polarization.'* The twisting of the two wires around the capillary was intended to separate them from each other and make them independent. HH&K tested whether this was the case and found that changing the voltage of one did indeed slightly affect the voltage of the other. (Although not stated here, these changes are much smaller than the electrode polarization that occurs in the single-electrode voltage clamp.) Nevertheless, HH&K simply report the error and proceed. Later, the effects of mutual polarization will be considered. Thus, while the recognition of this error shapes the conclusions, it does not impede the progress of experimentation. This approach, of proceeding with an understanding of the experiment's limitations, turns out to be characteristic throughout the papers (see note 6).

inner electrodes (c) and (d) were made from 0·5 mm. silver wire. Exposed portions of the electrodes were coated electrolytically with chloride and the wires connecting the electrodes with external terminals were insulated with shellac.

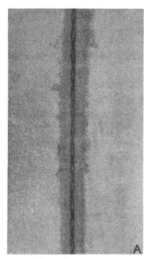 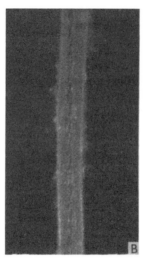

Fig. 5. Photomicrographs of giant axon and internal electrode. A, transmitted light; B, dark ground. The axon diameter was about 600 μ. The glass rod supporting the wires is not clearly seen.

Feed-back amplifier

A simplified diagram of the feed-back amplifier is shown in Fig. 6. It consisted of a differential d.c. amplifier with cathode follower input and output. The output of the amplifier was coupled to the input in such a way that negative feed-back was employed. This meant that any spontaneous change in membrane potential caused an output current to flow in a direction which restored the membrane potential to its original value. The level at which the potential was held constant was   18 determined by the bias voltage v_3 and the control voltage v_4. v_3 was set so that no current passed   19 through the nerve in the resting condition. This preliminary operation was carried out with the protective resistance R_f at its maximum value. This was important since an incorrect setting would   20 otherwise have caused a large current to flow through the membrane. R_f was gradually reduced to zero; at the same time v_3 was adjusted to keep the membrane current zero. In order to change the membrane potential a rectangular pulse $\pm v_4$ was fed into the second stage of the amplifier. A large current then flowed into the membrane and changed its potential abruptly to a new level determined by

$$v_1 - v_2 = \beta v_4, \tag{2}$$

where v_1 and v_2 are the two input voltages and v_4 is the control voltage; β is a constant determined   21 by resistance values and valve characteristics. Its value was of the order of 0·001. Any tendency to depart from Equation 2 was neutralized by a large output current which promptly restored the equilibrium condition defined by this relation.

In the majority of the experiments the slider of the potentiometer P was set to zero. Under   22 these conditions the potential difference between the internal and external recording electrodes ($v_b - v_c$) was directly proportional to ($v_1 - v_2$). If α is the voltage gain of the cathode followers (about 0·9), then   23

$$v_b - v_c = \frac{2}{\alpha}(v_1 - v_2) = \frac{2\beta v_4}{\alpha}. \tag{3}$$

The performance of the feed-back amplifier was tested in each experiment by recording the time course of the potential difference between the internal and external electrodes. This showed that the recorded potential followed the control voltage with a time lag of about 1 μsec. and an   24

18. *'feed-back amplifier. . . . meant that any spontaneous change in membrane potential caused an output current to flow in a direction which restored the membrane potential to its original value.'* The final phrase defines voltage clamp, which was referred to at the time simply as 'negative feedback' (see Historical Background). Without the voltage clamp, a 'spontaneous change in membrane potential' would arise from ionic currents flowing across the membrane. The feedback circuit of the amplifier holds the membrane potential at the desired value by withdrawing an identical current through the current electrode. The amount of current withdrawn provides a direct measure of ionic current. The response time of the amplifier (on the order of microseconds) is orders of magnitude faster than the intrinsic changes in membrane potential caused by changes in membrane conductance. Consequently, the voltage is always held to within a fraction of a millivolt of the command voltage. The circuit diagram of the amplifier, which is referred to in notes 19–24, is discussed in detail in note 25, opposite *Figure 6*.

19. *'the bias voltage v_3 and the control voltage v_4.'* The bias voltage, v_3, which is equivalent to the 'junction null' in modern amplifiers, is adjusted to compensate for the combined junction potentials of the voltage-sensing and ground electrodes, and subtracts out the axonal resting potential. With v_3 properly set, a control voltage, v_4, of zero will hold the transmembrane potential at its resting potential, operationally defined as the voltage at which the membrane current is zero. The control voltage, v_4, or command voltage (V_{cmd}) in modern terminology, is thus the independent variable in the experiments. Because the absolute resting potential varied from axon to axon and the voltage-clamp amplifier could not measure absolute voltage, HH&K defined the resting potential as 0 mV and measured voltage relative to this value.

20. *'protective resistance R_f at its maximum value.'* R_f is a variable resistor that sets the effective gain of the feedback amplifier, limiting the amount of current delivered when the two input voltages, v_1 and v_2, differ. R_f is initially made large (making the amplifier gain low) while the bias voltage, v_3, is being adjusted in the initial set up, to prevent the amplifier from inadvertently passing large currents that can destroy the axon.

21. *'where v_1 and v_2 are the two input voltages and v_4 is the control voltage' Equation 2* is related to the overall gain of the amplification stages, vacuum tubes T_4 & T_5 and T_6 & T_7 (see note 25). With $\beta \approx 0.001$, a command voltage of about 60 volts is required to displace the voltage $v_1 - v_2$ by approximately 60 mV. $v_1 - v_2$ is, in turn, directly proportional to the membrane potential, $v_b - v_c$, after series resistance compensation (see note 23).

22. *'the potentiometer P was set to zero.'* Setting P to zero means adjusting the connection from P to v_1 in *Figure 6* so that the arrow would point between the two 5K (kΩ) resistors. In this case, the voltage dividers (see Appendix 2.2) on the outputs of T_1 and T_2 are identical and consequently no series resistance compensation is applied (see note 26).

23. *'α is the voltage gain of the cathode follower (about 0.9)'* The cathode followers, or voltage followers, T_1 and T_2, are vacuum tubes with low gain (0.9). The purpose of these followers is therefore not to amplify the voltage but to present a high impedance to the electrodes, thereby drawing little current and permitting measurements without interfering with the system being measured (see Appendix 3.1). The 5K (kΩ) voltage dividers on the outputs of T_1 and T_2 (with P set to zero) further attenuate the signals from electrodes b and c by a factor of two, such that $v_1 - v_2 = 0.9(v_b - v_c)/2$. Substituting in βv_4 for $v_1 - v_2$ (*Equation 2*) yields *Equation 3*.

24. *'time lag of about 1 μsec.'* This time lag is very brief in comparison to the time course of the ionic currents being measured, and it is comparable to or better than most modern patch-clamp amplifiers.

accuracy of 1–2%. It is therefore unnecessary to discuss the numerous approximations which have to be made in order to derive Equation 2.

The voltage gain of the feed-back amplifier and cathode followers was about 400 in the steady state. At high frequencies the gain was about 1200, since the condenser C_1 increased the gain under transient conditions. The mutual conductance of the feed-back system $\left[\dfrac{\partial i}{\partial\left(v_b - v_c\right)}\right]$ was about 1 mho in the steady state and 3 mhos at high frequencies. The maximum current that the amplifier could deliver was about 5 mA.

The method described would be entirely satisfactory if there were no resistance, apart from that of the membrane, between internal and external electrodes. In practice there was a small series

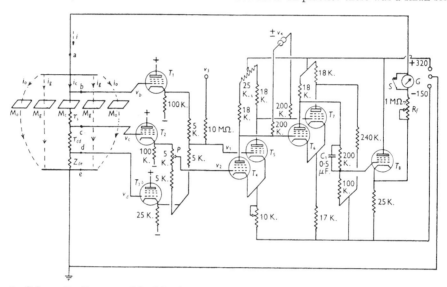

Fig. 6. Schematic diagram of feed-back amplifier. Screen resistances, grid stoppers and other   25 minor circuit elements have been omitted. T_1, T_2, T_3 and T_8 are cathode followers; T_4, T_5, T_6 and T_7 are d.c. amplifiers. All valves were 6 AK 5 except T_1 and T_2 which were 1223. G is a microammeter used in setting-up. S is a switch for short-circuiting G. M_c is the membrane in the central section of the guard system. M_g, membrane in guard channels. M_o, membrane outside guard system. i_c, i_g and i_o are currents through these elements. r_{cd}, fluid-resistance used to measure current (74 Ω. at 20° C.). r_s, resistance in series with membrane (about 52 Ω. at 20° C.). z_{de}, impedance of large earthed electrode and sea water between d and e. Potentials are given with respect to earth.

resistance, represented by r_s in Fig. 6 and discussed further on p. 444. This meant that the true   26 membrane potential was in error by the quantity $r_s i_c$. Thus

$$v_i - v_o = v_b - v_c - r_s i_c = 2\beta v_4/\alpha - r_s i_c, \qquad (4)$$

where $v_i - v_o$ is the potential difference between the inner and outer surfaces of the membrane, r_s is the resistance in series with the membrane and i_c is the current flowing through the central area of membrane.

In principle the error introduced by r_s can be abolished by setting the potentiometer P to an appropriate value. All three cathode followers (T_1, T_2, T_3) had the same gain so that v_1 and v_2 were determined by the following equations:

$$v_1 = \tfrac{1}{2}\alpha\left(v_b + v_d\right), \qquad (5)$$

$$v_2 = \tfrac{1}{2}\alpha\left[v_c + v_d + p\left(v_c - v_d\right)\right], \qquad (6)$$

and
$$v_1 - v_2 = \tfrac{1}{2}\alpha\left[v_b - v_c - p\left(v_c - v_d\right)\right], \qquad (7)$$

25. ***Figure 6.*** In broad outline, the circuit consists of a set of high-impedance cathode followers (see note 23 and Appendix 3.1) followed by two stages of amplification (gain). The central elements of the circuit are vacuum tubes, which are voltage amplifiers, illustrated by circles and labeled T_n. The vacuum tubes T_1, T_2, and T_3, each configured as a cathode follower, convert the high-impedance inputs from electrodes *b*, *c*, and *d* into low-impedance outputs. These signals are fed through a network of resistors (for series resistance compensation; see below), and the outputs, v_1 and v_2, are the inputs to the gain stages of the amplifier (see Appendix 3.2). The two pairs of vacuum tubes, T_4 & T_5, and T_6 & T_7, form a cascade of two differential amplifiers, which amplify the difference between v_1 and v_2 by a large factor. T_8 (also configured as a cathode follower), converts the differential output of the amplification stages into a low-impedance signal relative to ground, which is fed back into the axon through R_f and into the current-passing electrode (*a*). The remaining elements of the circuit perform the additional functions necessary for the experiments.

Series resistance compensation. Since T_1 reports v_b and T_2 reports v_c, the difference between these outputs, $v_2 - v_1$, gives a measure of the transmembrane potential. This value would be accurate if the membrane were the only resistance between internal electrode *b* and external electrode *c*. In reality, the residual connective tissue and the seawater surrounding the axon introduce additional resistance, called the series resistance, r_s (see note 72). When the amplifier supplies a current, *i*, to clamp the membrane potential, the voltage $v_b - v_c$ is shifted by ir_s, the voltage drop owing to the series resistance. HH&K reduce the series resistance to the extent possible by partial removal of connective tissue around the axon (***Figure 5***). Even so, the voltage error can be significant. To compensate for this error, the outputs of T_1 and T_2 are connected through separate pairs of resistors to the output of T_3 (which reports v_d). This part of the circuit subtracts a signal from $v_1 - v_2$ that is exactly proportional to the injected current reported by v_d (***Equations 5–10***). The constant of proportionality is set by the variable resistor *P* in the circuit.

Voltage command and current offset. Each stage of the amplifier includes one 'offset' input, which adds a fixed signal to its output. The bias voltage, v_3, which nulls the junction potentials and resting potential (see note 19), is applied to the input to T_5, one arm of the first differential amplifier; the control voltage, v_4, is applied across the inputs to T_6 & T_7, the second differential amplifier.

Current limiter. Any negative feedback system can generate damaging, large-amplitude oscillations when the gain of the amplifier is high and the time lag of its response is non-zero. Therefore, during the early phases of an experiment, the variable resistor, R_f, is set high to limit the output current and prevent oscillations.

26. **'small series resistance, represented by r_s'** The series resistance is mainly a result of the narrow clefts between the Schwann cells that surround the axon; these clefts provide the only electrical path between the axonal membrane and the external solution (Hodgkin & Frankenhaeuser 1956). The series resistance also includes the resistance from the residual connective tissue and the seawater between the membrane and electrode *c* (see note 25). HH&K measure this resistance to be 74 Ω at 20°C. (In modern patch-clamp recordings, the primary series resistance is usually at the pipette tip and is on the order of a few MΩ.) HH&K note that setting the variable resistor, *P*, to a value of 0.7 subtracts a signal exactly equal to ir_s, effectively nullifying the error introduced by series resistance. Series resistance compensation, however, is a form of *positive* feedback, whereby a signal in proportion to the output of the amplifier is fed back into its inputs with net positive polarity. If the compensation signal grows too large (i.e., if *P* is increased too much), the overall gain of the amplifier will no longer provide *negative* feedback, and instead will produce uncontrolled output currents that oscillate at high amplitude, which can damage the axon. For this reason, HH&K set *P* slightly below the value at which it would exactly counteract the effects of series resistance. In many cases, they do not apply compensation at all, but only after using the circuit to determine that the error can be safely ignored (see note 74).

where p is proportional to the setting of the potentiometer P and varied between extremes of 0 and 1 and v_d is the potential of electrode d.

From Ohm's law

$$v_c - v_d = r_{cd} i_c, \tag{8}$$

where r_{cd} is the resistance of the central channel between electrodes c and d.

From Equations 4, 7 and 8

$$v_1 - v_2 = \tfrac{1}{2}\alpha \left[v_i - v_o + i_c \left(r_s - p r_{cd} \right) \right]. \tag{9}$$

If $p = r_s/r_{cd}$

$$v_i - v_o = \frac{2}{\alpha} (v_1 - v_2) = \frac{2\beta v_4}{\alpha}. \tag{10}$$

The ratio r_s/r_{cd} was found to be about 0·7 and subsequent trials showed that a setting of $p = 0·6$ could be used with safety. This procedure, which will be called compensated feed-back, was used successfully in seven of the later experiments. It had to be employed with considerable caution since a system of this type is liable to oscillate. Another difficulty is that if p is inadvertently made greater than r_s/r_{cd} the overall feed-back becomes positive and there is a strong probability that the membrane will be destroyed by the very large currents which the amplifier is capable of producing.

Auxiliary equipment

In addition to the feed-back amplifier we employed the following additional units: (1) A d.c. amplifier and cathode-ray oscillograph for recording membrane current and potential. (2) A voltage calibrator, with a built-in standard cell, giving ± 110 mV. in steps of 1 mV. (3) A time calibrator consisting either of an electrically maintained 1 kcyc./sec. tuning fork, or a 4 kcyc./sec. fork with circuits to give pulses at 4, 2, 1 or 0·5 kcyc./sec. (4) Two units for producing rectangular pulses. These pulses were of variable amplitude (0–100 V.) and the circuits were arranged in such

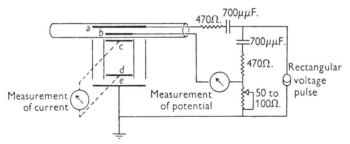

Fig. 7. Diagram of arrangement for recording response of membrane to short shock.

a way that the outputs of each generator were symmetrical with respect to earth. A single pulse generator was used in the early experiments, and its output was applied to the feed-back amplifier in the manner shown in Fig. 6. When required, the output of a second pulse generator was applied in parallel through a second pair of resistances. (5) An electrically operated refrigerator unit for cooling the preparation. All these items were of conventional design and need no detailed description.   27

Stimulation with brief currents

In the early stages of the work it was important to prove that the membrane was capable of giving an action potential of normal size. For this purpose we disconnected the feed-back 28 amplifier and employed the arrangement shown in Fig. 7. A rectangular voltage step v_4 was 29 applied to one internal electrode through a 700 $\mu\mu$F. condenser. The total area of membrane 30 exposed to current flow from the electrode was about 0·3 cm.^{2} (1·5 cm. $\times \pi \times 0·06$ cm.). It therefore had a capacity of about 0·3 μF. When v_4 was suddenly changed by 10 V. the membrane potential 31 was displaced by about 23 mV. (10 V. $\times 700/300{,}000$). With this arrangement the membrane current consisted of very brief currents at the beginning and end of the voltage step. The size of the current could be varied by altering the size of the step, while the membrane current in the central channel of the guard-system could be measured by recording the potential difference

27. **'All these items . . . need no detailed description.'** The d.c. amplifier amplifies the input signal and is used to measure absolute membrane potential only in a subset of experiments that do not require the voltage clamp. The cathode-ray oscillograph is an oscilloscope. The voltage calibrator produces voltage pulses of known amplitude, and the time calibrator uses a tuning fork to generate a signal of known frequency; these are used to calibrate the magnitude and duration of the measurements. In many figures, one can see the top of the sine wave from the tuning fork signal, which defines the time base. As necessary, the voltage values are scaled to report current (by dividing by the relevant resistance), and/or the image is projected onto graph paper for specific measurements to be made (see note 37). The rectangular voltage pulse generator, connected to v_4 in the clamp circuit, serves to step the voltage of the voltage clamp. The refrigerator cools the preparation, in order to delay the deterioration of the tissue and/or to slow the kinetics for precise measurements of fast phenomena.

28. **'capable of giving an action potential of normal size.'** HH&K prepare to test whether the axon can fire a *nonpropagating* action potential when it is placed in the guard system and threaded by the two-electrode assembly to keep the central stretch of membrane isopotential. They will also evaluate the extent to which that action potential resembles the normal propagating action potential (see note 39).

29. **Figure 7.** This figure includes a simplified illustration of the axon in the guard system (note the axon and the five electrodes). For these experiments, the feedback circuit within the amplifier is disconnected, such that the axon is not voltage clamped. Instead a short shock is applied (as a rectangular voltage pulse at v_4) so that the membrane potential can be transiently depolarized or hyperpolarized. Transmembrane voltage is measured from b to c, and membrane current from c to d.

30. **'700 µµF. condenser.'** Micromicrofarads ($10^{-6} \times 10^{-6}$ F) are equivalent to picofarads (10^{-12} F) or pF. Condenser is a synonym for capacitor.

31. **'0.3 µF.'** The voltage-clamped region of the axon has a capacitance of 300,000 pF, more than four orders of magnitude greater than the cell body of most mammalian neurons. It is indeed a giant axon, and the currents will be correspondingly giant.

between electrodes c and d. The potential difference between the voltage wire (b) and the external electrode (c) was equal to the sum of the membrane potential and the potential difference across the ohmic resistance in series with the membrane. The second component was eliminated by the bridge circuit illustrated in Fig. 7.

Experimental procedure

Giant axons with a diameter of 400–800 μ. were obtained from the hindmost stellar nerve of   32
Loligo forbesi and freed from all adherent tissue. Careful cleaning was important since the guard system did not operate satisfactorily if the axon was left with small nerve fibres attached to it. A further advantage in using cleaned axons was that the time required for equilibration in a test solution was greatly reduced by removing adherent tissue.   33

The axon was cannulated and mounted in the same type of cell as that described by Hodgkin & Huxley (1945) and Hodgkin & Katz (1949a). A conventional type of internal electrode, consisting of a long glass capillary, was thrust down the axon for a distance of 25–30 mm. This was then removed and the double wire electrode inserted in its place. Action potentials and resting potentials were recorded from the first electrode and the axon was rejected if these were not reasonably uniform over a distance of 20 mm. Another reason for starting with a conventional   34 type of electrode was that the double wire electrode, in spite of the rigidity of its glass support, could not be inserted without buckling unless the axon had first been drilled with the glass capillary.   35

When the wire electrode was in position the guard system was brought into place by means of a micromanipulator. This operation was observed through a binocular microscope and care was taken to ensure that the central channel coincided exactly with the exposed portion of the internal voltage wire. The front of the guard-ring box was gently pressed into position and finally sealed by firm pressure with a pair of forceps. Before applying the front, spots of a vaseline-oil mixture were placed in such a position that they completed the seal round the axon when the front was pressed home (Figs. 1 and 3).

After the axon was sealed into position cold sea water (3–11° C.) was run into the cell and this temperature was maintained by means of a cooling coil which dipped into the cell. Air was bubbled through the cell in order to stir the contents and obtain a uniform temperature.

Before proceeding to study the behaviour of the axon under conditions of constant voltage its response to a short shock was observed. The experiment was discontinued if the action potential recorded in this way was less than about 85 mV. If the axon passed this test it was connected to   36 the feed-back amplifier in the manner described previously.

Solutions were changed by running all the fluid from the cell and removing it from the guard-ring assembly with the aid of a curved capillary attached to a suction pump. A new solution was then run into the cell and was drawn into the guard rings by applying suction at appropriate places.

Calibration

The amplifier was calibrated at the end of each experiment, and all photographic records were analysed by projecting them on to a calibration grid. The readings obtained in this way were converted into current by dividing the potential difference between the two external electrodes c and d by the resistance between these electrodes (r_{cd}). This resistance was determined by blocking   37 up the outer compartments of the guard-ring assembly and filling the central channel with sea water or with one of the standard test fluids. A silver wire was coated with silver chloride and inserted into the position normally occupied by the axon (Fig. 1). A known current was applied between the central wire and the outer electrode (e). The resistance between the two external electrodes c and d could then be obtained by measuring the change in potential difference resulting from a given application of current. It was found that the resistance between these electrodes was 74 Ω. when the central chamber was filled with sea water at 20° C. This value was close to that calculated from the dimensions of the system.

Membrane currents were converted to current densities by dividing them by the area of mem-   38 brane exposed to current flow in the central compartment. The area was calculated from the measured axon diameter and the distance between the partitions A_2 and A_3 (Fig. 1).

32. *'400–800 μ.'* The giant axon has a diameter of about one-half millimeter, approximately 500 times that of a mammalian central neuron.

33. *'time required for equilibration . . . was greatly reduced by removing adherent tissue.'* HH&K recognized the importance of ensuring that the extracellular solution had access to the membrane without the diffusion barrier presented by adherent tissue. They therefore dissected away this tissue to the extent possible, as shown in *Figure 5*. One of the principal objections to the 'sodium hypothesis'—that an influx of sodium was responsible for the upstroke of the action potential—stemmed from reports that other excitable cells could continue to generate electrical signals after removal of sodium from the external solution. It later became clear that the low-sodium solutions had effectively failed to reach the axon membrane owing to diffusion barriers from surrounding connective tissue (see Historical Background).

34. *'rejected if these were not reasonably uniform over a distance of 20 mm.'* HH&K report their exclusion criteria.

35. *'had first been drilled with the glass capillary.'* The axoplasm is gel-like. The first capillary makes an opening for the insertion of the delicate double-wire electrode assembly.

36. *'less than about 85 mV.'* Again, exclusion criteria are reported. Note that the magnitude is given as the change in voltage relative to rest. Assuming a resting potential of −60 mV, 85 mV corresponds to an overshoot to an absolute potential of about +25 mV.

37. *'projecting them onto a calibrating grid . . . converted into current by . . . dividing by the resistance between these electrodes (r_{cd}).'* Oscilloscope traces were recorded by photographing the oscilloscope face. The camera shutter was opened just before the start and closed just after the end of a trace. Once the film was developed, each frame was projected onto graph paper. The amplitude and time course of the voltage were then measured point by point, by hand. Current was calculated as the voltage measured between electrodes c and d divided by the resistance of the solution between the two electrodes.

38. *'current densities'* Dividing the measured current by the area of membrane serves to normalize the data across different size axons. The current magnitude per unit area is also more likely to be a biological variable, independent of the technical variables such as the length of axon that is clamped. All currents are therefore quantified in units of mA/cm^2.

RESULTS

Stimulation with brief currents

Before investigating the effect of a constant voltage it was important to establish that the membranes studied were capable of giving normal action potentials. This was done by applying a brief shock to one internal electrode and recording changes in membrane potential with the other. Details of the method are given on p. 431; typical results are shown in Fig. 8 (23° C.) and Fig. 9 (6° C.).

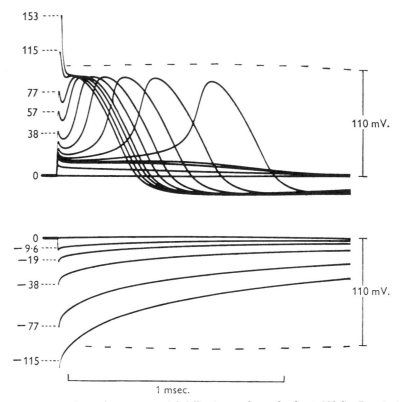

Fig. 8. Time course of membrane potential following a short shock at 23° C. Depolarizations shown upwards. Axon 18. The numbers attached to the curves give the strength of shock in mμcoulomb/cm.². Shock strengths for unlabelled curves are 29, 23, 19·2, 17·3, 16·7, 15·3, 9·6.

The shock used to displace the membrane potential was calibrated by recording the membrane current in the central channel (Fig. 7). This test showed that the current pulse consisted of a brief surge which was 95% complete in about 8μsec. and reached a peak amplitude of about 50 mA./cm.² at the highest strengths. The total quantity of current passing through the central channel was evaluated by integrating the current record and was used to define the strength of the shock. The numbers attached to the records in Figs. 8 and 9 give the charge applied per unit area in mμcoulomb/cm.². It

39. ***Figure 8.*** This figure shows the responses to brief shocks that serve to depolarize or hyperpolarize the axon. This experiment is performed by delivering a short pulse to the voltage command input (v_4) of the amplifier with the feedback disconnected, so that the membrane potential is displaced from rest but is allowed to vary freely afterward. The primary observation of this experiment is that a stretch of membrane can produce an action potential with the normal characteristics of a threshold, overshoot, and afterhyperpolarization, even when it is held isopotential (space-clamped). Therefore, these characteristics do not depend on propagation *per se*, but must be intrinsic to the membrane. The applied current density is converted to charge by integrating over time and the values written on the figure are given as mμcoulomb/cm^2 (millimicrocoulomb/cm^2, equivalent to nanocoulomb/cm^2). Negative charge applied through the internal current-passing electrode brings the membrane potential to a more hyperpolarized level, after which the voltage decays passively back to rest. Positive charge depolarizes the membrane: small positive shocks elicit subthreshold responses (see notes 42 and 47; see also Historical Background), whereas large ones evoke action potentials. The larger the depolarization, the shorter the latency to the action potential. Note the time base: squid action potentials are extremely brief, only about a third of a millisecond at 23°C.

will be seen that the initial displacement of membrane potential was proportional to the charge applied and that it corresponded to a membrane capacity of about $0.9 \mu F./cm.^2$. Values obtained by this method are given in Table 1.   40

Although the initial charging process was linear, the subsequent behaviour of the membrane potential varied with the strength of shock in a characteristic

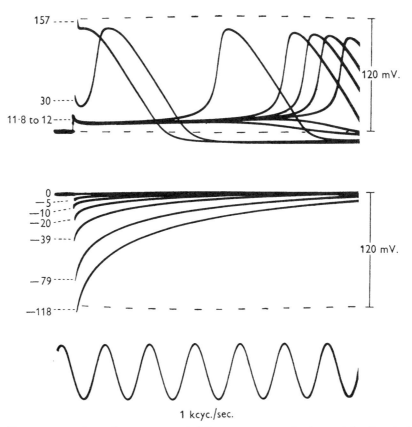

Fig. 9. Time course of membrane potential following a short shock at 6° C. Depolarization shown upwards. Axon 17. The numbers attached to the curves give the strength of shock in $m\mu coulomb/cm.^2$. The initial displacement in the case of the uppermost curve cannot be seen; its value was about 200 mV.

manner. All the anodal records had the same general shape, but depolarizations of more than a few millivolts gave non-linear responses. If the depolariza-   41 tion was more than 15 mV. (Fig. 8) or 12 mV. (Fig. 9) the response became regenerative and produced an action potential of about 100 mV. If it was less than 12 or 15 mV. it was followed by a subthreshold response similar to that seen in most excitable tissues. If the potential was displaced to the threshold level it might remain in a state of unstable equilibrium for considerable periods   42 of time. This is illustrated by Fig. 9 which shows the effect of a small variation of shock strength in the region of threshold.

40. *'initial displacement of membrane potential was proportional to the charge applied and . . . it corresponded to a membrane capacity of about 0.9 µF/cm².'* This statement says that, within the brief time scale of the shocks, before the ionic currents begin to respond to the change in membrane potential, voltage is proportional to charge, or $V \propto Q$. The squid axon therefore obeys a law that governs a simple physical system, namely a capacitor. More exactly, because $V = Q/C$, the proportionality constant must be $1/C$. Thus, HH&K deduce the 'biological constant' of specific membrane capacitance (see Historical Background). Measurements of 'membrane capacity' are given in **Table 1** (see also note 67).

41. *'anodal records . . . same general shape . . . depolarizations . . . non-linear responses.'* 'Anodal records' correspond to hyperpolarizations. The similarity of their shape is referred to at different points as *passive* and *linear*: the voltage change is proportional to the applied charge. These passive responses contrast with the *active, non-linear* responses to depolarizations. The ability to make such an active voltage response (an action potential) is the definition of an excitable tissue.

42. *'unstable equilibrium'* After small depolarizations (11.8 to 12 in **Figure 9**), instead of decaying exponentially back to rest, the voltage hovers at a fixed value for several milliseconds before falling back to rest (hence, the apparent 'equilibrium' is 'unstable'). Later experiments, culminating in the simulations of Paper 5, will reveal that this unstable equilibrium occurs near threshold where the opposing effects of inward sodium current and outward potassium current are nearly equal; in these examples, the sodium current eventually inactivates and the potassium current brings the voltage back to rest (see notes 39 and 47).

Records such as those in Fig. 9 may be used to estimate the relation between membrane potential and ionic current. The total membrane current density (I) is negligible at times greater than $200\,\mu$sec. after application of the short

TABLE 1. Membrane capacities

Axon no.	Diameter (μ.)	Temper- ature (° C.)	Change in potential (mV.)	Membrane capacity (μF./cm.²)		R_s (Ω.cm.²)	r_s (Ω.)	r_s/r_{cd}
				Anodic	Cathodic			
			A. Voltage clamp					
13	520	9	+36 − 36	0·76	0·83	8·2	72	0·77
			+56 − 56	0·83	0·90			
			+98 − 98	0·83	0·96			
14	430	9	+36 − 34	0·81	0·83	5·8	61	0·65
17	588	7	+31 − 32	0·72	0·76	8·3	64	0·65
18	605	21	+30 − 31	0·92	0·91	5·5	41	0·57
19	515	8	+43 − 45	0·93	0·90	7·8	69	0·73
20	545	6	+42 − 43	0·88	0·86	9·1	76	0·77
21	533	9	+42 − 44	0·98	1·01	9·1	78	0·84
22	542	23	+40 − 41	1·01	1·03	4·0	34	0·50
25	603	8	+39 − 41	0·88	0·86	7·0	53	0·57
25*	603	7	+39 − 41	0·84*	0·82*	8·8*	66*	0·55*
26	675	20	+40 − 42	0·97	0·93	7·7	52	0·70
Average	—	—	—	0·88	0·90	7·3	60	0·68
				0·89				
			B. Short shock					
13	520	9	+58 − 50	1·07	1·11	—	—	—
17†	588	6	—	0·79†	0·74†	—	—	—
18†	605	23	—	0·85†	0·88†	—	—	—
Average	—	—	—	0·90	0·91	—	—	—
				0·91				
			C. Constant current					
29	540	21	—	—	1·49	6·4	42	0·57
41	585	4	—	—	0·78	11·9	92	0·88
Average	—	—	—	—	1·13	9·2	67	0·73
				1·13				
Average by all methods	—	—	—	0·91		7·6	61	0·68

* In this experiment choline was substituted for sodium in the external solution. The values obtained are excluded from the averages.

† In these experiments the shock strength was not measured directly but was obtained from the calibration for axon 13. The values for C_M are means obtained from a wide range of shock strengths.

shock. This means that the ionic current density (I_i) must be equal to the product of the membrane capacity per unit area (C_M) and the rate of depolarization. Thus if $I = 0$, Equation 1 becomes

$$I_i = -C_M \frac{\partial V}{\partial t}. \tag{11}$$

43

44

Fig. 10 illustrates the relation between membrane potential and ionic current at a fixed time ($290\,\mu$sec.) after application of the stimulus. It shows that

43. *'after application of the short shock.'* By 200 µs after the onset of the brief depolarizing (or hyper-polarizing) pulse, the applied current has decayed to zero. Therefore, any further changes in voltage must be due to rearrangements of charge on the membrane capacitor resulting from the flow of currents that are intrinsic to the membrane.

44. *'$I_i = -C_M(\partial V/\partial t)$'* With no externally applied current and no longitudinal current flow along the axon, all charges that flow through the isopotential portion of the membrane as ionic current, I_i, are deposited directly onto the membrane capacitor. For all these charges deposited on the membrane, countercharges are displaced on the opposite side, forming the capacity current, I_c. This redistribution of charge changes the transmembrane potential according to $\Delta V = \Delta Q/C$. Thus, the capacity current is equal and opposite to the ionic current (see note 13). HH&K can therefore use a measure of capacity current, $-C_M(dV/dt)$, as a proxy for ionic current, even without imposing a voltage clamp onto the membrane. Accordingly, from **Figures 8** and **9**, which illustrate the transmembrane voltage with a variety of shock strengths, they calculate dV/dt at a fixed time, $t = 290$ µs, well after the applied current has fallen to zero. Any currents flowing are therefore intrinsic to the axon. Scaled by C_M, the slope gives I_c, which is equal and opposite to *the total ionic current flowing at any voltage*. In this way, HH&K extract an estimate of current as a function of voltage (called a 'current-voltage relation' or '*I-V* curve') from the non-voltage-clamped action potential, illustrated in **Figure 10**.

ionic current and membrane potential are related by a continuous curve which   45
crosses the zero current axis at $V = 0$, $V = -12$ mV. and $V = -110$ mV. Ionic
current is inward over the regions -110 mV. $< V < -12$ mV. and $V > 0$, and
is outward for $V < -110$ mV. and -12 mV. $< V < 0$. $\partial I_i / \partial V$ is negative for
-76 mV. $< V < -6$ mV. and is positive elsewhere.

A curve of this type can be used to describe most of the initial effects seen
in Figs. 8 and 9. When the membrane potential is increased by anodal shocks
the ionic current associated with the change in potential is in the inward
direction. This means that the original membrane potential must be restored
by an inward transfer of positive charge through the membrane. If the

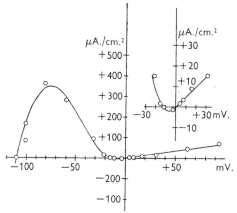

Fig. 10. Relation between ionic current density (I_i) and displacement of membrane potential
(V). Abscissa: displacement of membrane potential from its resting value (anodal displacement shown positive). Ordinate: ionic current density obtained from $-C_M \dfrac{dV}{dt}$ (inward
current shown positive). Inset: curve in region of origin drawn with tenfold increase in
vertical scale. Axon 17; $C_M = 0.74 \,\mu\text{F./cm.}^2$; temperature 6·3° C. Measurements made
0·29 msec. after application of shock.

membrane potential is depolarized by less than 12 mV., ionic current is outward and again restores the resting condition by repolarizing the membrane
capacity. At $V = -12$ mV., I_i is zero so that the membrane potential can   46
remain in a state of unstable equilibrium. Between $V = -110$ mV. and   47
$V = -12$ mV., I_i is inward so that the membrane continues to depolarize
until it reaches $V = -110$ mV. If the initial depolarization is greater than
110 mV. I_i is outward which means that it will repolarize the membrane
towards $V = -110$ mV. These effects are clearly seen in Figs. 8 and 9.   48

Membrane current under conditions of controlled potential

General description

The behaviour of the membrane under a 'voltage clamp' is illustrated by   49
the pair of records in Fig. 11. These show the membrane current which flowed
as a result of a sudden displacement of the potential from its resting value to

45. **'ionic current and membrane potential are related by a continuous curve'** The graph in **Figure 10** resembles, but is not identical to, the relation that will later become recognizable as an *I-V* curve for isolated voltage-gated sodium current (see note 46). HH&K draw attention to the three points at which the current is zero. These correspond to the resting potential, around −60 mV *abs.*; threshold, around −48 mV *abs.*; and the peak of the action potential, around +50 mV *abs.* At all these points, the voltage is not changing (dV/dt = 0), meaning that the inward and outward currents are equal in amplitude. The word 'continuous' is especially meaningful here. The action potential is an all-or-none phenomenon—one that is nonlinear and discontinuous, and hence it appears to be intractable to simple mathematical relations. But the continuity of this curve gives them an initial piece of evidence that the underlying mechanisms may be mathematically tractable.

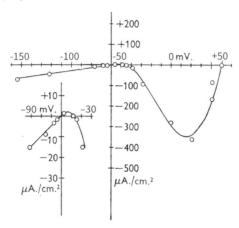

46. **'restores the resting condition by repolarizing the membrane capacity.'** In the voltage range from just below threshold to well below rest (more hyperpolarized than about −58 mV *abs.*, magnified in the inset to **Figure 10**), the relationship of current to voltage is linear with a positive slope. Hence, after the membrane is briefly hyperpolarized from rest, current flows inward to depolarize the membrane back toward the resting potential; conversely, after the resting membrane is slightly depolarized, current flows outward to repolarize the membrane to the resting potential. Such responses are expected from a passive, ohmic system with a fixed resistance ($I = V/R$). In the analysis that follows, these currents will ultimately be referred to as 'leak currents,' which dominate at potentials too hyperpolarized to activate voltage-gated conductances. The leak accounts for the region of the graph that differs from an *I-V* curve for an isolated voltage-gated sodium current, which appears in **Figure 13** (see note 60).

47. **'unstable equilibrium.'** At voltages just hyperpolarized to −12 mV (−48 mV *abs.*), ionic current flows outward and further hyperpolarizes the membrane, and at voltages just depolarized to it, current flows inward and further depolarizes the membrane. Hence, small deviations from −12 mV in either direction make current flow to bring the voltage *away* from this potential. This equilibrium point, which corresponds to the threshold of the action potential, is therefore unstable (see notes 39 and 42).

48. **'towards V = −110 mV.'** In the range above threshold, the graph of **Figure 10** resembles an *I-V* curve for voltage-gated sodium current. HH&K note that above threshold, the ionic current is no longer ohmic (linear). The current flows *inward* with depolarization, which does not 'restore the resting condition' but depolarizes it further, evidence of an 'active' process. This process apparently pushes the membrane potential toward a different equilibrium, the one at the peak of the action potential, about −110 mV (+50 mV *abs.*). At voltages more hyperpolarized than the peak value, current is inward; at more depolarized voltages, current is outward. This peak voltage would be a stable equilibrium if the magnitudes of the underlying currents did not change rapidly with time.

49. **'behaviour of the membrane under a "voltage clamp" . . . is illustrated.'** Here, HH&K illustrate the first voltage-clamped currents recorded with high-quality space clamp (see Historical Background). Note that 'voltage clamp' is in quotation marks as the term is sufficiently novel that it requires introduction!

a new level at which it was held constant by electronic feed-back. In the upper record the membrane potential was increased by 65 mV.; in the lower record it was decreased by the same amount. The amplification was the same in both cases. 50

The first event in both records is a slight gap, caused by the surge of 'capacity current' which flowed when the membrane potential was altered suddenly. The surge was too rapid to be visible on these records, but was examined in other experiments in which low gain and high time base speed were employed (see p. 442). The ionic current during the period of constant potential was small when the membrane potential was displaced in the anodal direction, and is barely visible with the amplification used in Fig. 11. The top record in Fig. 12 gives the same current at higher amplification and shows 51

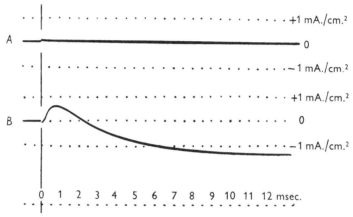

Fig. 11. Records of membrane current under a voltage clamp. At zero time the membrane potential was increased by 65 mV. (record *A*) or decreased by 65 mV. (record *B*); this level was then maintained constant throughout the record. Inward current is shown as an upward deflexion. Axon 41; diameter 585 μ. Temperature 3·8° C. Compensated feed-back.

that an increase of 65 mV. in the membrane potential was associated with an inward ionic current of about 30 μA./cm.² which did not vary markedly with time. The sequence of events was entirely different when the membrane potential was reduced by 65 mV. (Fig. 11 *B*). In this case the current changed sign during the course of the record and reached maximum amplitudes of +600 and −1300 μA./cm.². The initial phase of ionic current was inward and was therefore in the opposite direction to that expected in a stable system. If it had not been drawn off by the feed-back amplifier it would have continued to depolarize the membrane at a rate given by Equation 11. In this experiment C_M was 0·8 μF./cm.² and I_i had a maximum value of 600 μA./cm.². The rate of depolarization in the absence of feed-back would therefore have been 750 V./sec., which is of the same general order as the maximum rate of rise of the spike (Hodgkin & Katz, 1949 *a*, *b*). The phase of inward current was not maintained but changed fairly rapidly into a prolonged period of outward 52 53 54 55

50. **'increased by 65 mV . . . decreased by the same amount.'** Recall that in HH&K's terminology 'increased' means 'hyperpolarized' (enlarges the magnitude of the transmembrane voltage) and 'decreased' means 'depolarized' (reduces the magnitude of the transmembrane voltage).

51. **'"capacity current" which flowed when the membrane potential was altered suddenly.'** 'Capacity current' is also in quotation marks because it is shown here for the first time. This current is the physical illustration of $I_c = C_M(dV/dt)$: I_c is nonzero when the membrane potential changes but falls to zero when the potential is held constant.

52. **'did not vary markedly with time.'** Hyperpolarizations relative to the resting potential evoked a small, steady current, which will later be termed a 'leak' current. It is inward below rest, and linear (ohmic) with voltage, consistent with the data graphed in **Figure 10**. The leak current is described in reference to **Figure 11A** but is more visible in **Figure 12** (top 4 traces of their records).

53. **'The sequence of events was entirely different . . . to that expected in a stable system.'** These three sentences, which describe **Figure 11B**, revolutionized neurophysiology. In modern terms, depolarization evokes a time-dependent current that flows first inward and then outward. Such a pattern of ionic flux predicts that depolarization of a non-voltage-clamped membrane would induce an inward current that would generate a *further* depolarization, suggestive of a nonstable, positive feedback mechanism. Afterward, the outward phase of current would elicit a repolarization of the membrane. Thus, the measured currents can qualitatively account for the action potential.

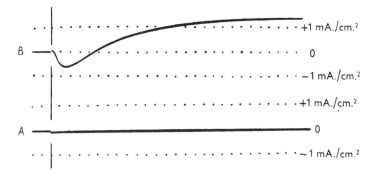

54. **'If it had not been drawn off by the feedback amplifier'** This statement refers to the action of the voltage clamp, which counteracts the ionic current flowing through the membrane. In HH&K's conventions, when positive ions flow into the axon, the amplifier 'draws off' an *identical* positive current, thereby preventing any change in membrane potential. It has become common, however, to describe the action of the amplifier as injecting a current that is *'equal and opposite'* to the ionic current: an inward flux of positive ions into the axon is balanced by an equivalent injection of negative ions (Cl⁻) into the axon. This equivalent depiction is concordant with modern conventions of illustrating an inward positive current as a downward deflection. This 'negative' current can be viewed as the current applied by the amplifier to neutralize the inward flux of positive ions. HH&K point out that if the clamp had not effectively held the potential constant, the membrane would have depolarized further. Such depolarization can indeed occur when the quality of the voltage clamp is inadequate, a phenomenon now described by the term 'voltage escape.'

55. **'same general order as the maximum rate of rise of the spike.'** HH&K use the measured inward ionic current to calculate the predicted rate of voltage change, from $I_i = -I_c = -C_M(dV/dt)$. The match between the calculated value and measured maximum rate of rise of the spike suggests that the current that they measured under voltage clamp is indeed *sufficient* to account for the change in membrane potential during the action potential (see Historical Background).

current. In the absence of feed-back this current would have repolarized the membrane at a rate substantially greater than that observed during the falling phase of the spike. The outward current appeared to be maintained for an indefinite period if the membrane was not depolarized by more than 30 mV. With greater depolarization it declined slowly as a result of a polarization effect discussed on p. 445.

56

57

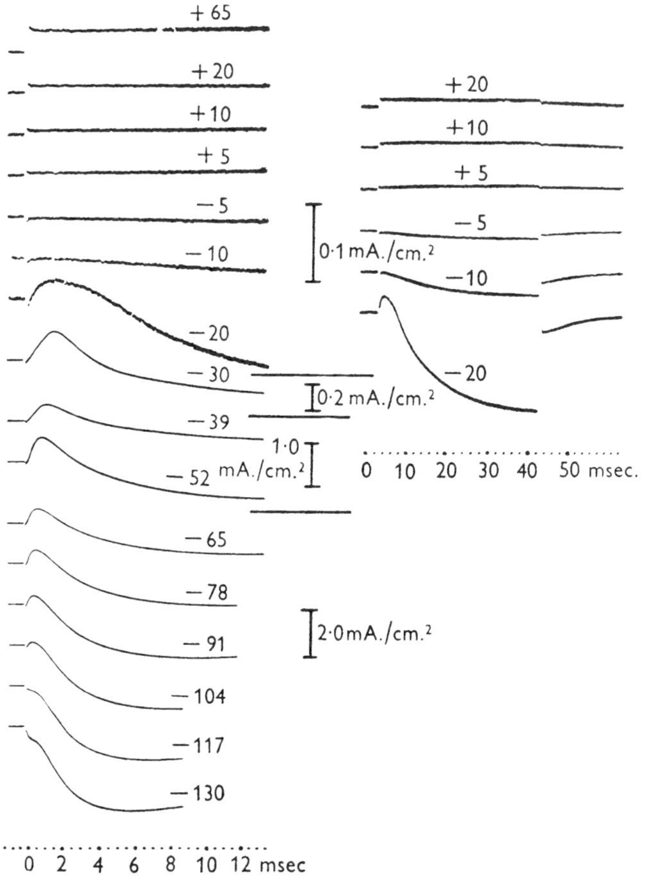

Fig. 12. Records of membrane current under a voltage clamp. The displacement of membrane potential (V) is given in millivolts by the number attached to each record. Inward current is shown as an upward deflexion. Six records at a lower time base speed are given in the right-hand column. Experimental details as in Fig. 11.

The features illustrated in Fig. 11 B were found over a wide range of voltages as may be seen from the complete family of curves in Fig. 12. An initial phase of inward current was conspicuous with depolarizations of 20–100 mV., while the delayed rise in outward current was present in all cathodal records. A convenient way of examining these curves is to plot ionic current density against membrane potential. This has been done in Fig. 13, in which the abscissa gives the displacement of membrane potential and the ordinate gives

58

56. **'would have repolarized . . . rate substantially greater than that observed during the falling phase of the spike.'** The outward current measured under voltage clamp may underlie repolarization, although HH&K note that its magnitude after 12 msec at +5 mV *abs.* is larger than necessary to account for the downstroke of the action potential. Consistent with this observation, the data in Paper 5 show that this outward current does not reach its full magnitude during the brief depolarization of the action potential.

57. **'polarization effect'** HH&K were concerned about electrode polarization because they were aware that such polarization could distort the current record (see note 75; see also Historical Background).

58. **'inward current was conspicuous with depolarizations of 20–100 mV . . . delayed rise in outward current . . . all cathodal records.'** The inward current has a fixed range, as shown in the complete family of traces of **Figure 12**; with depolarizations of 117 mV or more (≥+57 mV *abs.*) it is no longer inward, much like the current calculated from $C_M(dV/dt)$ in **Figure 10**. The late outward current, however, is outward for all depolarizations. Note the different scaling on different sets of illustrated currents.

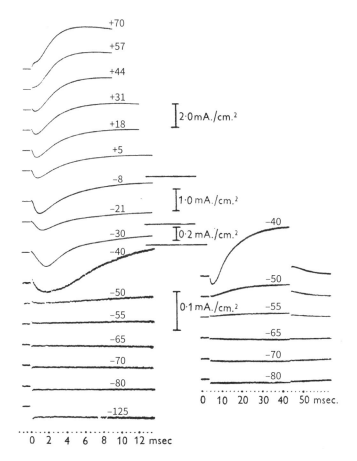

the ionic current density at a short time (curve A) and in the 'steady state' (curve B). It will be seen that there is a continuous relation over the whole range, but that small changes in membrane potential are associated with large changes in current. At short times the relation between ionic current density   59

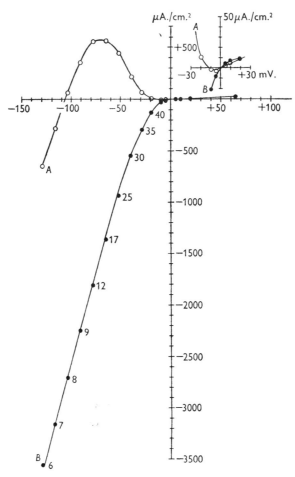

Fig. 13. Relation between membrane current density and membrane potential. Abscissa: displacement of membrane potential from its resting value in mV. Ordinate: membrane current density at 0·63 msec. after beginning of voltage step (curve A) and in 'steady state' (curve B). The numbers attached to curve B indicate the times in msec. at which the measurements were made. Inset: curves in region of origin drawn with a tenfold increase in the vertical scale. Inward current density is taken as positive and the membrane potential is given in the sense external potential minus the internal potential. Measurements were made from the records reproduced in Fig. 12 (3·8° C.).

and membrane potential is qualitatively similar to that obtained indirectly in Fig. 10. Ionic current is inward over the region $-106 \text{ mV.} < V < -12 \text{ mV.}$ and   60 for $V > 0$; it is outward for $V < -106 \text{ mV.}$ and for $-12 \text{ mV.} < V < 0$. $\partial I_i / \partial V$ is negative for $-70 \text{ mV.} < V < -7 \text{ mV.}$ and is positive elsewhere. More

59. *'a continuous relation over the whole range . . . large changes in current.' Figure 13* plots the first true current versus voltage relation (often called an *I-V* curve) for the early transient current and the late sustained current, which are later identified as sodium current and potassium current, respectively. Note that the data points for the early current are not peak values, but are the current amplitudes measured at a fixed time (isochronal) of 0.63 msec after depolarization. The late current is measured at the steady state, at the times indicated on the curve. HH&K again emphasize that the relationships are continuous rather than discontinuous, and thus lack a discrete threshold. This idea is fundamental as it provides evidence for the ultimate conclusion that each conductance has a graded dependence on voltage, which can be described by simple physical equations.

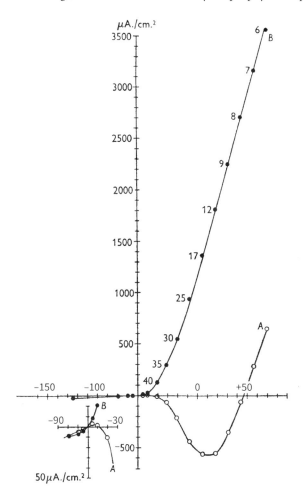

60. *'qualitatively similar to that obtained indirectly in Figure 10.'* The similarity demonstrates that voltage-clamping does not distort the properties of the axonal membrane and justifies exploring the possibility that the currents measured under voltage clamp *are* the underlying source of the voltage changes seen in the unclamped action potential. The specific values reveal that inward sodium current becomes appreciable near threshold, about 12 mV depolarized from the resting potential (−48 mV *abs.*).

quantitative comparisons are invalidated by the fact that the ionic current is a function of time as well as of membrane potential. At long times depolarization is invariably associated with an outward current and $\partial I_i/\partial V$ is always positive.

The electrical resistance of the membrane varied markedly with membrane potential. In Fig. 13, $\partial V/\partial I_i$ is about $2500\,\Omega.\,\text{cm.}^2$ for $V > 30\,\text{mV}$. For 61 $V = -110\,\text{mV.}$ it is $35\,\Omega.\,\text{cm.}^2$ (curve A) or $30\,\Omega.\,\text{cm.}^2$ (curve B). At $V = 0$, $\partial V/\partial I_i$ is $2300\,\Omega.\,\text{cm.}^2$ at short times and $650\,\Omega.\,\text{cm.}^2$ in the steady state. These results are comparable with those obtained by other methods (Cole & Curtis, 1939; Cole & Hodgkin, 1939).

62

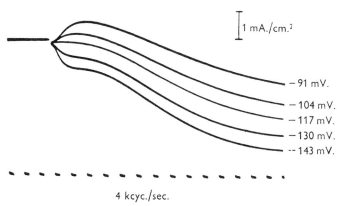

Fig. 14. Time course of membrane current during large depolarizations. Abscissa: time. Ordinate: inward current density. The numbers attached to the records give the displacement of membrane potential from its resting value. Axon 41; temperature 3·5° C. Compensated feed-back.

Fig. 14 illustrates the initial phase of ionic current at large depolarizations in greater detail. These records were obtained from the same axon as Fig. 12 but at an earlier stage of the experiment. They show that the initial 'hump' of ionic current changed sign at a potential of $-117\,\text{mV}$. At $-130\,\text{mV.}$ the 63 initial hump consists of outward current while it is plainly inward at $-104\,\text{mV}$. The curve at $-117\,\text{mV.}$ satisfies the condition that $\partial I_i/\partial t = 0$ at short times and has no sign of the initial hump seen in the other records. It will be shown later that this potential probably corresponds to the equilibrium potential for sodium and that it varies with the concentration of sodium in the external medium (Hodgkin & Huxley, 1952a).

The effect of temperature

The influence of temperature on the ionic currents under a voltage clamp is illustrated by the records in Fig. 15. These were obtained from a pair of axons from the same squid. The first axon isolated was examined at 6° C. and gave the series of records shown in the left-hand column. About 5 hr. later the second axon, which had been kept at 5° C. in order to retard deterioration, was

61. **'resistance . . . varied markedly with membrane potential.'** For passive devices that obey Ohm's law ($V = IR$, or $I = V/R$), a graph of I as a function of V will be a straight line through the origin with slope $1/R$ (as in $y = ax$). The curves in **Figure 13** include linear regions, but they also include non-linear, smoothly curving stretches. HH&K examine the slopes in different voltage ranges as well as at different times (curve A, measured early, versus curve B, measured late), and see that the slope, which theoretically should represent the inverse of the resistance is not constant, and instead varies over two orders of magnitude. Therefore, if the axonal membrane does obey Ohm's law, R must be a variable resistor, having values that vary with voltage and time. Note that the high resistance at the most hyperpolarized potentials, 2500 Ω cm^2, is what we would now call the 'input resistance,' R_{input}. The value that HH&K estimate from squid would correspond to 125 MΩ for a 20 pF spherical neuronal cell body, which is on the order of magnitude of that seen in many mammalian neurons.

62. **'Cole & Curtis 1939; Cole & Hodgkin 1939'** Cole and Curtis used a Wheatstone bridge (see Appendix 3.4) to measure the resistance of the axonal membrane. Their results had provided evidence for a 30- to 40-fold increase in membrane conductance during an action potential. (See Historical Background.)

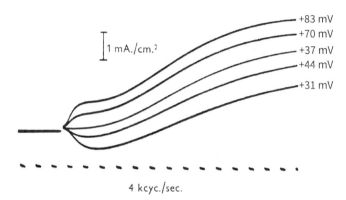

63. **'the initial "hump" of ionic current changed sign at . . . –117 mV.' Figure 14** documents the reversal of the early transient current at about +57 mV *abs.*: at more hyperpolarized potentials, the current is inward; at more depolarized potentials, the current is outward, hence the current has a 'reversal potential' of +57 mV. HH&K express this idea by saying that the *change in current with time* (the slope) is zero at this potential. They foreshadow that the experimentally measured reversal potential will be equivalent to the calculated equilibrium potential for sodium ions.

examined in a similar manner at 22° C. Its physiological condition is likely to have been less normal than that of the first but the difference is not thought to be large since the two axons gave propagated action potentials of amplitude 107 and 103 mV. respectively, both measured at 22° C. The resting potentials were 55 mV. in both cases. The results obtained with the second axon are given 64 in the right-hand column of Fig. 15. It will be seen that the general form and

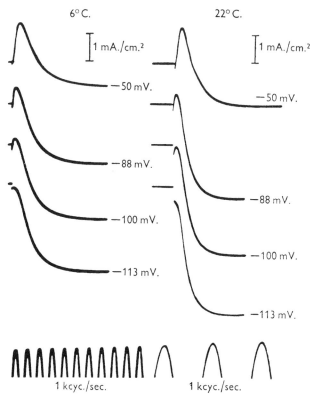

Fig. 15. Membrane currents at different temperatures. Axons 17 (6° C.) and 18 (22° C.), from the same squid. Inward current is shown as an upward deflexion. The numbers attached to each curve give the displacement of membrane potential. Uncompensated feed-back was employed.

amplitude of the two sets of records are similar but that the rate at which the ionic current changes with time was increased about sixfold by the rise in temperature of 16° C. It was found that the two families could be roughly superposed by assuming a Q_{10} of 3 for the rate at which ionic current changes 65 with time. Values between 2·7 and 3·5 were found by analysing a number of experimental records obtained under similar conditions, but with different axons at different temperatures. In the absence of more precise information we shall use a temperature coefficient of 3 when it is necessary to compare rates measured at different temperatures.

64. *'The resting potentials were 55 mV. in both cases.'* This is one of the few instances when the absolute transmembrane voltages (relative to ground) are given, although the sign is still opposite to modern conventions.

65. *'rate at which the ionic current changes with time increased about sixfold by the rise in temperature of 16° C. . . . a Q_{10} of 3'* The Q_{10}, or temperature coefficient, indicates the fold change in rate per 10 degrees Celsius. Electrochemical processes, such as diffusion, are not greatly accelerated by increases in temperature and have Q_{10}'s of 1–1.5. Biochemical processes that involve proteins, such as enzymatic reactions, have considerably higher temperature sensitivity, with Q_{10}'s of 2–4. The observation that the Q_{10} of the kinetics of the currents is near 3 gives HH&K a hint that the *temporal* changes in current (as distinct from the magnitude) may result from a biochemical process (see note 66).

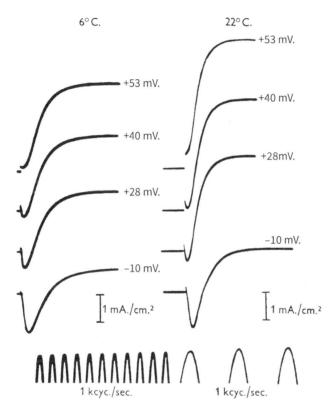

The absolute magnitude of the current attained at any voltage probably varies with temperature, but much less than the time scale. In the experiment   66 of Fig. 15 a rise of 16° C. increased the outward current about 1·5-fold, while the inward current at −50 mV. was approximately the same in the two records. Since the initial phase of inward current declined relatively rapidly as the axon deteriorated it is possible that a temperature coefficient of about 1·3 per 10° C. applies to both components of the current. Temperature coefficients of the order of 1·0–1·5 were also obtained by examining a number of results obtained with other axons.

The capacity current

The surge of current associated with the sudden change in membrane potential was examined by taking records at high time-base speed and low amplification. Tracings of a typical result are shown in Fig. 16. It will be seen

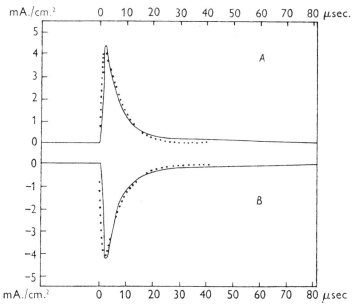

Fig. 16. Current through capacitative element of membrane during a voltage clamp. Abscissa: time in μsec. Ordinate: membrane current density (mA./cm.^{2}) with inward current taken as positive. At $t=0$ the potential difference between external and internal electrodes was displaced $+40$ mV. in curve A or -40 mV. in curve B. The continuous curves were traced from experimental records. The dotted curves were calculated according to the equation

$$I^* = 6\cdot8\left[\exp\left(-0\cdot159t\right) - \exp\left(-t\right)\right],$$

where I^* is the current in mA./cm.^{2} and t is time in μsec. This follows from the assumptions given in the text. Axon 25; temperature 8° C.

that the current records for anodal and cathodal changes are almost symmetrical and that the charging process is virtually complete in $50\,\mu$sec. In the anodal record the observed current declined from a peak of 4·5 mA./cm.^{2} to a steady level of about 0·04 mA./cm.^{2}. The steady current is barely visible in the tracing

66. *'absolute magnitude of the current . . . varies with temperature . . . less than the time scale.'* The Q_{10} for the amplitude of the currents at a fixed time is only ~1.3, distinct from the Q_{10} for kinetics. This observation suggests that current *magnitude* may depend on a nonbiochemical process, such as diffusion (see note 65).

but could be seen more clearly at higher amplification and lower time-base speed; records taken under these conditions were similar to those in Fig. 12.

The membrane capacity was obtained from the change in potential and the area under the curves. A small correction was made for ionic current but the resting membrane conductance was sufficiently low for uncertainties here to be unimportant. In the experiment illustrated by Fig. 16 A the charge entering the membrane capacity in 60 μsec. was 35 mμcoulomb/cm.^{2}, while the change in potential was 40 mV. Hence the membrane capacity per unit area was about 0·9 μF./cm.^{2}. Table 1 (p. 435) gives the results of other experiments of this kind. It also shows that replacement of all the sodium in sea water by choline had little effect on the membrane capacity and that there was no large change of capacity with temperature.

If a perfect condenser is short-circuited through zero resistance it loses its charge instantaneously. Fig. 16 suggests that the nerve capacity charged or discharged with a time constant of about 6 μsec. under a 'voltage clamp'. This raises the question whether the finite time of discharge was due to an imperfection in the membrane capacity or whether it arose from an imperfection in the method of holding the membrane potential constant. The records in Fig. 16 were obtained with uncompensated feed-back, which means that there was a small resistance in series with the capacitative element of the membrane. This clearly reduces the rate at which the membrane capacity can be discharged, and must be allowed for. It is also necessary to take account of the finite time constant of the recording amplifier (about 1 μsec. at this gain). Both effects have been considered in calculating the dotted lines in Fig. 16. These were drawn on the assumption that a 0·9 μF. condenser was charged to $\pm$ 40 mV. through a resistance of 7 Ω. and that the resulting pulses of current were recorded by an amplifier with an exponential time lag of 1 μsec. It will be seen that there is good agreement between the amplitude and general form of the two pairs of curves. Deviations at short times are not considered important since there was some uncertainty in the correction for amplifier delay.

At relatively long times (> 25 μsec.) the current record shows a 'tail' which is not explained by the presence of a series resistance. This effect was present in all records and was larger at higher temperatures. It can be explained by supposing that the membrane capacity was not perfect but behaved in the manner described by Curtis & Cole (1938). The records in Fig. 16 are roughly consistent with a constant phase angle of 80°, while those at higher temperatures require somewhat lower values. These statements must be regarded as tentative since our experiments were not designed to measure the phase angle and do not give good data for quantitative analysis. For the time being the principal point to be emphasized is that the surge associated with a sudden

67

68

67. *'membrane capacity was obtained from the change in potential and the area under the curves.'* *Figure 16* shows the transient capacity current that flows upon depolarization or hyperpolarization of the membrane. Note the time scale: the capacity currents decay within tens of microseconds, well before the ionic currents change noticeably in response to a voltage step. (The small, steady current is the ionic leak current.) Since I_c is dQ/dt, the integral under the curve (current over time) gives Q. HH&K integrate the records by measuring the area under the curves. (The method is not specified but techniques in use at the time involved tracing the records on graph paper and either counting squares or cutting out the areas of interest and weighing them.) Since the change in potential is known (40 mV), the capacitance can be calculated as $C = Q/V$ (see note 40).

68. *'a perfect condenser . . . short-circuited through zero resistance . . . loses its charge instantaneously.'* In a perfect condenser (capacitor) with no series resistance, the charge required to produce a step change in voltage would be transferred onto the capacitor instantaneously by an infinite current of infinitesimal duration (a delta function). In the squid axon, the series resistance through which the capacitor is charged is not zero, but is closer to 7 Ω. With a series resistance, the current in response to a voltage step will follow an exponential decay, with a time constant $\tau = RC$. The recorded charging curve, however, is slightly distorted by the amplifier 'lag,' which is the time between the change of voltage at the input and the resulting change at the output. HH&K mention that this lag changed as a function of the gain (amplification factor) but is about 1 μsec at the gain used in *Figure 16*. When both the series resistance and amplifier lag are applied to a theoretically perfect condenser of 0.9 μF/cm², the predicted current (dotted curve) provides a good fit to the recorded current (solid curve). At longer times, small deviations might reflect imperfections in the capacitor (e.g., a phase angle of 80° rather than 90°, as reported by Cole; see Historical Background). HH&K recognize this possibility but emphasize that the data can be 'adequately described' by treating the membrane as a perfect capacitor in which the only deviations are introduced by the amplifier and series resistance.

HH&K treat the 1-μs amplifier lag as an exponential decay, rather than a simple time shift. Their approach is based on the approximation that, when a step function is applied at the input, the amplifier's output voltage is proportional to the sum of the input voltage at all previous times, weighted by an exponential decay. The equation that describes the dotted lines in *Figure 16*, $I^\star = 6.8[(e^{-0.159t} - e^{-t})]$, therefore, is derived using a Green's Function, convolving the exponential decay of the response of the membrane capacitor ($\tau = 6$ μsec) with that of the impulse response of the amplifier ($\tau = 1$ μsec).

change in potential is adequately described by assuming that the membrane has a capacity of about $1\,\mu$F./cm.^{2} and a series resistance of about $7\,\Omega$.cm.^{2}.

The surge of capacity current was larger in amplitude and occupied a shorter time when compensated feed-back was employed. These experiments were not suitable for analysis, since the charging current was oscillatory and could not be adequately recorded by the camera employed. All that could be seen in records of ionic current is a gap, as in Fig. 14.

Our values for the membrane capacity are in reasonable agreement with those obtained previously. Using transverse electrodes 5·6 mm. in length, Curtis & Cole (1938) obtained the following values in twenty-two experiments: average membrane capacity at 1 kcyc./sec., $1\cdot1\,\mu$F./cm.^{2}, range $0\cdot66\,\mu$F./cm.^{2} to $1\cdot60\,\mu$F./cm.^{2}; average phase angle, 76°, range 64–85°.

The values for membrane capacity in the upper part of Table 1 were obtained by integrating the initial surge of current over a total time of about $50\,\mu$sec. If the phase angle is assumed to be 76° at all frequencies the average value of $0\cdot89\,\mu$F./cm.^{2} obtained by this method is equivalent to one of $1\cdot03\,\mu$F./cm.^{2} at 1 kcyc./sec. This is clearly in good agreement with the figures given by Curtis & Cole (1938), but is substantially less than the value of $1\cdot8\,\mu$F./cm.^{2} mentioned in a later paper (Cole & Curtis, 1939). However, as Cole & Curtis point out, the second measurement is likely to be too large since the electrode length was only 0·57 mm. and no allowance was made for end-effects.

The series resistance

Between the internal and external electrodes there is a membrane with a resting resistance of about $1000\,\Omega$.cm.^{2}. This resistance is shunted by a condenser with a capacity of about $1\,\mu$F./cm.^{2}.   69
In series with the condenser, and presumably in series with the membrane as a whole, there is a small resistance which, in the experiment illustrated by Fig. 16, had an approximate value of $7\,\Omega$.cm.^{2}. This 'series resistance' can be estimated without fitting the complete theoretical curve shown in that figure. A satisfactory approximation is to divide the time constant determining the decline of the capacitative curve by the measured value of the membrane capacity. This procedure   70
was followed in calculating the values for the series resistance given in the upper part of Table 1. The symbol r_s gives the actual resistance in series with the central area of membrane, while R_s is the same quantity multiplied by the area of membrane exposed to current flow in the central channel of the guard system. The last column gives the ratio of r_s to the resistance (r_{cd}) between the current measuring electrodes, c and d. This ratio is of interest since it determined the potentiometer setting required to give fully compensated feed-back (pp. 430–1).

Another method of measuring the series resistance was to apply a rectangular pulse of current to the nerve and to record the potential difference (v_{bc}) between the internal electrode b and the external electrode c as a function of time. The current in the central channel of the guard system was also obtained by recording the potential difference (v_{cd}) between the external electrodes c and d. The two records were rounded to the same extent by amplifier delay so that the series resistance and the membrane capacity could be determined by fitting the record obtained from the internal electrode by the following equation

$$v_{bc} = \frac{r_s}{r_{cd}}\,v_{cd} + \frac{1}{r_{cd}c}\int_0^t v_{cd}\,\mathrm{d}t,$$

where c is the capacity of the area of membrane exposed to current flow in the central channel.

69. *'a membrane with resting resistance . . . shunted by a condenser with a capacity of about 1 μF/cm².'*
This statement describes the membrane as a parallel *RC* circuit (see Appendix 2.3) between electrode *b* and *c*. HH&K point out, however, that another resistance, r_s, is presented by the seawater between the membrane and electrode *c*. This resistance also lies between electrodes *b* and *c*, in series with the *RC* circuit of the membrane, hence the term 'series resistance.'

70. *'divide the time constant determining the decline of the capacitative curve by the measured value of the membrane capacity.'* Since the time constant τ is equal to *RC*, the magnitude of the series resistance, r_s, can be estimated as τ/C. The alternative approach described in the next paragraph by HH&K involves calculating the series resistance r_s from the time course of the voltage between electrodes *b* and *c*, v_{bc}, and between *c* and *d*, v_{cd}. The calculated series resistance is 61 Ω (see note 26).

This analysis was made with two axons and gave satisfactory agreement between observed and calculated values of v_{bc}, with values of r_s/r_{cd} and C_M which were similar to those obtained by the voltage clamp method (see Table 1).

The observed value of the series resistance ($r_s = 61\,\Omega$.) cannot be explained solely by convergence of current between the electrodes used to measure membrane potential. Only about 30% was due to convergence of current between electrode c and the surface of the nerve, while convergence between the membrane and the internal electrode should not account for more than 25%, unless the specific resistance of axoplasm was much greater than that found by Cole & Hodgkin (1939). The axons used in the present work were surrounded by a dense layer of connective tissue, 5–20μ. in thickness, which adheres tightly and cannot easily be removed by dissection. According to Bear, Schmitt & Young (1937) the inner layer of this sheath has special optical properties and may be lipoid in nature. It seems reasonable to suppose that one or other of these external sheaths may have sufficient resistance to account for 45% of the series resistance. There was, in fact, some evidence that the greater part of the series resistance was external to the main barrier to ionic movement. Substitution of choline sea water for normal sea water increased r_{cd} by 23%, but it reduced r_s/r_{cd} by only 3·5% (Table 1, axon 25). This suggests that about 80% of r_s varied directly with the specific resistance of the external medium. Since the composition of axoplasm probably does not change when choline is substituted for sodium (Keynes & Lewis, 1951) it seems likely that most of the series resistance is located outside the main barrier to ionic movement. Further experiments are needed to establish this point and also to determine whether   71 the resistive layer has any measurable capacity.

Accuracy of method

The effect of the series resistance. The error introduced by the series resistance (r_s) was discussed on p. 430. Its magnitude was assessed by comparing records obtained with uncompensated feed-back ($p = 0$) with those obtained with compensated feed-back ($p = 0·6$). The effect of compensation was most conspicuous with a depolarization of about 30 mV. Fig. 17 shows typical curves in this   72 region. A, B and C were obtained with uncompensated feed-back; α, β and γ with compensated feed-back. A gives the potential difference between external and internal electrodes. B is the   73 potential difference between the external electrodes used to measure current and is equal to the product of the membrane current and the resistance r_{cd}. The true membrane potential differs from A by the voltage drop across r_s which is equal to (r_s/r_{cd}) B. C shows the membrane potential calculated on the assumption that r_s/r_{cd} had its average value of 0·68. α, β and γ were obtained in exactly the same manner as A, B and C, except that the potentiometer setting (p) was increased from 0 to 0·6. It will be seen that this altered the form of the upper record in a manner which compensates for the effect of current flow. The error in C is about 20%, while γ deviates by only about 2·5%. Hence any error present in β is likely to be increased eightfold in B. Since the two records are not grossly different, β may be taken as a reasonably faithful record of membrane current under a voltage clamp.

Experiments of this type indicated that use of uncompensated feed-back introduced errors but that it did not alter the general form of the current record. Since the method of compensated feed-   74 back was liable to damage axons it was not employed in experiments in which the preparation had to be kept in good condition for long periods of time.

Polarization effects. The outward current associated with a large and prolonged reduction of membrane potential was not maintained, but declined slowly as a result of a 'polarization effect'. The beginning of this decline can be seen in the lower records in Fig. 12. It occurred under conditions which had little physiological significance, for an axon does not normally remain with a membrane potential of − 100 mV. for more than 1 msec. Nor does the total current through the membrane approach that in Fig. 12.

In order to explain the effect it may be supposed either: (1) that 'mutual polarization' of the   75 electrode (p. 428) is substantially greater inside the axon than in sea water; (2) that currents may cause appreciable changes in ionic concentration near the membrane; (3) that some structure in series with the membrane may undergo a slow polarization. We were unable to distinguish

71. *'most of the series resistance is located outside the main barrier to ionic movement.'* The main barrier to ionic movement is the membrane. Changing the external solution from sodium-based seawater to a choline-based solution raises the resistance between electrodes c and d, r_{cd}, while leaving the ratio r_s/r_{cd} about the same, suggesting that r_s also depends on the ionic makeup of the bath solution. The other likely contributor to the series resistance that HH&K cite is the residual connective tissue that adheres to the axonal membrane after dissection and cleaning (see Historical Background and note 26).

72. *'The effect of compensation was most conspicuous with a depolarization of about 30 mV.'* The series resistance introduces an extra voltage drop when current flows across it, so that the voltage across the membrane is not equal to the command voltage whenever current is applied by the voltage clamp. The lower the membrane resistance, the greater the proportionate effect of series resistance (see Appendix 3.1); consequently, when large currents are flowing (during the periods of low membrane resistance), the voltage error is greatest. HH&K compensate for the series resistance by increasing the setting on the potentiometer, P, in **Figure 6** (see note 25), which subtracts a signal proportional to the error, pI_iR_s, from the recorded membrane potential. As p approaches 1, the compensation becomes more accurate, but the tendency of the amplifier to oscillate and kill the axon also increases. HH&K note that the effect of adding series resistance compensation is greatest with depolarizations of about 30 mV, which is consistent with the sodium current peak amplitude being greatest at about -30 mV *abs*.

73. *'with compensated feed-back.'* **Figure 17** compares traces in the region of maximal loss of clamp owing to series resistance error, with and without compensation (see note 76).

74. *'did not alter the general form of the current record.'* HH&K recognize and quantify the error introduced by series resistance, but they note that recordings of current are not greatly distorted by the series resistance. In the interest of preserving the axon from the damage that can occur by applying too much compensation, they choose to conduct most of their experiments without compensation. This instance is another example of how HH&K were undeterred by the imperfections of the technique; they evaluated the magnitude of the error, acknowledged it, and proceeded.

75. *'to explain the effect'* With prolonged depolarization, the outward current decayed slowly. HH&K consider various bases for this decay, and their terminology favors the idea that their measurement of voltage slowly changed, rather than the underlying current itself. The possibilities include that the two electrodes influenced each other; that potassium accumulated externally after flowing out of the cell (later supported by Hodgkin and Frankenhaeuser 1956), which would depolarize the equilibrium potential for potassium, decrease the driving force, and reduce the current; and that yet-unknown resistive elements might introduce a slowly changing voltage difference, altering the potential across the membrane. A final possibility, not raised by HH&K, is that the potassium conductance indeed decays with prolonged periods of depolarization, which turns out to be the case for many kinds of potassium channels.

between these suggestions, but it was clear that the 'polarization effect' had little to do with the active changes, since it was also present in moribund axons and was little affected by temperature.

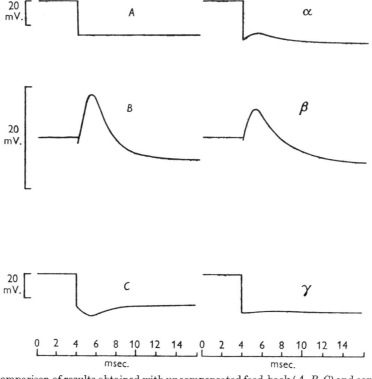

Fig. 17. Comparison of results obtained with uncompensated feed-back (A, B, C) and compensated 76 feed-back (α, β, γ). A, α: potential difference between external and internal electrodes ($v_c - v_b$). B, β: potential difference between current measuring electrodes ($v_d - v_c$). C, γ: membrane potential calculated as $C = A - 0\cdot68\,B$, or $\gamma = \alpha - 0\cdot68\,\beta$. Records B and β may be converted into membrane current density by dividing by $11\cdot9\,\Omega.\text{cm.}^2$. Temperature 4° C. Axon 34.

SUMMARY   77

1. An experimental method for controlling membrane potential in the giant axon of *Loligo* is described. This depended on the use of an internal electrode consisting of two silver wires, a guard system for measuring membrane current and a 'feed-back' amplifier for clamping the membrane potential at any desired level.

2. Axons impaled with the double electrode gave 'all-or-nothing' action potentials of about 100 mV. when stimulated with a brief shock. The action potential had a well-defined threshold at a critical depolarization of about 15 mV. Depolarizations less than 10–15 mV. gave graded responses similar to those seen in other excitable tissues.

3. The feed-back amplifier was arranged to make the membrane potential undergo a sudden displacement to a new level at which it was held constant for

76. ***Figure 17.*** The uncompensated records in the left column show the command potential (*A*), the early inward transient current (*B*), and the actual voltage across the membrane illustrating the voltage 'escape,' evident as the bump to a more depolarized potential (*C*). The records on the right show that the application of series resistance compensation briefly reduces the depolarization across the membrane and series resistance at the time of greatest current flow (α). Consequently, the peak current amplitude is smaller (β), as the membrane potential is well-clamped to a fixed potential (γ) rather than depolarizing excessively during the step.

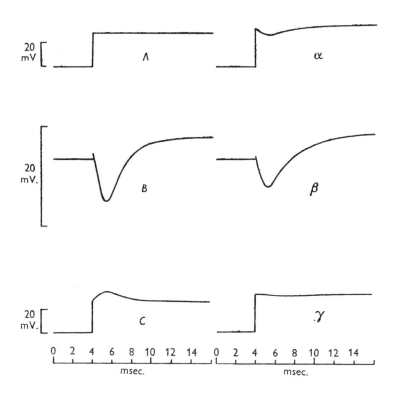

77. **'*Summary*'** At the conclusion of Paper 1, HH&K have accomplished the following: (1) introduced the squid axon preparation, (2) confirmed that it can generate a stationary action potential, (3) extracted the underlying currents from the action potential waveform, (4) built and applied the two-electrode voltage-clamp amplifier, (5) distinguished ionic and capacity current and the relationships between them, (6) identified an early inward transient current and late outward sustained current that flow upon depolarization of the membrane, (7) measured the current versus voltage relations for these ionic currents, (8) identified these curves as continuous and similar to those calculated from the action potential, (9) defined a strong temperature sensitivity of the kinetics but not of the amplitude of these currents, and (10) assessed the influence of various errors, including series resistance, on the measurements.

10–50 msec. Under these conditions the membrane current consisted of a brief surge of capacity current, associated with the sudden change in potential, and an ionic current during the period of maintained potential. The brief surge was proportional to the displacement of membrane potential and corresponded to the charging of a membrane with an average capacity of $0.9\,\mu\text{F./cm.}^2$. The sign and time course of the ionic current varied markedly with the membrane potential. Anodal displacements gave small currents which were always inward in direction. Depolarizations of less than 15 mV. gave outward currents which were small initially but increased markedly with time. Depolarizations of 15–110 mV. gave an initial phase of inward current which changed fairly rapidly into a large and prolonged outward current. The phase of inward current disappeared at about 110 mV. and was replaced by one of outward current. There was a continuous relation between ionic current and membrane potential. At short times this relation was similar to that derived from the rising phase of the action potential.

4. The maximum inward and outward ionic currents were little altered by temperature, but the rate at which the ionic current changed with time was increased about threefold for a rise of $10°$ C.

5. There was evidence of a small resistance in series with the capacitative element of the membrane. Errors introduced by this resistance were reduced by the use of compensated feed-back.

We wish to thank the Rockefeller Foundation for financial aid and the Director and staff of the Marine Biological Association for assistance at all stages of the experimental work.

REFERENCES

BEAR, R. S., SCHMITT, F. O. & YOUNG, J. Z. (1937). The sheath components of the giant nerve fibres of the squid. *Proc. Roy. Soc.* B, **123**, 496–529.

COLE, K. S. (1949). Dynamic electrical characteristics of the squid axon membrane. *Arch. Sci. physiol.* **3**, 253–258.

COLE, K. S. & CURTIS, H. J. (1939). Electric impedance of the squid giant axon during activity. *J. gen. Physiol.* **22**, 649–670.

COLE, K. S. & HODGKIN, A. L. (1939). Membrane and protoplasm resistance in the squid giant axon. *J. gen. Physiol.* **22**, 671–687.

CURTIS, H. J. & COLE, K. S. (1938). Transverse electric impedance of the squid giant axon. *J. gen. Physiol.* **21**, 757–765.

HODGKIN, A. L. & HUXLEY, A. F. (1945). Resting and action potentials in single nerve fibres. *J. Physiol.* **104**, 176–195.

HODGKIN, A. L. & HUXLEY, A. F. (1952a). Currents carried by sodium and potassium ions through the membrane of the giant axon of *Loligo*. *J. Physiol.* **116**, 449–472.

HODGKIN, A. L. & HUXLEY, A. F. (1952b). The components of membrane conductance in the giant axon of *Loligo*. *J. Physiol.* **116**, 473–496.

HODGKIN, A. L. & HUXLEY, A. F. (1952c). The dual effect of membrane potential on sodium conductance in the giant axon of *Loligo*. *J. Physiol.* **116**, 497–506.

HODGKIN, A. L. & HUXLEY, A. F. (1952d). A quantitative description of membrane current and its application to conduction and excitation in nerve. *J. Physiol.* (in the press).

HODGKIN, A. L., HUXLEY, A. F. & KATZ, B. (1949). Ionic currents underlying activity in the giant axon of the squid. *Arch. Sci. physiol.* **3**, 129–150.

HODGKIN, A. L. & KATZ, B. (1949a). The effect of sodium ions on the electrical activity of the giant axon of the squid. *J. Physiol.* **108**, 37–77.

No annotations to this page.

HODGKIN, A. L. & KATZ, B. (1949b). The effect of temperature on the electrical activity of the giant axon of the squid. *J. Physiol.* **109**, 240–249.

HUXLEY, A. F. & STÄMPFLI, R. (1951). Effect of potassium and sodium on resting and action potentials of single myelinated nerve fibres. *J. Physiol.* **112**, 496–508.

KEYNES, R. D. & LEWIS, P. R. (1951). The sodium and potassium content of cephalod nerve fibres. *J. Physiol.* **114**, 151–182.

MARMONT, G. (1949). Studies on the axon membrane. *J. cell. comp. Physiol.* **34**, 351–382.

ROTHENBERG, M. A. (1950). Studies on permeability in relation to nerve function. II. Ionic movements across axonal membranes. *Biochim. biophys. acta*, **4**, 96–114.

Paper 2

Hodgkin, A. L. and Huxley, A. F. (1952a) Currents carried by sodium and potassium ions through the membrane of the giant axon of *Loligo*. *Journal of Physiology* 116:449–472.

Even before these experiments were conducted, Hodgkin and Huxley (abbreviated H&H) had acquired evidence that an influx of sodium ions contributed to the action potential upstroke and that an efflux of potassium ions contributed to the downstroke, referred to as the 'sodium hypothesis' (see Historical Background). First, their intracellular recordings of the action potential in the squid giant axon had shown that the absolute magnitude of 'the action potential exceeded the resting potential.' The former was about 100 mV peak-to-peak, and the latter was about 50 mV. The action potential therefore surpassed or 'overshot' 0 mV, indicating that the electrical impulse was not a simple breakdown in membrane selective permeability but that it involved an 'active' process (Hodgkin & Huxley 1939; Hodgkin & Huxley 1945). Second, a substance 'like potassium' leaked from crab nerves during prolonged electrical activity, suggesting that potassium efflux repolarized the membrane after the active depolarization phase (Hodgkin & Huxley 1947). Third, intracellular recordings from squid axons demonstrated that reducing the concentration of sodium ions in the bathing solution slowed the upstroke of the action potential, for which the simplest explanation was that the voltage change was generated by sodium influx (Hodgkin & Katz 1949). Additionally, radioactive tracer data provided supporting evidence for sodium crossing the membrane (Keynes 1949), and work by Cole and Curtis (1939) was suggestive of a conductance change during the action potential. Nevertheless, other competing hypotheses remained. Direct recordings of ionic currents were lacking; the factors that initiated and regulated any such currents were unknown; and the relationship between a theoretical sodium and potassium flux was entirely unexplored. The second of the 1952 papers unequivocally identifies sodium as the ion entering the axon upon depolarization, separates the total current into the parallel components carried by sodium and potassium, quantifies the magnitudes and time courses of the currents they carry, demonstrates the mutual independence of the two ionic fluxes, and provides evidence that membrane potential is the variable controlling them.

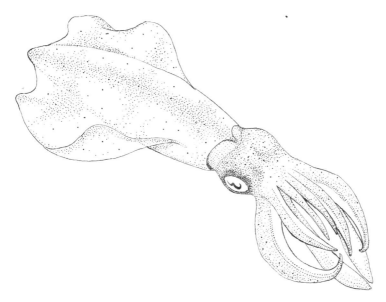

CURRENTS CARRIED BY SODIUM AND POTASSIUM IONS THROUGH THE MEMBRANE OF THE GIANT AXON OF *LOLIGO*

By A. L. HODGKIN and A. F. HUXLEY

From the Laboratory of the Marine Biological Association, Plymouth, and the Physiological Laboratory, University of Cambridge

(*Received* 24 *October* 1951)

In the preceding paper (Hodgkin, Huxley & Katz, 1952) we gave a general description of the time course of the current which flows through the membrane of the squid giant axon when the potential difference across the membrane is suddenly changed from its resting value, and held at the new level by a feed-back circuit ('voltage clamp' procedure). This article is chiefly concerned with the identity of the ions which carry the various phases of the membrane current.   1

One of the most striking features of the records of membrane current obtained under these conditions was that when the membrane potential was lowered from its resting value by an amount between about 10 and 100 mV. the initial current (after completion of the quick pulse through the membrane capacity) was in the inward direction, that is to say, the reverse of the direction of the current which the same voltage change would have caused to flow in an ohmic resistance. The inward current was of the right order of magnitude, and   2 occurred over the right range of membrane potentials, to be the current responsible for charging the membrane capacity during the rising phase of an action potential. This suggested that the phase of inward current in the voltage   3 clamp records might be carried by sodium ions, since there is much evidence (reviewed by Hodgkin, 1951) that the rising phase of the action potential is caused by the entry of these ions, moving under the influence of concentration and potential differences. To investigate this possibility, we carried out voltage   4 clamp runs with the axon surrounded by solutions with reduced sodium concentration. Choline was used as an inert cation since replacement of sodium   5 with this ion makes the squid axon completely inexcitable, but does not reduce the resting potential (Hodgkin & Katz, 1949; Hodgkin, Huxley & Katz, 1949).

1. **'identity of the ions which carry . . . membrane current.'** Although the results of the previous paper was consistent with the inference that the early inward current was carried by sodium, the present paper explores and tests this hypothesis quantitatively.

2. **'reverse of the direction of the current which the same voltage change would have caused to flow in an ohmic resistance.'** For an 'ohmic' resistance, a depolarization from rest would generate an outward positive current, which would tend to restore a non-voltage-clamped membrane to its resting value. H&H draw attention to the observation that the current elicited by 'lowering' the membrane potential (depolarizing it) was not ohmic but instead was inward, which would amplify the depolarization.

3. **'right order of magnitude . . . right range of membrane potentials . . . for . . . the rising phase of an action potential.'** Note the reasoning applied here: H&H define the attributes of the phenomenon being sought and then undertake an investigation of something with those attributes.

4. **'sodium ions . . . moving under the influence of concentration and potential differences.'** H&H begin with the hypothesis that the current *is* carried by sodium ions, and then seek to falsify the hypothesis (although none appear). They will make and test predictions based on this assumption, until an unfulfilled prediction disproves the hypothesis. A series of fulfilled predictions will provide multiple lines of convergent evidence that permit the hypothesis to be accepted, though always provisionally.

5. **'To investigate this possibility . . . solutions with reduced sodium concentration.'** The experiments in the previous paper were more *observational*, in that HH&K applied a voltage clamp in control solutions and observed the transient inward and sustained outward current. The present experiments are more *investigative*, in that they make a manipulation, perturbing a variable (the sodium concentration) that is predicted by the sodium hypothesis to have specific effects, and testing whether those predictions are fulfilled. These experiments introduce the use of ion substitution to identify the ionic species carrying voltage-clamped currents, an approach that is still in use some seven decades later.

METHOD

The apparatus and experimental procedure are fully described in the preceding paper (Hodgkin *et al.* 1952). 'Uncompensated feed-back' was employed.

Sea water was used as a normal solution. Sodium-deficient solutions were made by mixing sea water in varying proportions with isotonic 'choline sea water' of the following composition:

Ion	g. ions/kg. H_2O	Ion	g. ions/kg. H_2O
Choline$^+$	484	Mg^{++}	54
K$^+$	10	Cl$^-$	621
Ca^{++}	11	HCO$_3^-$	3

The mixtures are referred to by their sodium content, expressed as a percentage of that in sea water (30% Na sea water, etc.).

RESULTS

Voltage clamps in sodium-free solution

Fig. 1 shows the main differences between voltage clamp records taken with the axon surrounded by sea water and by a sodium-free solution. Each record gives the current which crossed the membrane when it was depolarized by 65 mV. After the top record was made, the sea water surrounding the axon was replaced by choline sea water, and the middle record was taken. The fluid was again changed to sea water, and the bottom record taken. The amplifier gain was the same in all three records, but a given deflexion represents a smaller current in the choline solution, since the current was detected by the potential drop along a channel filled with the fluid which surrounded the axon, and the specific resistance of the choline sea water was about 23% higher than that of ordinary sea water.

The most important features shown in Fig. 1 are the following: (1) When the external sodium concentration was reduced to zero, the inward current disappeared and was replaced by an early hump in the outward current. (2) The late outward current was only slightly altered, the steady level being 15–20% less in the sodium-free solution. (3) The changes were reversed when sea water was replaced. The currents are slightly smaller in the bottom record than in the top one, but the change is not attributable to an action of the choline since a similar drop occurred when an axon was kept in sea water for an equal length of time.

A series of similar records with different strengths of depolarization is shown in Fig. 2. The features described in connexion with Fig. 1 are seen at all strengths between −28 and −84 mV. At the weakest depolarization (−14 mV.) the early phase of outward current in the sodium-free record is too small to be detected. At the highest strengths the early current is outward even in sea water, and is then increased in the sodium-free solution.

These results are in qualitative agreement with the hypothesis that the inward current is carried by sodium ions which, as an early result of the decrease in membrane potential, are permitted to cross the membrane in both

6

7

8

9

6. *'top record . . . middle record . . . bottom record taken.'* By recording first in seawater, then in sodium-free (choline) seawater, and then in seawater again, H&H introduce the approach of making a control recording, a test recording, and a reversal (or wash) recording. Time is almost always a variable that affects electrophysiological recordings, frequently owing to deterioration of the preparation (generically termed 'rundown'). This experimental design serves to control for any changes that might result simply from the passage of time rather than the manipulation itself.

7. *'specific resistance of the choline sea water was about 23% higher than that of ordinary sea water.'* Recall that the voltage-clamped current is measured as the voltage difference between electrodes *c* and *d,* divided by the resistance between them ($I = V/R$). Because choline ions are larger than sodium ions, they have a lower mobility in solution, which reduces the conductance (increases the resistance) of that solution. The same measured voltage change between current-sensing electrodes *c* and *d* will therefore represent a smaller current in choline solution.

8. *'inward current disappeared . . . outward current was only slightly altered . . . changes were reversed when sea water was replaced.'* The observation that removing sodium from the external medium reversibly eliminates the inward current without affecting the outward current is strong evidence for the 'sodium hypothesis.' It further provides the first suggestion that the inward and outward currents may be independent of one another. The latter idea is particularly significant, since no previous data offered insight into the relationship between the theoretical sodium and potassium fluxes, and the idea that one *current* might trigger another *current* was still in vogue (see Historical Background). As H&H note, however, both conclusions are qualitative rather than quantitative.

9. *'early current is outward even in sea water, and is then increased in the sodium-free solution.'* With this first 'family of currents' recorded under voltage clamp, H&H draw attention to the conditions under which the early current is outward both in control solutions (with large depolarizations) and in sodium-free solutions (always), which will provide the stepping-stone for the quantitative analyses of reversal potentials.

directions under a driving force which is the resultant of the effects of the concentration difference and the electrical potential difference across the membrane. When the axon is in sea water, the concentration of sodium out- 10 side the membrane $[\text{Na}]_o$ is 5–10 times greater than that inside, $[\text{Na}]_i$. This tends to make the inward flux exceed the outward. The electrical potential

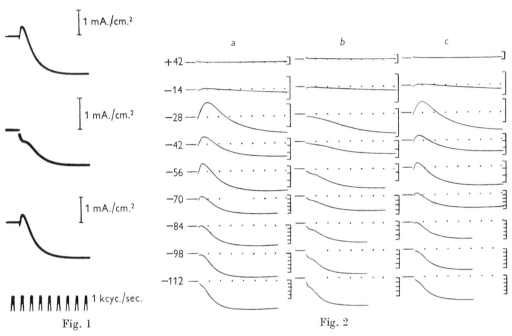

Fig. 1.   Fig. 2.

Fig. 1. Records of membrane current during 'voltage clamps' in which membrane potential was lowered by 65 mV. Top record: axon in sea water. Centre record: axon in choline sea water. Bottom record: after replacing sea water. Axon no. 15; temperature 11° C. Inward current is shown upwards in this and all other figures.

Fig. 2. Records of membrane current during 'voltage clamps'. *a*, axon in sea water; *b*, axon in choline sea water; *c*, after replacing sea water. Displacement of membrane potential indicated in mV. Axon no. 21; temperature 8·5° C. Vertical scale: 1 division is 0·5 mA./cm.². Horizontal scale: interval between dots is 1 msec.

difference E also helps the inward and hinders the outward flux so long as it is positive, i.e. in the same direction as the resting potential. The net current carried by the positive charge of the sodium ions is therefore inward unless the depolarization is strong enough to bring E to a sufficiently large negative value to overcome the effect of the concentration difference. The critical value 11 of E at which the fluxes are equal, and the net sodium current is therefore zero, will be called the 'sodium potential', E_{Na}. Its value should be given by 12 the Nernst equation

$$E_{\text{Na}} = \frac{RT}{F} \log_e \frac{[\text{Na}]_i}{[\text{Na}]_o}. \tag{1}$$

10. *'sodium ions . . . cross the membrane in both directions under a driving force which is the resultant of the effects of the concentration difference and the electrical potential difference across the membrane.'* This is a verbal rendering of the equations stating that ionic current is proportional to driving force, which depends on the transmembrane electrical potential *and* the chemical equilibrium potential of the ion, the latter being set by concentration gradient of that ion. Eventually, these ideas will be formulated (in modern conventions) as $I \propto (V - E_{\text{Na}})$ and $E_{\text{Na}} \propto \log ([\text{Na}]_{\text{out}} / [\text{Na}]_{\text{in}})$.

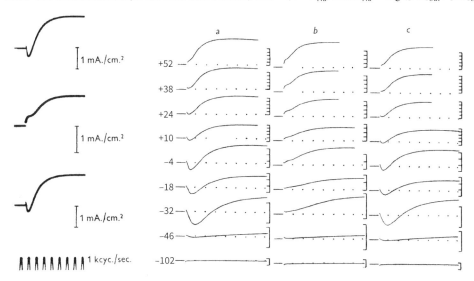

11. *'The net current carried by . . . sodium ions is therefore inward unless the depolarization is strong enough . . . to overcome the effect of the concentration difference.'* Current carried by positive ions is *always inward* at voltages more hyperpolarized than the equilibrium potential, and *always outward* at voltages more depolarized than the equilibrium potential. In **Figure 2**, the initial current in control solutions is inward at voltages more hyperpolarized than +38 mV *abs.*, but is outward at +52 mV *abs.* In choline seawater, the initial phase of current is always outward.

12. *'called the "sodium potential," E_{Na}.'* Here the membrane currents are related to known physical phenomena and the equations that describe them. If the assumption of a relation between the biology and the physical chemistry is correct, the value of E_{Na} *should* be given by the Nernst equation. Note that the expression of the Nernst equation given in **Equation 1** has the intracellular concentration in the numerator and extracellular concentration in the denominator, which is inverted relative to modern conventions. R is the ideal gas constant (8.314 Joules/degree Kelvin/mole), T is temperature (in degrees Kelvin), and F is the Faraday constant (96,485 coulombs/mole).

With values of E more negative than this, the net sodium flux is outward, causing the early phase of the outward current seen in the lowest record of the first and third columns of Fig. 2, where the axon was in sea water and was depolarized by 112 mV. A family of voltage clamp records which shows particularly well this transition from an initial rise to an initial fall as the strength of depolarization is increased is reproduced as Fig. 14 of the preceding paper.

When the axon is placed in a sodium-free medium, such as the 'choline sea water', there can be no inward flux of sodium, and the sodium current must always be outward. This will account for the early hump on the outward current which is seen at all but the lowest strength of depolarization in the   13 centre column of Fig. 2.

Voltage clamps with reduced sodium concentration

The results of reducing the sodium concentration to 30 and 10% of the value in sea water are shown in Figs. 3 and 4 respectively. These figures do   14 not show actual records of current through the membrane. The curves are graphs of ionic current against time, obtained by subtracting the current through the capacity from the recorded total current. The initial surge in an   15 anodal record was assumed to consist only of capacity current, and the capacity current at other strengths was estimated by scaling this in proportion to the amplitude of the applied voltage change.

As would be expected, the results are intermediate between those shown in Fig. 2 for an axon in sea water and in choline sea water. Inward current is present, but only over a range of membrane potentials which decreases with the sodium concentration, and within that range, the strength of the current is reduced. A definite sodium potential still exists beyond which the early hump of ionic current is outward, but the strength of depolarization required to reach it decreases with the sodium concentration. Thus, in the first column   16 of Fig. 3, with the axon in 30% sodium sea water, the sodium potential is almost exactly reached by a depolarization of 79 mV. In the second column, with sea water surrounding the axon, the sodium potential is just exceeded by a depolarization of 108 mV. In column 3, after re-introducing 30% sodium sea water, the sodium potential is slightly exceeded by a depolarization of 79 mV. Similarly, in Fig. 4, the sodium potentials are almost exactly reached by depolarizations of 105, 49 and 98 mV. in the three columns, the axon being in sea water, 10% sodium sea water and sea water respectively. The sequence of changes in the form of the curves as the sodium potential is passed is remarkably similar in all cases.

The external sodium concentration and the 'sodium potential'

Estimation of the 'sodium potential' in solutions with different sodium concentrations is of particular importance because it leads to a quantitative

13. *'This will account for the early hump on the outward current'* The hypothesis, grounded in the Nernst equation, has 'explanatory power,' since it predicts the shift in E_{Na} in choline seawater that accounts for the direction of current flow. H&H formulate the argument that the early inward hump seen in control solutions at all but the most depolarized potentials is converted into the early outward hump in choline. Although *all* current is outward in sodium-free medium, it has two phases, the early hump and the later sustained current; H&H suggest that this early outward current hump is distinct from the late outward sustained current and comes from an *efflux* of sodium. This conclusion rejects the alternative possibility that early inward and outward currents might arise from two different mechanisms, e.g., if certain ions can only cross the membrane in one direction.

14. *'reducing the sodium concentration to 30 and 10% of the value in sea water'* These experiments make the data progressively more quantitative. Instead of having only two categories, of 100% and 0% of normal sodium, the intermediate values set up for curve fitting, estimation of specific values, and other mathematical analyses.

15. *'subtracting the current through the capacity from the recorded total current.'* In modern terminology, the current was 'capacitance subtracted.' In voltage clamp, the ionic and capacity current are independent of one another and sum linearly: $I_{total} = I_c + I_i$. The capacity current scales linearly with the size of the voltage change: $I_c \propto dV/dt$. It can therefore be measured in isolation by hyperpolarizing the membrane to voltages at which little ionic current flows. The result can then be inverted and scaled to match the size of the voltage step at which ionic current is recorded, and finally subtracted from the I_{total} trace to isolate the ionic current.

16. *'A definite sodium potential . . . exists . . . but the strength of depolarization required to reach it decreases with the sodium concentration.'* The reversal potential for the early current hump has a defined value that varies systematically with $[Na]_{out}$. When one variable changes systematically with respect to another, it becomes possible to determine a quantitative relationship between them.

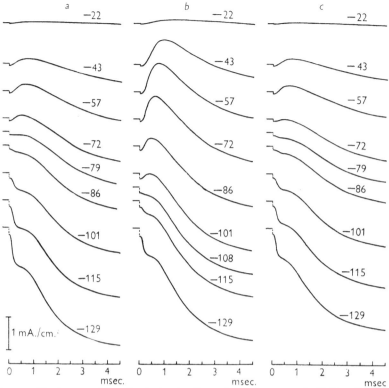

Fig. 3. Curves of ionic current density during 'voltage clamps'. *a*, axon in 30 % sodium sea water;   17
b, axon in sea water; *c*, after replacing 30 % sodium sea water. Displacement of membrane
potential indicated in millivolts. Axon no. 20; temperature 6·3° C.

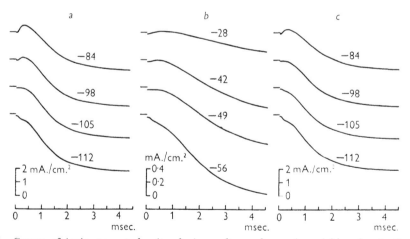

Fig. 4. Curves of ionic current density during voltage clamps in neighbourhood of sodium
potential. *a*, axon in sea water; *b*, axon in 10 % sodium sea water; *c*, after replacing sea water.
Note that ordinate scale is larger in *b* than in *a* and *c*. Displacement of membrane potential
in millivolts indicated for each curve. Axon no. 21; temperature 8·5° C.

17. **Figure 3 and Figure 4.** These two figures permit a close look at the reversal of the early transient current in control and reduced-sodium seawater. **Figure 3** shows the changes in currents with a switch from 30% sodium (a) to 100% sodium (b), which changes the reversal potential from +19 *abs.* to +48 mV *abs.* **Figure 4** switches the recordings from 100% sodium (a) to 10% sodium (b), which changes the reversal potential from +45 *abs.* to −11 mV *abs.* In both experiments, the test solution (b) is followed by a return to the initial solution (c), to control for time-dependent changes ('rundown').

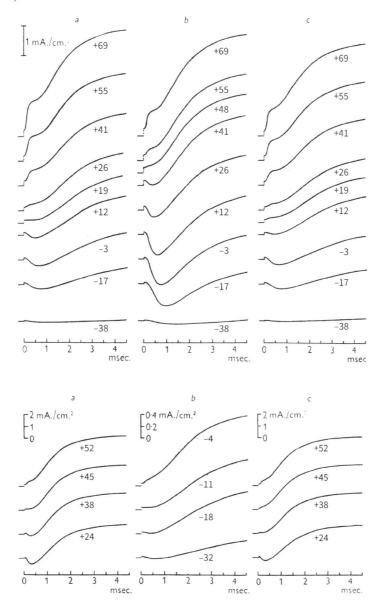

test of our hypothesis. Equation (1) gives the sodium potential in sea water [18] (E_{Na}), and the corresponding quantity (E'_{Na}) when the external sodium concentration is reduced to $[Na]'_o$ is given by

$$E'_{Na} = \frac{RT}{F} \log_e \frac{[Na]_i}{[Na]'_o}.$$

Hence $E'_{Na} - E_{Na} = \frac{RT}{F} \left\{ \log_e \frac{[Na]_i}{[Na]_o} - \log_e \frac{[Na]_i}{[Na]'_o} \right\} = \frac{RT}{F} \log_e \frac{[Na]_o}{[Na]'_o}.$ (2)

The displacements of membrane potential, V, corresponding to these values are $V_{Na} = E_{Na} - E_r$ and $V'_{Na} = E'_{Na} - E'_r$, where E_r and E'_r are the values of the resting potential in sea water and in the test solution respectively. Hence

$$(V'_{Na} - V_{Na}) + (E'_r - E_r) = \frac{RT}{F} \log_e \frac{[Na]_o}{[Na]'_o}.$$ (3)

Each term in this equation can be determined experimentally, and data [19] were obtained in four experiments on two axons. The results are given in Table 1, where the observed shift in sodium potential is compared with that predicted from the change in sodium concentration by Equation (3). It will be

TABLE 1. Comparison of observed and theoretical change in sodium potential when the fluid [20] surrounding an axon is changed from sea water to a low sodium solution. Observed change:

$$E'_{Na} - E_{Na} = (V'_{Na} - V_{Na}) + (E'_r - E_r); \text{ theoretical change} = \frac{RT}{F} \log_e \frac{[Na]_o}{[Na]'_o}.$$

| | | | | | | Sodium potential shift | |
Axon no.	Temp. (°C.)	$\frac{[Na]'_o}{[Na]_o}$	V_{Na} (mV.)	V'_{Na} (mV.)	$(E'_r - E_r)$ (mV.)	Observed (mV.)	Theoretical (mV.)
20	6·3	0·3	−105	−78	+3	+30	+28·9
20	6·3	0·1	−96	−45	+4	+55	+55·3
21	8·5	0·1	−100	−48	+4	+56	+55·6
21	8·5	0·1	−95	−45	+4	+54	+55·6

seen that there is good agreement, providing strong evidence that the early rise or fall in the recorded ionic current is carried by sodium ions, moving under the influence of their concentration difference and of the electric potential difference across the membrane.

Details of the estimation of the quantities which enter into Equation (3) are given in the following paragraphs.

Determination of V_{Na}. At the sodium potential there is neither inward sodium current, shown by an initial rise in the ionic current, nor outward sodium current, shown by an early hump in the outward current. It was found that these two criteria did in fact define the sodium potential very sharply, i.e. a hump appeared as soon as the ionic current showed an initial fall. It was therefore permissible to take as V_{Na} the strength of depolarization which gave an ionic current curve which started horizontally. This criterion was much more convenient to apply than the [21] absence of a hump, since records were taken at fairly wide intervals of V (usually 7 mV.) and an interpolation procedure was necessary in order to estimate V_{Na} to the nearest 0·5 mV.

18. **'quantitative test of our hypothesis.'** In the quantitative version of the sodium hypothesis, the Nernst equation should predict the values of the reversal potential of the early inward current in normal and low concentrations of sodium. Calculating these potentials, however, requires knowing the concentration of intracellular sodium, which H&H cannot measure directly. To address this limitation, H&H rearrange the Nernst equation to express the predicted *change* in reversal potential, $E'_{Na} - E_{Na}$, from the known *change* in extracellular concentration. In essence, they solve two equations for one unknown.

19. **'Each term in this equation can be determined experimentally'** **Equation 2** expresses the changes in the reversal potential in terms of absolute transmembrane voltage, E. Recall, however, that in H&H's measurements, the resting potential is set to be 0 mV when they begin each experiment, so the voltage where the current reverses, V_{Na}, is the depolarization from rest necessary to reach the sodium potential. **Equation 3** takes into account the effect of external sodium on both the reversal and the resting potential.

The powerful aspect of rearranging the Nernst equation into **Equation 3**, as H&H indicate, is that they can obtain *measurements* of (1) the change in resting potential, (2) the change in reversal potential, and (3) the external sodium concentration in control and low-sodium solutions. They can then compare the observed shift in reversal based on recordings of currents (the left side of the equation) to the expected shift based on the changes in sodium concentration (right side of the equation). If both sides are indeed equal, then it will demonstrate that the currents can be directly accounted for by a transient selective permeability of the membrane to sodium ions.

20. **Table 1.** The 'Theoretical' sodium potential shift predicted by the Nernst equation is compared to the 'Observed' sodium potential shift. Given the close agreement between theory and experiment—to within less than 2 mV—H&H refer to the early current as 'sodium current' after this point.

21. **'started horizontally.'** With a voltage step to the sodium reversal potential, a small change in leak current occurs instantaneously, but because the net sodium flux is zero, the 'early hump' is absent, making the current trace extend horizontally for a fraction of a millisecond (e.g., **Figure 3**, column 1, −79 mV trace, or +19 mV *abs.* in redrawn figure).

Change in resting potential. Experiments with ordinary capillary internal electrodes showed that the resting potential increased on the average by 4 mV. when the sea water surrounding the axon was replaced by choline sea water (a correction of 1·5 mV. for junction potentials in the external solutions is included in this figure). With intermediate sodium concentrations, the change in resting potential was assumed to be proportional to the change in sodium concentration. For instance, the resting potential in 30 % sodium sea water was taken as 2·8 mV. higher than that in sea water.   22

Slow change in condition of axon. When an axon is kept in sea water, its sodium content rises (Steinbach & Spiegelman, 1943; Keynes & Lewis, 1951) and its resting potential falls. Both of these effects bring E_r and E_{Na} closer together, diminishing the absolute magnitude of V_{Na}. In comparing V_{Na} in two solutions, it was therefore necessary to determine V_{Na} first in one solution, then in the other and finally in the first solution again. The second value of V_{Na} was then compared with the mean of the first and third.   23

The internal sodium concentration and the sodium potential

In freshly mounted fibres the average difference between the sodium potential and the resting potential was found to be -109 mV. (ten axons with a range of -95 to -119 mV. at an average temperature of $8°$ C.). The average resting potential in these fibres was 56 mV. when measured with a micro-electrode containing sea water. By the time the sodium potential was measured the resting potential had probably declined by a few millivolts and may be taken as 50 mV. Allowing 10–15 mV. for the junction potential between sea water and axoplasm (Curtis & Cole, 1942; Hodgkin & Katz, 1949) this gives an absolute resting potential of 60–65 mV. The absolute value of the sodium potential would then be -45 to -50 mV. The sodium concentration in sea water is about 460 m.mol./kg. H_2O (Webb, 1939, 1940) so that the internal concentration of sodium would have to be 60–70 m.mol./kg. H_2O in order to satisfy Equation 1. This seems to be a very reasonable estimate since the sodium concentration in freshly dissected axons is about 50 m.mol./kg. H_2O while that in axons kept for 2 or 3 hr. is about 100 m.mol./kg. H_2O (Steinbach & Spiegelman, 1943; Keynes & Lewis, 1951; Manery, 1939, for fraction of water in axoplasm).   24

Outward currents at long times

So far, this paper has been concerned with the earliest phases of the membrane current that flows during a voltage clamp. The only current which has the opposite sign from the applied voltage pulse is the inward current which occurs over a certain range of depolarizations when the surrounding medium contains sodium ions. This inward current is always transient, passing over into outward current after a time which depends on the strength of depolarization and on the temperature. The current at long times resembles that in an ohmic resistance in having the same sign as the applied voltage change, but differs in that the outward current due to depolarization rises with a delay to a density which may be 50–100 times greater than that associated with a similar increase in membrane potential. Figs. 1–3 show that this late current   25

22. *'resting potential increased . . . by 4 mV . . . choline sea water'* The resting membrane potential hyperpolarizes in choline, possibly due to a loss of a small contribution of sodium to the leak conductance. The mobility of choline is also lower than sodium, which introduces a junction potential, for which H&H correct (see note 7).

23. *'its resting potential falls.'* As the preparation deteriorates, the ionic gradients of the axon run down and the absolute value of the resting potential decreases. Since V_{Na} is the driving force measured relative to rest, V_{Na} is also reduced. H&H therefore take the average of V_{Na} in the initial and the wash solution.

24. *'resting potential was 56 mV. when measured with a microelectrode containing seawater.'* Recall again that at the beginning of each experiment, H&H set the voltage-clamp amplifier to read 0 mV at the resting membrane potential, so their estimate of sodium potentials is relative to rest, rather than an absolute potential. To obtain an absolute measurement, they use a d.c. amplifier (see Paper 1, note 27) and a glass *microelectrode* filled with seawater. H&H use the microelectrode to record the absolute membrane potential to be −56 mV (negative in modern conventions; positive for H&H), which they estimate is reduced to −50 mV by rundown at the time they measure the sodium reversal potential. Because a junction potential of 10–15 mV arises between the seawater in the microelectrode and the axoplasm, the actual transmembrane potential must be −60 to −65 mV. This estimate serves to calibrate the rest of the measurements. With a difference of 109 mV between the resting potential and the sodium equilibrium potential, E_{Na} can be taken as +45 to +50 mV *abs*. A check on this value is obtained by calculating the *intracellular* sodium concentration required to give an equilibrium potential in this range. The calculated concentration is consistent with previous estimates. Thus, H&H obtain an absolute measure of V_{rest} and E_{Na}, as well as another estimate of the sodium concentration in axoplasm.

25. *'current at long times resembles that in an ohmic resistance . . . but . . . outward current due to depolarization rises with a delay to . . . be 50–100 times greater than that associated with an increase in membrane potential.'* The outward current flows in the direction that would restore the membrane potential to its initial value (if the axon were not voltage clamped), which is characteristic of an ohmic response, but the current is distinguished in two ways. First, it is time dependent, increasing over the course of a few milliseconds. Second, it is much larger than that seen with hyperpolarizations of equivalent magnitude. Therefore, in addition to any passive leak currents, the outward current appears to include an actively voltage-gated, time-dependent current, an idea that H&H first proposed in their 1947 paper on potassium leakage from the nerve during activity.

is not greatly affected by the concentration of sodium in the fluid surrounding the axon.

An outward current which arises with a delay after a fall in the membrane potential is clearly what is required in order to explain the falling phase of the action potential. The outward currents reached in a voltage clamp may con-   26
siderably exceed the maximum which occurs in an action potential; this may   27
well be because the duration of an action potential is not sufficient to allow the outward current to reach its maximum value. These facts suggest that the outward current associated with prolonged depolarization is the same current which causes the falling phase of the action potential. The evidence (reviewed by Hodgkin, 1951) that the latter is caused by potassium ions leaving the axon is therefore a suggestion that the former is also carried by potassium ions. Direct evidence that such long-continued and outwardly directed membrane currents are carried by potassium ions has now been obtained in *Sepia* axons by a tracer technique (unpublished experiments). We shall therefore assume that this delayed outward current is carried by potassium ions, and we shall   28
refer to it as 'potassium current', I_K. Since it is outward, it is not appreciably affected by the external potassium concentration, and evidence for or against   29
potassium being the carrier cannot easily be obtained by means of experiments analogous to those which have just been described with altered external sodium concentration.

I_K *in sea water and choline.* As has been mentioned, the later part of the current record during a constant depolarization is much the same whether the axon is surrounded by sea water or by one of the solutions with reduced sodium concentration. There are, however, certain differences. For a given strength of depolarization, the maximum outward current is smaller by some 10 or 20% in the low-sodium solution, and at the higher strengths where the outward current is not fully maintained, the maximum occurs earlier in the low-sodium solution. Part of the difference in amplitude is explained by the difference of resting potential. Since the resting potential is greater in the   30
low-sodium medium, a higher strength of depolarization is needed to reach a given membrane potential during the voltage clamp. This difference can be allowed for by interpolation between the actual strengths employed in one of the solutions. In most cases, this procedure did not entirely remove the difference between the amplitudes. There are, however, two other effects which are likely to contribute. In the first place, the effect of not using 'compensated feed-back' is probably greater in the low-sodium solution (see   31
preceding paper, p. 445). This further reduces the amplitude of the voltage change which actually occurs across the membrane. In the second place, the fact that the current reached its maximum earlier suggests that 'polarization' (preceding paper, p. 445) had a greater effect in the low-sodium solution. We   32
do not know enough about either of these effects to estimate the amount by

26. **'An outward current which rises with a delay . . . is required . . . to explain the falling phase of the action potential.'** Here the logical link is made between the necessary attributes of any mechanism for repolarization and the observed properties of the voltage-clamped outward current that is being recorded for the first time.

27. **'outward currents . . . in a voltage clamp may considerably exceed the maximum which occurs in an action potential'** H&H recognize that the sustained nature of the outward current is a consequence of the sustained depolarization induced by a voltage step. With a real action potential, the depolarization is brief, so the outward current might likewise be relatively short. Additionally, because of its delayed rise, the outward current may not reach its peak value during an action potential (see Paper 1, note 56). Both these ideas will be supported by modeling in the fifth paper, and were confirmed in later decades with measurements of voltage-clamped potassium currents during command potentials with the shape of an action potential, as opposed to a voltage step.

28. **'We shall therefore assume that this delayed outward current is carried by potassium ions'** As stated in note 1, and summarized in this paragraph, H&H had considerable evidence that potassium was responsible for the downstroke of the action potential. The question here is whether the sustained outward current under voltage clamp is indeed produced by potassium ions flowing out of the cell. They will proceed on this assumption until an observation to the contrary is made or until multiple lines of evidence substantiate the idea beyond reasonable doubt.

29. **'Since it is outward, it is not appreciably affected by the external potassium concentration'** Small changes in extracellular potassium concentration would indeed shift the equilibrium potential, but as long as the concentration remains relatively low, the current will not reverse direction but will remain outward with all step depolarizations. This issue is addressed in Paper 3.

30. **'outward current is smaller . . . in the low-sodium solution. . . . Part of the difference . . . is explained by the difference of resting potential.'** The observation that the outward current is reduced in low-sodium solution raises the question of whether the early and late currents are indeed independent and/or whether the late current might be influenced or even partly carried by sodium ions. H&H point out the additional variable of the resting potential, which hyperpolarizes in low-sodium solution. Recall that the voltage clamp depolarizes the membrane by the command voltage, V, *relative* to the resting potential, which is defined as zero. If the absolute resting potential itself becomes more hyperpolarized, as it does in choline seawater, then the final absolute potential reached during a step, V, is correspondingly more hyperpolarized. If the magnitude of potassium current is sensitive to *absolute* membrane potential, it will therefore be smaller in choline seawater.

31. **'effect of not using "compensated feedback" is probably greater in low-sodium solution'** The choline solution has a lower conductance, so the series-resistance error is greater that in control solutions.

32. **'"polarization" . . . had a greater effect in the low-sodium solution.'** The polarization of the voltage-sensing electrode by current in the current-passing electrode was thought to be a technical artifact (see Paper 1, notes 17 and 75). The effect is small, but is exacerbated in the low-sodium external solution, which has lower conductance than seawater because of the lower mobility of choline.

which they may have reduced the potassium current. It does seem at least possible that they account for the whole of the discrepancy, and we therefore assume provisionally that substituting choline sea water for sea water has no direct effect on the potassium current.   33

Separation of ionic current into I_{Na} and I_K

The results so far described suggest that the ionic current during a depolarization consists of two more or less independent components in parallel, an early transient phase of current carried by sodium ions, and a delayed long-lasting phase of current carried by potassium ions. In each case, the direction 34 of the current is determined by the gradient of the electro-chemical potential of the ion concerned. It will clearly be of great interest if it is possible to estimate separately the time courses of these two components. There is enough information for doing this in data such as are presented in Fig. 2, if we make certain assumptions about the effect of changing the solution around the axon. If we compare the currents when the axon is in the low-sodium solution with those in sea water, the membrane potential during the voltage clamp being the same in both cases, then our assumptions are: 35

(1) The time course of the potassium current is the same in both cases.

(2) The time course of the sodium current is similar in the two cases, the amplitude and sometimes the direction being changed, but not the time scale or the form of the time course.

(3) $\dfrac{dI_K}{dt} = 0$ initially for a period about one-third of that taken by I_{Na} to reach its maximum.

The first two of these assumptions are the simplest that can be made, and do not conflict with any of the results we have described, while the third is strongly suggested by the form of records near the sodium potential, as pointed out on p. 454. These points are sufficient reason for trying this set of assumptions first, but their justification can only come from the consistency of the results to which they lead. The differences between the effects of lack of com- 36 pensation, and of the polarization phenomenon, in the two solutions, referred to at the end of the last section, will of course lead to certain errors in the 37 analysis in the later stages of the ionic current.

The procedure by which we carried out this analysis was as follows:

(1) Three series of voltage clamp records at a range of strengths were taken, the first with the axon in one of the solutions chosen for the comparison, the second with the axon in the other solution, and the third with the first solution again. Such a set of records is reproduced in Fig. 2.

(2) Each record was projected on to a grid in which the lines corresponded to equal intervals of time and current, and the current was measured at a series 38 of time intervals after the beginning of the voltage change.

33. *'assume provisionally that substituting choline . . . has no direct effect on the potassium current.'*
The key word here is 'direct,' as H&H's interest is in finding out whether the early and late currents are independent. Again, the idea is taken as true until falsified.

34. *'more or less independent components in parallel . . . carried by sodium . . . and . . . potassium ions.'* H&H explicitly state the hypothesis. The next goal is to describe each of these currents mathematically, as a function of voltage to define its *amplitude* and as a function of time to define its *kinetics*. If the assumption that the currents are independent is valid, they should be linearly separable, and so they should sum to give the total ionic current (see Appendix 2.1). Therefore, the first step involves separating, or isolating, the currents.

35. *'our assumptions are'* The assumptions are systematically laid out so that if falsification of the hypothesis occurs, it will be possible to go back and reject the false assumption. The three points relate to what would change and what would remain the same in control and low-sodium solutions if the currents were indeed independent: First, the potassium current would be the same. Second, the sodium current time course (kinetics) would be the same, whereas the magnitude (amplitude), including sign, may differ. Third, since the potassium current is delayed, a short period in which potassium current remains zero ($dI_K/dt = 0$) would exist at the beginning of the depolarizing voltage step. This final assumption is validated by inspection of the recordings made at the reversal potential for the sodium current.

36. *'These points are sufficient reason for trying this set of assumptions first, but their justification can only come from the consistency of the results to which they lead.'* This is a profoundly important direct statement of the logical underpinnings of the entire body of work. This study does not consist of a single 'smoking gun' experiment. Neither does each experiment demonstrate a single point conclusively on which the next experiment builds serially. It is therefore not possible to present a 'story' in which *one* thing leads to *another*. Instead, *multiple* (indeed, numerous) lines of evidence, taken together, support the ultimate conclusions. Since not all the evidence can be given at once, the ideas are stated provisionally, and their validity is tested as the experimental narrative proceeds.

37. *'will of course lead to certain errors'* H&H again acknowledge the inevitability of experimental errors. These may be technical errors, such as imprecise measurements because of imperfect devices, or they may be biological errors, which would reflect mistakes in the assumptions about what the axon is really doing. H&H will discuss these errors as they arise.

38. *'Each record was projected on to a grid . . . current was measured at . . . time intervals'* Because the records are photographs of the oscilloscope screen, H&H must manually digitize the records by making point-by-point measurements on graph paper of the time values, $x_1 . . . x_n$, and the corresponding current values, $y_1 . . . y_n$, for every trace, in each condition (initial, test, and wash solutions). Note that the initial and wash solutions are later averaged to control for rundown.

(3) The time course of the initial pulse of current through the membrane capacity was determined from anodal records as described on p. 452 above, and subtracted from the measured total currents. Different corrections were 39 needed in the two solutions, because the capacity current had a slower time course in the low sodium solutions, perhaps as a result of their lower conductivity. This procedure yielded a family of curves of ionic current against time such as is shown in Fig. 3.

(4) Each pair of curves in the first and third series at the same strength was averaged, in order to allow for the slow deterioration in the condition of the axon that took place during the experiment.

(5) The difference in resting potential was allowed for by interpolating between consecutive curves in either the second series or the series of averaged 40 curves.

(6) We have now obtained curves of ionic current against time in the two solutions, with strengths of depolarization which reach the same membrane potential during the voltage clamp. The ionic current will be called I_i in sea water and I_i' in the low-sodium solution. The components carried by sodium and potassium in the two cases will be called I_{Na}, I_{Na}', I_K and I_K' respectively. The next step was to plot I_i' against I_i, and to measure the initial slope k of 41 the resulting graph (corresponding to the beginning of the voltage clamp).

Since we assume that initially $dI_K/dt = 0$, and that I_{Na} and I_{Na}' have similar time courses, $k = I_{Na}'/I_{Na}$. Further, since we assume that $I_K = I_K'$,

$$I_i - I_i' = I_{Na} - I_{Na}' = I_{Na}(1-k).$$

Hence
$$I_{Na} = (I_i - I_i')/(1-k), \tag{4}$$

$$I_{Na}' = k(I_i - I_i')/(1-k), \tag{5}$$

and
$$I_K = I_K' = I_i - I_{Na} = (I_i' - kI_i)/(1-k). \tag{6}$$

These equations give the values of the component currents at any time in terms of the known quantities I_i and I_i' at that time. Curves of I_{Na} and I_K against time could therefore be constructed by means of these equations.

This procedure is illustrated in Fig. 5, which shows two pairs of ionic current curves together with the deduced curves of I_{Na}, I_{Na}' and I_K against time. The complete family of I_K curves from this experiment is shown in Fig. 6b, while Fig. 6a shows the family derived by the same procedure from another experiment. A satisfactory feature of these curves, which is to some extent a check on the validity of the assumptions, is that the general shape is the same at all strengths. If the time courses of I_{Na} and I_{Na}' had not been of similar form, 42 Equation (6) would not have removed sodium current correctly. It would then have been unlikely that the curve of potassium current at a potential away from the sodium potential would have been similar to that at the sodium potential, where the sodium current is zero and Equation (6) reduces to $I_K = I_i$ because $k = \infty$.

39. *'time course of the initial pulse . . . was . . . subtracted'* The capacity current was manually subtracted to isolate the ionic current (see note 15).

40. *'difference in resting potential was allowed for by interpolating between consecutive curves'* For the subtraction process to be accurate, the voltages for any pair of current records must match. Therefore, if recordings were made at −50 and −40 mV *abs.* in control but at −53 and −43 mV *abs.* in choline, H&H must correct the recordings in choline by interpolation.

41. *'to plot I_i' against I_i and to measure the initial slope k'* This section begins a more quantitative test of the hypothesis that sodium and potassium currents are independent and sum linearly. If so, then total ionic current, I_i, will equal $I_{Na} + I_K$ (or in choline seawater, $I_i' = I_{Na}' + I_K$). If the currents in both solutions were identical, then plotting I_i' versus I_i would give a straight line with a slope of 1. Deviations from unity will show how the currents have changed. Since I_K is assumed to be unaffected by choline substitution, the deviations will represent changes in I_{Na}. This idea is most useful at the earliest time points, when I_K is 0 and I_{Na} is smallest. These points fall at the base of the I_i' versus I_i plot. The slope in this initial region, k, gives the fractional reduction of I_{Na} in choline seawater. The subsequent algebraic equations give the method of deducing the *isolated* I_{Na}, I_{Na}', or I_K in terms of the measured I_i, I_i', and estimated k.

42. *'to some extent a check on the validity of the assumptions, is that the general shape is the same at all strengths.'* **Figure 6** shows the potassium current from pairs of recordings done in seawater and either 30% or 10% seawater as in **Figure 5**. The sodium current reversed at different voltages in the two solutions, and so the I_{Na}' waveforms at each potential would have been different. Nevertheless, when these distinct sodium currents were subtracted from the total ionic current in the two conditions, the shape of the resulting isolated potassium current was similar, as predicted by the hypothesis of independent currents.

On the other hand, it is clearly inconsistent that the I_{Na} and I'_{Na} curves in
the lower part of Fig. 5 reverse their direction at 2 msec. after the beginning   43
of the pulse. This is a direct consequence of the fact, discussed on p. 456 above,
that the late outward current is somewhat greater in sea water than in the low-
sodium solutions, even when allowance is made for the resting potential shift.

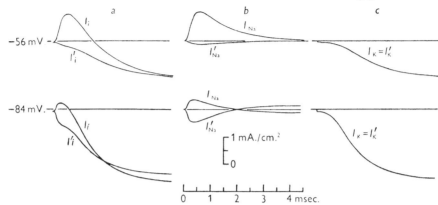

Fig. 5. Curves illustrating separation of ionic current into I_{Na} and I_{K}. Upper part of figure.   44
a, ionic currents: I_i, axon in sea water, membrane potential lowered by 56 mV.; I'_i, axon in
10% sodium sea water, membrane potential lowered by 60 mV. (average of curves taken
before and after I_i). b, sodium currents: I_{Na}, sodium current in sea water; I'_{Na}, sodium
current in 10% sodium sea water. c, potassium current, same in both solutions. Lower part
of figure. Same, but membrane potential lowered by 84 mV. in sea water and 88 mV.
in 10% sodium sea water. Current and time scales same for all curves. Axon no. 21;
temperature 8·5° C.

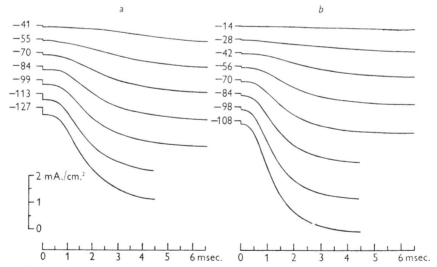

Fig. 6. Curves of potassium current against time for various strengths of depolarization. Dis-   45
placement of membrane potential when axon is in sea water is indicated for each curve, in
millivolts. a, derived from voltage clamps with axon in 30% sodium sea water, sea water
and 30% sodium sea water. Axon no. 20; temperature 6·3° C. b, derived from voltage clamps
with axon in 10% sodium sea water, sea water and 10% sodium sea water. Axon no. 21;
temperature 8·5° C.

43. **'inconsistent that the I_{Na} and I'_{Na} curves . . . reverse their direction'** The apparent change in direction of current flow is evident at the 2-msec point in the lower records of **Figure 5b**. A switch from inward to outward flux of sodium during the voltage step is kinetically unlikely as it would require an extreme change in the sodium equilibrium potential over the course of two milliseconds. The apparent reversal can be accounted for by the point raised earlier, namely, that the resting potential hyperpolarized in choline, with the outcome that slightly more potassium current was activated in control relative to choline seawater. If a little extra outward current were subtracted to obtain the control sodium currents, the isolated 'sodium' current would appear to reverse direction.

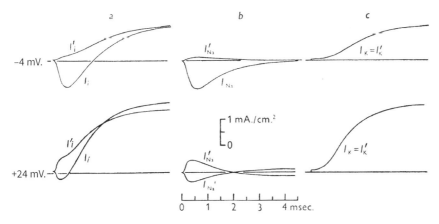

44. **Figure 5.** The total current (*a*), isolated sodium current (*b*), and isolated potassium current (c) are shown for normal (*I*) and low-sodium (*I'*) solutions. Note that these traces provide the first clear evidence that the sodium current itself decays back to baseline during a maintained depolarization, rather than simply being overwhelmed by a large flux in the opposite direction. This result was unexpected (see notes 46 and 47).

45. **Figure 6.** After the sodium current is subtracted, the remaining currents include the leak current as well as the voltage-gated potassium current. The leak current is evident as a small step increase in outward current at the onset of the voltage step, which scales with the magnitude of the depolarization (see note 49).

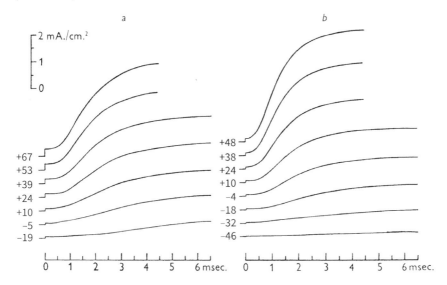

It was pointed out there that the difference may well be due to lack of compensation and to 'polarization'. Until these effects can be eliminated, estimates of sodium current at the longer times will be quite unreliable, and the corresponding estimates of potassium current will be somewhat reduced by these errors.

All the sodium current curves agree in showing that I_{Na} rises to a peak and then falls. With weak depolarizations (less than 40 mV.) the steady state value is definitely in the same direction as the peak, but at higher strengths the measured I_{Na} tends to a value which may have either direction. Since the sources of error mentioned in the last paragraph can cause an apparent reversal of I_{Na} during the pulse, it is possible that if these errors were larger than we suppose the whole of the apparent drop of I_{Na} from its peak value might also be spurious. At the time that an account of preliminary work with this technique was published (Hodgkin *et al.* 1949) we were unable to decide this point, and assumed provisionally that I_{Na} did not fall after reaching its maximum value. We are now convinced that this fall is genuine: (1) because of improvements in technique; (2) because of further experiments of other kinds

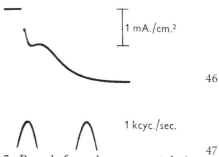

46

47

Fig. 7. Record of membrane current during a voltage clamp with axon in choline sea water, showing early maximum of outward current. Displacement of membrane potential during clamp $= -84$ mV. Axon no. 24; temperature 20° C.

which are described in the next two papers of this series (Hodgkin & Huxley, 1952 *a*, *b*); and (3) because we occasionally observed records of the kind shown in Fig. 7. This is a record of membrane current during a voltage clamp in which an axon in choline sea water was depolarized by 84 mV. It will be seen that the early hump of outward current (due to sodium ions) was so marked that the total current reached a maximum at about 0·2 msec. and then fell before finally rising to the plateau attributable to movement of potassium ions. Unless we make the quite unwarrantable assumption that I_K itself has this double-humped form, this curve can only be explained by supposing that I_{Na} (outward in this case) falls after passing through a maximum value. The 48 fact that such a clear maximum was not regularly observed was no doubt due to I_{Na} usually being smaller in relation to I_K than in this case.

We do not present a family of I_{Na} curves here, because the sequence of the curves is interrupted at the sodium potential. For this reason, the information is better given in the curves of 'sodium conductance' which are derived later in this paper (pp. 461–2 and Fig. 8). The variation of peak sodium current with strength of depolarization is shown in Fig. 13 for axons both in sea water and in low-sodium solutions.

46. *'the whole of the apparent drop of I_{Na} from its peak value might also be spurious.'* The observation is that sodium current rises to a peak and then decays, but H&H have pointed out that the apparent change in direction of I_{Na} (from inward early to outward late in the trace, or vice versa, as in the lower records in *Figure 5b*) arises because their measurement exaggerates the decay in I_{Na}. They indicate the reasons why they think the error is small, but they consider the possibility that the error might be large, and that sodium current may not in fact decay at all. If this were the case, I_{Na} might actually be a sustained current that is distorted by technical limitations.

47. *'we were unable to decide this point. . . . We are now convinced that this fall is genuine'* What is striking here is the direct manner in which H&H acknowledge how hard this question was to resolve. They offer three bases for their change of mind to thinking that the decay, which is now recognized as inactivation, is real. Again, they rely on multiple lines of evidence. Not only have the technical concerns been mitigated, but the whole of Paper 4 will also provide evidence for inactivation of sodium current. For now, they focus on the evidence provided by the curve in *Figure 7* (see note 48).

48. *'the quite unwarrantable assumption that I_K itself has this double-humped form, . . . I_{Na} . . . falls after passing through a maximum value.'* This experiment provides an example of a single, high-resolution observation that nonetheless requires an explanation. The situation is distinct from the type of experiment in which many measurements must be gathered to obtain an estimate of a mean value of a variable. Here, the observation that the outward current (evoked by depolarization in choline seawater) rises, falls, and rises again suggests one of two possibilities. Either the triphasic response comes entirely from the late potassium current (superimposed on a small sustained outward sodium current), or the total current has two components: a biphasic sodium current that rises and falls, and a slowly rising sustained potassium current that contributes the third phase. The former possibility is kinetically unlikely (in H&H's words, 'unwarrantable'): A single stimulus (a change in voltage) is not expected to trigger a process—here, one that permits potassium current to flow—to turn on, off, and on again. The latter possibility is therefore more likely.

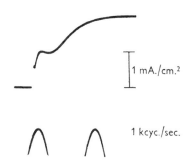

1 mA./cm.²

1 kcyc./sec.

Current carried by other ions. It seems to be possible to account for the variation of current with time during the voltage clamp by variations in the currents carried across the membrane by two ions, namely sodium and potassium. If, however, the membrane allowed constant fluxes of one or more other ion species, the current carried by these would form part of the 'I_K' which is deduced by our procedure, since this current would be independent 49 both of time and of sodium concentration, and I_K is defined by its satisfying these criteria during the earliest part of the pulse. Reasons will be given in the next paper (Hodgkin & Huxley, 1952a) for supposing that the current carried by other ions is appreciable, though not of great importance except when the membrane potential is near to or above its resting value. Each of the I_K curves in Figs. 5 and 6 therefore includes a small constant component carried by other ions. This component probably accounts for much of the step in 'I_K' at the beginning of the voltage pulse.

Expression of ionic currents in terms of conductances

General considerations. The preceding sections have shown that the ionic current through the membrane is chiefly carried by sodium and potassium ions, moving in each case under a driving force which is the resultant of the concentration difference of the ion on the two sides of the membrane, and of the electrical potential difference across the membrane. This driving force 50 alone determines the direction of the current carried by each ionic species, 51 but the magnitude of the current depends also on the freedom with which the membrane allows the ions to pass. This last factor is a true measure of the 'permeability' of the membrane to the ion species in question. As pointed out 52 by Teorell (1949a), a definition of permeability which takes no account of electrical forces is meaningless in connexion with the movements of ions, though it may well be appropriate for uncharged solutes.

The driving force for a particular ion species is clearly zero at the equilibrium potential for that ion. The driving force may therefore be measured as the difference between the membrane potential and the equilibrium potential. Using the same symbols as in Equations (1)–(3), the driving force for sodium ions will be $(E - E_{Na})$, which is also equal to $(V - V_{Na})$. The permeability of the membrane to sodium ions may therefore be measured by $I_{Na}/(E - E_{Na})$. This quotient, which we denote by g_{Na}, has the dimensions of a conductance (cur- 53 rent divided by potential difference), and will therefore be referred to as the sodium conductance of the membrane. Similarly, the permeability of the membrane to potassium ions is measured by the potassium conductance g_K, which is defined as $I_K/(E - E_K)$. Conductances defined in this way may be called chord conductances and must be distinguished from slope conductances (G) defined as $\partial I/\partial E$. 54

These definitions are valid whatever the relation between I_{Na} and $(E - E_{Na})$,

49. **'fluxes of . . . other ion species . . . would form part of the "I_K" which is deduced by our proce-dure'** H&H point out that they subtracted I_{Na} from the total current and lumped the rest as I_K, but other currents besides potassium may be present. These turn out to be the leak currents, discussed later. As they indicate, leak current dominates the total ionic current only at membrane potentials near or more hyperpolarized than the resting potential.

50. **'ionic current . . . carried by sodium and potassium ions, moving in each case under a driving force . . . and of the electrical potential difference across the membrane.'** H&H reiterate the point made earlier (see note 10), which sets up the explicit discussion of the ionic flux and the equilibrium potentials in the context of Ohm's law.

51. **'driving force alone determines the direction of the current'** Any cationic current is inward at volt-ages more hyperpolarized than its equilibrium potential and outward at voltages more depolarized than its equilibrium potential. Note that since H&H's conventions set the resting potential at 0 mV, the term 'driving force'—the difference between the reversal potential and the membrane potential—at rest is synonymous with the equilibrium potential. (H&H mention this idea in the following paragraph.)

52. **'magnitude . . . depends also on the freedom with which the membrane allows ions to pass. This . . . factor is a true measure of the "permeability" of the membrane'** This statement is par-ticularly important. The idea of membrane *permeability* had been discussed for decades: the mem-brane was said to be 'selectively permeable' to potassium; the electrical signal of nerve was pro-posed to result from the membrane becoming 'permeable' to all ions. The equation that came to be known as the Goldman-Hodgkin-Katz (GHK) voltage equation defined resting potential in terms of the relative permeabilities to potassium, sodium, and chloride ions. Permeability, however, which has units of a rate (cm/sec), has no simple quantitative relationship to current, nor can it be measured directly. Here, H&H begin to consider the specific nature of the relationship between the magnitude of an ionic current and the electrical driving force on the ions, defining permeability in electrical terms for the first time.

53. **'has the dimensions of a conductance'** This key sentence introduces the idea that the phenomenon long described as 'ionic permeability' can be expressed as a conductance in the squid axonal mem-brane. What H&H propose is that each ionic current, I_{ion}, depends linearly on driving force, $(E - E_{ion})$, where E is the absolute membrane potential (often given as V_m in modern conventions), but also on a scaling factor that reflects the freedom of move-ment of ions. Thus, $I_{ion} \propto (E - E_{ion})$. According to Ohm's law ($I = V/R$), the factor that scales voltage to give current is the inverse of resistance ($1/R$), or the conductance, g.

54. **'Conductances defined in this way may be called chord conductances and must be distinguished from slope con-ductances'** A chord conductance is calculated from the slope of a line drawn on an *I-V* curve from the point of interest to the point at which the current reverses (*Figure A, dotted line*, modified from Figure 13 of Paper 1). The chord conductance, g, is then $(I_2 - I_1)/(V_2 - V_1)$ or $\Delta I / \Delta V$. The slope conductance, G, is given by the tangent to the *I-V* curve at a specific point (*Figure A, dashed line*), which is why H&H express it as a local derivative. Note that the slope conductance for the point in the figure is negative, reflecting the voltage dependence of the conductance. In contrast, chord conductances are never negative.

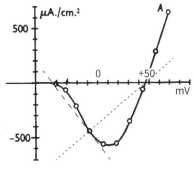

Figure A

or between I_K and $(E - E_K)$, but the usefulness of the definitions, and the degree to which they measure real properties of the membrane, will clearly be much increased if each of these relations is a direct proportionality, so that g_{Na} and g_K are independent of the strength of the driving force under which 55 they are measured. It will be shown in the next paper (Hodgkin & Huxley, 1952*a*) that this is the case, for both sodium and potassium currents, in an axon surrounded by sea water, when the measurement is made so rapidly that the condition of the membrane has no time to change.

Application to measured sodium and potassium currents. The determination of sodium current, potassium current and sodium potential have been described in earlier sections of the present paper. The method by which the potassium potential, E_K, was found is described in the next paper (Hodgkin & Huxley, 1952*a*), and the values used here are taken from that paper. We have 56

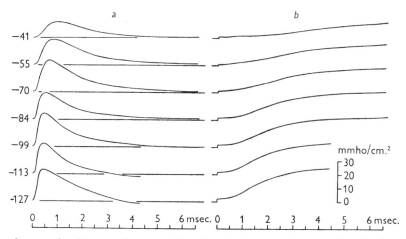

Fig. 8. Curves of sodium conductance (*a*) and potassium conductance (*b*). Displacement of 57 membrane potential (millivolts) when axon was in sea water is indicated on each curve. Curves of I_i and I_K in same experiment are shown in Figs. 3 and 6*a* respectively. Axon no. 20; temperature 6·3° C.

therefore sufficient data to estimate g_{Na} and g_K as functions of time during a voltage clamp. Families of g_{Na} and g_K curves, for various strengths of depolarization, are shown in Fig. 8. The sodium conductances are calculated from the sodium currents in sea water, divided by the difference between membrane potential and sodium potential in sea water. If the same pro- 58 cedure had been applied to the corresponding quantities in the low-sodium solution, a similar family would have been obtained, but the relative amplitudes of the members of the family would have been slightly different. The values 59 obtained from the sea water figures are the more interesting, both because they refer to a more normal condition, and because it is only in this case that the instantaneous relation between sodium current and voltage is linear

55. *'the usefulness of the definitions, and the degree to which they measure real properties of the membrane, will . . . be . . . increased if . . . g_{Na} and g_K are independent of the strength of the driving force'* This statement foreshadows Paper 3, on tail currents. If the membrane permeability behaves as a simple ohmic conductance, then current will scale linearly with voltage for a fixed state of the membrane, a relationship that has not been established at this point. It is interesting to note the formal scientific skepticism with which H&H introduce the idea that permeability may be represented as a conductance in an electrical sense. The link between permeability and conductance is now recognized so widely in electrophysiology that it can be difficult to grasp that it was a revolutionary concept that required a great deal of evidence, which is provided in these papers.

56. *'values used here are taken from that paper.'* E_K was found to be about 10 mV more hyperpolarized than rest, or −70 to −75 mV *abs*. This value is more depolarized than in most neurons, in which E_K is generally closer to −90 mV.

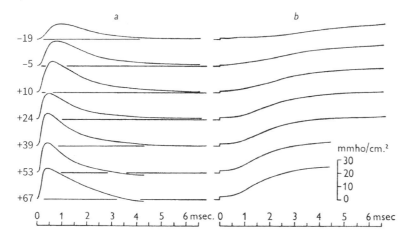

57. **Figure 8.** Note that these are *conductance* traces, unlike the *current* traces of **Figure 5** and **Figure 6**, and they have units of mhos per unit area. ('Mho' is 'ohm' spelled backward, reflecting the fact that conductance is the inverse of resistance. 'Mhos' has since been replaced by the metric unit, 'siemens'.) The conductance traces are calculated by manually dividing current traces point-by-point by the driving force. Note that upward deflections are positive and simply indicate that the conductance is active. Chord conductance is *never negative* (see note 54); the lowest value is zero, meaning that the membrane is nonconducting, or in modern terms, that all ion channels are closed.

These traces represent a significant experimental and conceptual step toward understanding the fundamental mechanisms underlying the action potential. Until this point, H&H have recorded and displayed currents, which depend both on the conductance of the membrane and the externally applied driving force (transmembrane potential). Here for the first time, the conductance is displayed independently of the driving force, as an isolated property of the membrane.

58. *'sodium conductances are calculated from . . . currents . . . divided by the difference between membrane potential and sodium potential'* This is a verbal rendering of the form of Ohm's law that is most relevant to studying membrane conductance: $g_{Na} = I_{Na}/(E - E_{Na})$.

59. *'in the low-sodium solution, a similar family would have been obtained, but the relative amplitudes . . . would have been slightly different.'* Here, H&H point out that the time course and voltage dependence of the conductance (*gating*) are independent of the flow of ions (*permeation*). In contrast, the magnitude of the conductance does depend on the concentration of ions available to flow across the membrane. Lower concentrations of permeant ions yield lower conductances (reducing the maximal value of g_{Na}), generally without changing kinetics; however, the current-voltage relationship will change, as shown in Paper 3.

(Hodgkin & Huxley, 1952a). The corresponding distinction does not arise with g_K, since both I_K and E_K are the same in both solutions.

The shapes of individual curves in Fig. 8 are of course similar to those of curves of I_{Na} or I_K, such as are shown in Figs. 5 and 6, since the driving force for each ion is constant during any one voltage clamp. The change of amplitude of the curves with strength of depolarization is, however, less marked than with the current curves. For potassium, this can be seen by comparing   60 Figs. 6a and 8b, which refer to the same experiment. For sodium, it is clear from Fig. 8a that the conductance curves undergo no marked change at the sodium potential, while the current curves reverse their direction at this point.   61

Membrane potential and magnitude of conductance. The effect of strength of depolarization on the magnitude of the conductances is shown in Figs. 9 and 10. For each experiment, the maximum values of g_{Na} and g_K reached in a voltage clamp of strength about 100 mV. are taken as unity, and the maximum values at other strengths are expressed in terms of these. Values of g_{Na} 62 are available only from the four experiments in which there were enough data in sea water and in a low-sodium solution for the complete analysis to be carried out. The maximum values of g_K were also estimated in two other experiments. This was possible without complete analysis because the late current was almost entirely carried by potassium when the axon was in choline sea water.

The two curves are very similar in shape. At high strengths they become flat, while at low strengths they approach straight lines. Since the ordinate is plotted on a logarithmic scale, this means that peak conductance increases exponentially with strength of depolarization. The asymptote approached by   63 the sodium conductances is probably steeper than that of the potassium data; the peak sodium conductances increase e-fold for an increase of 4 mV. in strength of depolarization; for potassium the corresponding figure is 5 mV.   64

The values of conductance at a depolarization of 100 mV., which are represented as unity in Figs. 9 and 10, are given in Table 2. In all these cases where enough measurements were made to construct a curve, the axon had been used for other observations before we took the records on which the analysis is based. In several cases, it was possible to estimate one or two values of sodium and potassium peak conductance at the beginning of the same experiment, and these were considerably higher than the corresponding values in Table 2, which must therefore be depressed by deterioration of the fibres.

More representative values of the peak g_K and g_{Na} at high strengths were estimated at the beginning of experiments on several fibres. Potassium current at long times can be estimated without difficulty, since I_{Na} is then negligible, especially at these depolarizations which are near the sodium potential. Nine fibres at 3–11° C. gave peak values of g_K at -100 mV. ranging from 22 to 41 m.mho/cm.^{2}, with a mean of 28; five fibres at 19–23° C. gave a range of

60. *'change of amplitude of the curves with strength of depolarization . . . less marked than with the current curves.'* For potassium *currents*, the amplitude increases with depolarization for two reasons: (1) because the conductance increases and (2) because the driving force increases. For potassium *conductance*, the driving force is not a factor, so the extent of increase is less for successive depolarizations.

61. *'conductance curves undergo no marked change at the sodium potential, while the current curves reverse their direction'* This point is extremely important, as it begins to establish that conductance, unlike current, is a monotonic function of voltage.

62. *'maximum values of g_{Na} and g_K . . . are taken as unity, and the maximum values at other strengths are expressed in terms of these.'* These data yield the first conductance-voltage or $g(V)$, curves, which plot conductance, g, normalized to a number between 0 and 1 by dividing by the maximal conductance, against depolarization relative to rest, V. Here the normalized conductance is plotted on a logarithmic y-axis.

63. *'peak conductance increases exponentially with strength of depolarization.'* On the semilog plot, the curves are linear at the more hyperpolarized potentials, indicating an exponential voltage dependence at the base of each curve. At the more depolarized potentials, the curves start to flatten, indicating saturation. The calculated conductances for sodium and for potassium define curves that differ primarily in their relative position on the x-axis and in the steepness of their rising phases. These are suggestive of common underlying mechanisms for the two conductances that differ only in the values of specific parameters.

64. *'peak sodium conductances increase e-fold for an increase of 4 mV . . . for potassium . . . 5 mV.'* Conductance-voltage curves like those in **Figures 9** and **10** are now more commonly plotted on linear (rather than semilog) axes, where they appear as S-shaped or 'saturating' curves. A simple function often used to fit saturating curves:

$$y = \frac{1}{1 + e^{-x}}$$

(see Appendix 1.4 and Paper 4, note 17). The rising phase of the curve defined by this equation can be made steeper or shallower by dividing x by a factor k, called 'the slope factor,' which here would indicate an e-fold increase in conductance for every k mV of depolarization; lower values of k therefore correspond to steeper curves and indicate a greater sensitivity to voltage. To a first approximation, the curves shown in **Figures 9** and **10** could have been fitted with a the two-state equation. Although H&H do not use this approximation (for reasons addressed in Paper 5), the slope factor, k, can be extracted from the slope of the straight line defined by the points at the base of the curves.

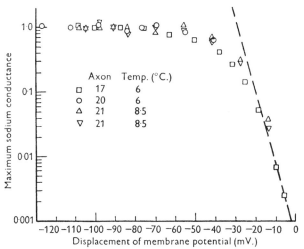

Fig. 9. Maximum sodium conductance reached during a voltage clamp. Ordinate: peak conduc-   65
tance relative to value reached with depolarization of 100 mV., logarithmic scale. Abscissa:
displacement of membrane potential from resting value (depolarization negative).

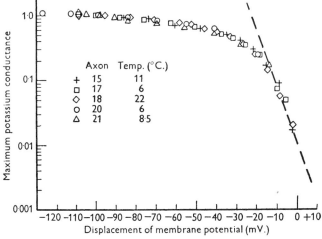

Fig. 10. Maximum potassium conductance reached during a voltage clamp. Ordinate: maximum   66
conductance relative to value reached with depolarization of 100 mV., logarithmic scale.
Abscissa: displacement of membrane potential from resting value (depolarization negative).

TABLE 2. Peak values of sodium and potassium conductance at a depolarization of 100 mV.   67
Same experiments as Figs. 9 and 10. In each case, the value given in this table is represented
as unity in Fig. 9 or Fig. 10.

| | | Peak conductances at − 100 mV. | |
| | | Sodium | Potassium |
Axon no.	Temp. (°C.)	(m.mho/cm.²)	(m.mho/cm.²)
15	11	—	21
17	6	18	20
18	21	—	28
20	6	22	23
21	8·5	23	31
21	8·5	17	—
	Mean	20	25

65. **Figure 9.** Note that half-maximal activation of sodium conductance occurs about 35 mV depolarized from rest, near −25 mV *abs*. The dashed line has a slope equal to $1/k$, where k, the slope factor, is 4 mV.

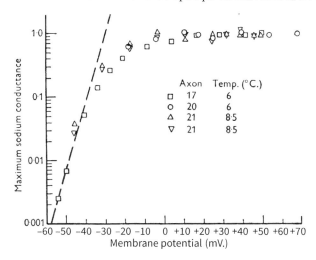

66. **Figure 10.** Note that half-maximal activation of potassium conductance occurs about 40 mV depolarized from rest, near −20 mV *abs*. The dashed line has a slope equal to $1/k$, where k, the slope factor, is 5 mV.

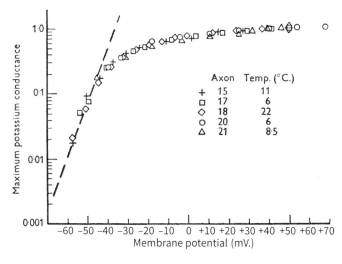

67. **Table 2.** Note that the mean maximal potassium conductance (measured at 100 mV more depolarized than rest, or at +40 mV *abs*.) is about 25% higher than the mean maximal sodium conductance. As H&H pointed out, however, the potassium conductance likely does not attain its peak during the brief depolarization associated with an action potential.

33–37 m.mho/cm.², mean 35. Values of the peak sodium conductance were obtained by measuring the peak inward current at a depolarization of about 60 mV. and dividing by the corresponding value of $(V - V_{Na})$. They are probably 10–20% low because current carried by potassium and other ions makes the peak inward current less than the peak sodium current, and because the peak conductance at 60 mV. depolarization is slightly less than that reached at 100 mV. Five fibres at 3–9° C. gave values ranging from 22 to 48 m.mho/cm.², mean 30; a single fibre at 22° C. gave 24 m.mho/cm.².

These results show that both g_K and g_{Na} can rise considerably higher than the values for the fully analysed experiments given in Table 2. They may be summarized by saying that on the average a freshly mounted fibre gives maximum conductances of about 30–35 m.mho/cm.² both for sodium and for potassium, corresponding to resistances of about 30 Ω. for 1 cm.² of membrane. 68 This value may be compared with the resting resistance of about 1000 Ω. cm.² (Cole & Hodgkin, 1939), and the resistance at the peak of an action potential, which is about 25 Ω. cm.² (Cole & Curtis, 1939).

Membrane potential and rate of rise of conductance. It is evident from Fig. 8 that the strength of depolarization affects not only the maximum values attained by g_{Na} and g_K during a voltage clamp, but also the rates at which these maxima are approached. This is well shown by plotting the maximum 69 rate of rise of conductance against displacement of membrane potential. This has been done for sodium conductance in Fig. 11 and for potassium conductance in Fig. 12. The data for g_K were taken from a fully analysed run, but in the case of sodium it is sufficient to take the maximum rate of rise of total ionic current, with the axon in sea water, and divide by $(V - V_{Na})$. The maximum rate of rise occurs so early that dI_K/dt is still practically zero, so that $dI_i/dt = dI_{Na}/dt$.

These graphs show that the rates of rise of both conductances continue to increase as the strength of the depolarization is increased, even beyond the point where the maximum values reached by the conductances themselves have become practically constant. 70

DISCUSSION

Only two aspects of the results described in this paper will be discussed at this stage. The first is the relationship between sodium current and external sodium concentration; the second is the application of the results to the interpretation 71 of the action potential. Further discussion will be reserved for the final paper of this series (Hodgkin & Huxley, 1952c).

Sodium current and external sodium concentration

General considerations and theory. We have shown in the earlier parts of this paper that there is good reason for believing that the component of membrane

68. *'maximum conductances of about 30–35 m.mho/cm² both for sodium and for potassium'* Translating into mammalian scale, a value of 30–35 m.mho/cm² would correspond to 60–70 nS (nanosiemens) in a 25 μm diameter cell body. At 1 μF/cm² (0.01 pF/μm²), such a cell would have a somatic capacitance of about 20 pF. Conductances of this order of magnitude are seen in mammalian neurons that fire spontaneously, in which voltage-gated sodium and potassium channels have a high density on the somatic membrane.

69. *'not only the maximum values . . . but also the rates at which these maxima are approached.'* Both the amplitude and the kinetics of the conductance changes are voltage dependent.

70. *'rates of rise . . . continue to increase . . . beyond the point where the maximum values of the conductances themselves have become practically constant.'* The larger the voltage step, the faster the conductance activates. Unlike the amplitudes, the kinetics do not saturate in the physiological voltage range.

71. *'sodium current and external sodium concentration'* This part of the discussion develops what is now referred to as the Goldman-Hodgkin-Katz *current* equation (see note 80).

current that we refer to as I_{Na} is carried by sodium ions which move down their own electrochemical gradient, the speed of their movement, and therefore the magnitude of the current, being also determined by changes in the freedom

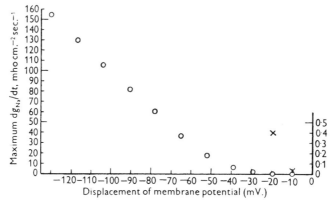

Fig. 11. Maximum rates of rise of sodium conductance during voltage clamps plotted against   72
displacement of membrane potential. Circles are to be read with the scale on the left-hand
side. The two lowest points are also re-plotted as crosses on 100 times the vertical scale, and
are to be read with the scale on the right-hand side. The peak sodium conductance reached
at high strengths of depolarization was 16 m.mho/cm.^{2}. Axon no. 41; temperature 3·5° C.
Compensated feed-back.

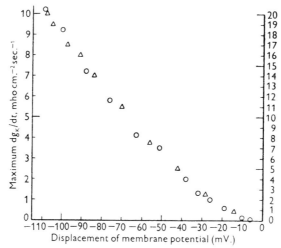

Fig. 12. Maximum rates of rise of potassium conductance during voltage clamps, plotted against   73
displacement of membrane potential. Circles, left-hand scale: axon no. 17; temperature
6° C. Triangles, right-hand scale: axon no. 21; temperature 8·5° C. At −100 mV. the maximum potassium conductance was 20 m.mho/cm.^{2} for axon no. 17 and 31 m.mho/cm.^{2} for
axon no. 21.

with which they are permitted to cross the membrane under this driving force.
If this is in fact the case, we should expect that sodium ions would cross the   74
membrane in both directions, the observed I_{Na} being the difference between
the opposing currents carried by these two fluxes. At the sodium potential

72. ***Figure 11.*** The maximal rate of rise in sodium conductance was calculated as the maximal slope of the conductance versus time curves in ***Figure 8a***. Recall from Paper 1 that action potential threshold was measured to be about 12 mV above rest. Here, a small, measurable conductance is activated 10 mV above rest (x's and right axis), and the activation rate accelerates at higher voltages, getting progressively faster up to and beyond the reversal potential.

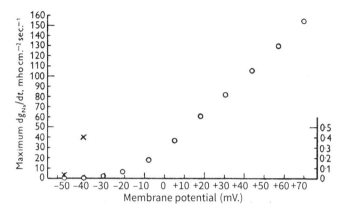

73. ***Figure 12.*** The maximal rates of rise for the potassium conductance are given for two different axons with different conductance densities (on different *y*-axis scales to account for the difference in temperature). The relative dependence on voltage is highly similar in both axons. As in the case of the sodium conductance, the activation rates do not saturate in the range of voltages tested.

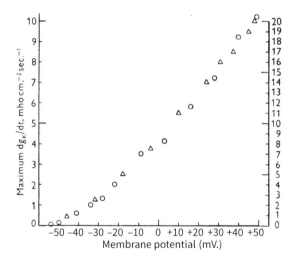

74. *'there is good reason for believing that . . . I_{Na} is carried by sodium ions. . . . If this is in fact the case, we should expect'* H&H begin their formal elaboration of what they term the 'independence principle.' This is a textbook example of the assembly and organization of observations, the formulation and statement of hypothesis, and the articulation of testable predictions.

the fluxes would be equal, making I_{Na} zero; as the membrane potential is increased from this value, the ratio of inward to outward flux would increase, making I_{Na} positive, and vice versa.

By making certain very general assumptions about the manner in which ions cross the membrane it is possible to derive an equation which predicts the effect of sodium concentration on sodium current. The theory on which this equation depends is closely connected with those of Behn (1897), Teorell (1949b) and Ussing (1949), but differs from them, both in the assumptions from which it is derived and in the range of cases to which it applies.

We assume only that the chance that any individual ion will cross the membrane in a specified interval of time is independent of the other ions which are present. The inward flux M_1 of any ion species will therefore be proportional to the concentration c_1 of that ion in the external fluid, and will not be affected by c_2, its concentration inside the axon. We may therefore write

$$M_1 = k_1 c_1, \tag{7}$$

where k_1 is a constant which depends on the condition of the membrane and on the potential difference across it. Similarly, the outward flux M_2 is given by

$$M_2 = k_2 c_2, \tag{8}$$

where k_2 is another constant, determined by the same factors as k_1 but in general different from it. Hence

$$M_1/M_2 = k_1 c_1 / k_2 c_2. \tag{9}$$

The condition for equilibrium is that $M_1 = M_2$, so that

$$k_2/k_1 = c_1^*/c_2,$$

where c_1^* is the external concentration that would be in equilibrium with the (fixed) internal concentration, under the existing value of E, the membrane potential.

Substituting for k_1/k_2 in (9), we have

$$M_1/M_2 = c_1/c_1^*. \tag{10}$$

Now $c_1^*/c_2 = \exp(-EF/RT)$ and $c_1/c_2 = \exp(-E^*F/RT)$, where E^* is the equilibrium potential for the ion under discussion, so that

$$c_1/c_1^* \quad = \exp(E - E^*)\,F/RT$$

and

$$M_1/M_2 = \exp(E - E^*)\,F/RT. \tag{11}$$

We now have in Equations (7), (8) and (11) three simple relations between M_1, M_2, c_1 and E. The effect of membrane potential on either of the fluxes alone is not specified by these equations, but is immaterial for our purpose.

If we wish to compare the sodium currents when the axon is immersed first in sea water, with sodium concentration $[Na]_o$, and then in a low-sodium

75. **'We assume only that the chance . . . is independent of the other ions'** The emphasis here is that the following derivation does not rely on any assumptions except that of independence; everything else is experimentally verified. Therefore, the hypothesis being tested is that the inward and outward flux of each ion is independent of the others, which H&H refer to as the independence principle.

76. **'The inward flux M_1 . . . will therefore be proportional to the concentration c_1'** The idea that flux depends on concentration can be intuitively grasped by recognizing that the number of people passing inward through an open doorway will be high with many people present outside the door, lower with few people outside, and zero with no one outside. In this simple analogy, the open door represents conductance, the number of people present represents the external concentration, and the number of people passing through represents the one-way current or flux.

77. **Equation 10.** At equilibrium, the ratio of inward flux to outward flux is the same as the ratio of external concentration of permeant ions to the inward concentration. The value c_1^* indicates the external concentration of permeant ion that *would have to be present* to set the reversal potential at the absolute membrane potential, E.

78. **'$c_1^*/c_2 = exp\ (-EF/RT)$ and $c_1/c_2 = exp\ (-E^*F/RT)$'** These expressions are rearrangements of the Nernst equation,

$$E = \left(\frac{RT}{F} \right) \ln \left(\frac{c_{in}}{c_{out}} \right).$$

They both give the absolute voltage in H&H's conventions, which are inverted relative to modern style (see note 18). The second equation is the more familiar form in which the equilibrium potential for an ion (e.g., E_{Na}) would be given in terms of the actual concentrations of the ion (e.g., sodium) inside (c_2) and outside (c_1) the axon. The rearrangement in Equation 11 is convenient from H&H's perspective because it converts the expressions from absolute voltages E and E^* into the differences between the voltage E and the equilibrium potential, which correspond to the values of V that they have measured.

solution with sodium concentration $[Na]'_o$, the membrane potential having the same value E in both cases, we have: 79

$$\frac{I'_{Na}}{I_{Na}} = \frac{M'_{Na_1} - M'_{Na_2}}{M_{Na_1} - M_{Na_2}}.$$

From (7), $M'_{Na_1}/M_{Na_1} = [Na]'_o/[Na]_o$ and from (8) $M'_{Na_2} = M_{Na_2}$. Using these relations and Equation (11)

$$\frac{I'_{Na}}{I_{Na}} = \frac{([Na]'_o/[Na]_o) \exp (E - E_{Na}) F/RT - 1}{\exp (E - E_{Na})F/RT - 1}.   \quad (12)   \quad 80$$

Strictly, activities should have been used instead of concentrations throughout. In the final Equation (12), however, concentrations appear only in the ratio of the sodium concentrations in sea water and the sodium-deficient solution. The total ionic strength was the same in these two solutions, so that the ratio of activities should be very close to the ratio of concentrations. The activity coefficient in axoplasm may well be different, but this does not affect Equation (12).

Equation (11) is equivalent to the relation deduced by Ussing (1949) and is a special case of the more general equation derived by Behn (1897) and Teorell (1949b). All these authors start from the assumption that each ion moves under the influence of an electric field, a concentration gradient and a frictional resistance proportional to the velocity of the ion in the membrane. This derivation is more general than ours in the respect that it is still applicable if, for instance, a change in c_1 alters the form of the electric field in the membrane and therefore alters M_2; in this case, Equations (8) and therefore (12) are not obeyed. On the other hand, it is more restricted than our derivation in that it specifies the nature of the resistance to movement of the ions.

Agreement with experimental results. Equation (12) is tested against experimental results in Fig. 13. Section (a) shows data from the experiment illus-   81
trated in Figs. 3 and 6a. The values of the sodium current in sea water (I_{Na}) and in 30% Na sea water (I'_{Na}) were derived by the procedure described in the 'Results' section. The crosses are the peak values of I_{Na}, plotted against V, the displacement of membrane potential during the voltage clamp. A smooth curve has been fitted to them by eye. V_{Na} was taken as the position   82
at which the axis of V was cut by this curve. Since

$$V = E - E_r, \quad (E - E_{Na}) = (V - V_{Na}),$$

and, for each point on the smoothed curve of I_{Na} against V, a corresponding value of I'_{Na} was calculated by means of Equation (12). These values are plotted as curve B. The experimentally determined peak values of I'_{Na} are shown as circles. These are seen to form a curve of shape similar to B, but of greater amplitude. They are well fitted by curve C, which was obtained from   83
B by multiplying all ordinates by the factor $1 \cdot 20$.

79. *'to compare the sodium currents . . . in sea water . . . and then in a low-sodium solution . . . we have:'* Note that the net current, I_{Na}, is the difference between the inward sodium flux, M_{Na_1}, and the outward sodium flux, M_{Na_2}.

80. **Equation 12.** This is one formulation of the Goldman-Hodgkin-Katz (GHK) current equation, which indicates that the ratio of currents in two different external solutions depends on the concentrations of permeant ions in each solution, the voltage at which the measurement is made, and the equilibrium potential.

81. *'Equation (12) is tested against experimental results in Fig. 13.'* As stated in note 74, the hypothesis that ionic current can be accounted for by flux governed by physical principles gives rise to testable predictions. Testability comes in the form of the mathematical 'model' of the GHK current equation. If the assumptions are correct, the model will generate a curve that overlaps well with the data. This scenario is different from *curve-fitting*, in which parameters are adjusted to replicate the pattern of the data. Curve-fitting is appropriate when the form of the function describing a phenomenon is known, and only the values of the parameters must be estimated for a specific case. Here, the question is whether the experimentally measured phenomena can be quantitatively predicted by an equation that emerges from a proposed theory, here, the independence principle.

82. *'A smooth curve has been fitted to them by eye.'* H&H plot the peak current-voltage relation from measurements of sodium current in seawater and draw a line *A* through the points, so that data can be estimated at all *x*-values. Next, from those values, they calculate what **Equation 12** predicts the current to be in low-sodium solution (curve *B*). Lastly, they compare the predicted values to those measured in low-sodium seawater.

83. *'a curve of shape similar to B, but of greater amplitude.'* The result is that **Equation 12** gives the correct *relative* voltage dependence, but the magnitude is too small by 20%.

Fig. 13b, c were obtained in the same way from experiments in which the low-sodium solutions were 10% Na sea water and choline sea water respectively. In each case, the peak values of I'_{Na} are well fitted by the values predicted by means of Equation (12), after multiplying by constant factors of 1·333 and 1·60 in (b) and (c) respectively.

These constant factors appear at first sight to indicate a disagreement with the theory, but they are explained quantitatively by an effect which is described in the fourth paper of this series (Hodgkin & Huxley, 1952b). The 84

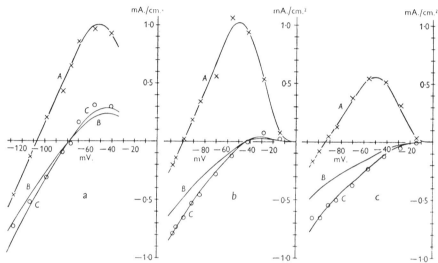

Fig. 13. Test of 'independence principle'. Three experiments. Crosses: peak sodium current 85
density during voltage clamp; axon in sea water. Curve A fitted by eye. Circles: peak
sodium current density during voltage clamp; axon in low sodium sea water. Curve B: peak
sodium current density in low sodium sea water, predicted from curve A by Equation (12).
Curve C: as curve B, but all ordinates multiplied by a factor f. Abscissa: membrane potential measured from resting potential in sea water. (a) Axon no. 20; temperature 6° C. Data
from voltage clamps in (1) 30% sodium sea water, (2) sea water, (3) 30% sodium sea water.
Circles are average values from runs (1) and (3). Value of V_{Na} inserted in Equation (12):
−106·8 mV. Factor f=1·20. (b) Axon no. 21; temperature 8·5° C. Data from voltage
clamps in (1) 10% sodium sea water, (2) sea water, (3) 10% sodium sea water. Circles are
average values from runs (1) and (3). V_{Na} = −98·8 mV., f=1·333. (c) Axon no. 21; temperature
8·5° C. Data from voltage clamps in (1) sea water, (2) choline sea water, (3) sea water, taken
later than (b). Crosses are average values from runs (1) and (3). V_{Na} = −93·8 mV., f=1·60.

resting potential was higher in the low-sodium solutions than in sea water, and it is shown in that paper that increasing the membrane potential by current flow allows a subsequent depolarization to produce greater sodium currents 86 than it would otherwise have done. The factor by which the sodium currents are thus increased is greater the lower the sodium concentration, and the poorer the condition of the fibre. The first of these effects explains why the factor is greater in (b) than in (a), while the second explains why it is greater in (c) than in (b). The experiments in Fig. 13b, c were performed on the same

84. *'appear at first sight to indicate a disagreement with the theory, but they are explained quantitatively by an effect described in the fourth paper'* Again, H&H acknowledge and then discuss the error. The 'effect' they will describe is inactivation of sodium currents.

85. ***Figure 13.*** As indicated in the text, the three predicted *I-V* curves in low-sodium seawater, labeled *B*, underestimates the measured current, but by a constant amount in each of three experiments. The measured current amplitudes can be matched by scaling the calculated values by 1.2–1.6 (a constant factor for each axon) to get three curves marked *C*, which match the data points well.

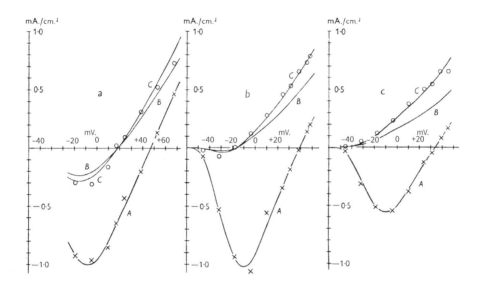

86. *'resting potential was higher in the low-sodium solutions . . . allows a subsequent depolarization to produce greater sodium currents'* The resting potential is hyperpolarized in choline, owing to the loss of a small depolarizing sodium flux. As discussed in Paper 4, the hyperpolarization of just a few millivolts allows the sodium conductance to recover partially from inactivation, which makes the currents evoked across all voltages increase by a fixed factor.

fibre, and the deterioration between the experiments is shown by the fact that the I_{Na} values are only about half as great in (c) as in (b).

We can therefore say that, within experimental error, the sodium currents in sea water and in low-sodium solutions are connected by Equation (12), suggesting that the 'independence principle' from which this equation was derived is applicable to the manner in which the ions cross the membrane. 87 This does not tell us much about the physical mechanism involved, since the 88 'independence' relations would be obeyed by several quite different systems. Examples are the 'constant field' system discussed by Goldman (1943), where the electric field through the membrane is assumed to be uniform and unaffected by the concentrations of ions present; and any system involving combination with carrier molecules in the membrane, so long as only a small proportion of the carrier is combined with the ion at any moment.

Origin of the action potential

The main conclusions that were drawn from the analysis presented in the 89 'Results' section of this paper may be summarized as follows. When the membrane potential is suddenly reduced (depolarization), the initial pulse of current through the capacity of the membrane is followed by large currents carried by ions (chiefly sodium and potassium), moving down their own electrochemical gradients. The current carried by sodium ions rises rapidly to a peak and then decays to a low value; that carried by potassium ions rises much more slowly along an S-shaped curve, reaching a plateau which is maintained with little change until the membrane potential is restored to its resting value.

These two components of the membrane current are enough to account qualitatively for the propagation of an action potential, the sequence of events 90 at each point on the nerve fibre being as follows: (1) Current from a neighbouring active region depolarizes the membrane by spread along the cable structure of the fibre ('local circuits'). (2) As a result of this depolarization, sodium current is allowed to flow. Since the external sodium concentration is several times greater than the internal, this current is directed inwards and depolarizes the membrane still further, until the membrane potential reverses its sign and approaches the value at which sodium ions are in equilibrium. (3) As a delayed result of the depolarization, the potassium current increases and the ability of the membrane to pass sodium current decreases. Since the internal potassium concentration is greater than the external, the potassium current is directed outwards. When it exceeds the sodium current, it repolarizes the membrane, raising the membrane potential to the neighbourhood of the resting potential, at which potassium ions inside and outside the fibre are near to equilibrium.

The further changes which restore the membrane to a condition in which it

87. *'within experimental error . . . the "independence principle" from which this equation was derived is applicable'* The good fit of the quantitative formulation of the hypothesis to the data supports the theory underlying it, after the factor relating to the inactivation of the sodium conductance is explained in Paper 4. The sodium and potassium ions indeed appear to flow independently of one another.

88. *'This does not tell us much about the physical mechanism involved'* Throughout the papers, H&H emphasize that the *mathematical* relations presented serve to (1) identify key variables and their relative dependence, (2) predict electrical behavior, and (3) provide evidence that the underlying physical structures must obey certain laws of electricity. H&H also stress, however, that these equations do not provide information about what the physical (molecular) referents might be. In later years, H&H stated that they assumed that the equations they derived would be a relatively short-lived transition state in the history of biophysics and would be supplanted by relations grounded in molecular biology.

89. *'The main conclusions'* This summary emphasizes (1) the distinction between capacity and ionic current, (2) the identification of sodium and potassium ions as the carriers of independent ionic currents, and (3) the description of the kinetics of each current.

90. *'enough to account qualitatively for the propagation of the action potential'* The experiments have been done with a 'stationary,' or 'nonpropagating,' action potential, elicited in a section of membrane held isopotential by a long internal wire. Recall that isopotential does not imply that that entire membrane is clamped at a fixed voltage over *time*, but that all points on the membrane across *space* vary together, including when an action potential is evoked. H&H indicate that they can use what has been learned to infer how an action potential is generated as well as how it propagates along an axon that is not isopotential. This idea will be revisited in Paper 5. Despite the quantitative support for the hypotheses, H&H call this stage of the work 'qualitative' because although they can quantitatively replicate the *I-V* curves from first principles, they have not yet replicated the action potential. This challenge will also be met in Paper 5.

can propagate another impulse have also been studied by the 'voltage clamp' technique and are described in subsequent papers (Hodgkin & Huxley, 1952 a, b). In the final paper of the series (Hodgkin & Huxley, 1952 c), we show that an action potential can be predicted quantitatively from the voltage clamp results, by carrying through numerically the procedure which has just been outlined.

SUMMARY   91

1. The effect of sodium ions on the current through the membrane of the giant axon of *Loligo* was investigated by the 'voltage-clamp' method.

2. The initial phase of inward current, normally associated with depolarizations of 10–100 mV., was reversed in sign by replacing the sodium in the external medium with choline.

3. Provided that sodium ions were present in the external medium it was possible to find a critical potential above which the initial phase of ionic current was inward and below which it was outward. This potential was normally reached by a depolarization of 110 mV., and varied with external sodium concentration in the same way as the potential of a sodium electrode.

4. These results support the view that depolarization leads to a rapid increase in permeability which allows sodium ions to move in either direction through the membrane. These movements carry the initial phase of ionic current, which may be inward or outward, according to the difference between the sodium concentration and the electrical potential of the inside and outside of the fibre.

5. The delayed outward current associated with prolonged depolarization was little affected by replacing sodium ions with choline ions. Reasons are given for supposing that this component of the current is largely carried by potassium ions.

6. By making certain simple assumptions it is possible to resolve the total ionic current into sodium and potassium currents. The time course of the sodium or potassium permeability when the axon is held in the depolarized condition is found by using conductance as a measure of permeability.

7. It is shown that the sodium conductance rises rapidly to a maximum and then declines along an approximately exponential curve. The potassium conductance rises more slowly along an **S**-shaped curve and is maintained at a high level for long periods of time. The maximum sodium and potassium conductances were normally of the order of 30 m.mho/cm.^{2} at a depolarization of 100 mV.

8. The relation between sodium concentration and sodium current agrees with a theoretical equation based on the assumption that ions cross the membrane independently of one another.

91. ***'Summary'*** By the end of Paper 2, H&H have applied the voltage clamp to measure and describe the ionic current flowing across the squid axon as a function of voltage and time. They have also separated the currents into components carried by sodium and (probably) potassium and provided evidence that the inward and outward fluxes of these ions are independent. Finally, they have introduced the concepts of conductance as a measure of ion permeability and of an inactivating phase to the sodium conductance. These latter two ideas will be examined further in Papers 3 and 4, respectively.

REFERENCES

BEHN, U. (1897). Ueber wechselseitige Diffusion von Elektrolyten in verdünnten wässerigen Lösungen, insbesondere über Diffusion gegen das Concentrationsgefälle. *Ann. Phys., Lpz.,* N.F. **62**, 54–67.

COLE, K. S. & CURTIS, H. J. (1939). Electric impedance of the squid giant axon during activity. *J. gen. Physiol.* **22**, 649–670.

COLE, K. S. & HODGKIN, A. L. (1939). Membrane and protoplasm resistance in the squid giant axon. *J. gen. Physiol.* **22**, 671–687.

CURTIS, H. J. & COLE, K. S. (1942). Membrane resting and action potentials from the squid giant axon. *J. cell. comp. Physiol.* **19**, 135–144.

GOLDMAN, D. E. (1943). Potential, impedance, and rectification in membranes. *J. gen. Physiol.* **27**, 37–60.

HODGKIN, A. L. (1951). The ionic basis of electrical activity in nerve and muscle. *Biol. Rev.* **26**, 339–409.

HODGKIN, A. L. & HUXLEY, A. F. (1952a). The components of membrane conductance in the giant axon of *Loligo. J. Physiol.* **116**, 473–496.

HODGKIN, A. L. & HUXLEY, A. F. (1952b). The dual effect of membrane potential on sodium conductance in the giant axon of *Loligo. J. Physiol.* **116**, 497–506.

HODGKIN, A. L. & HUXLEY, A. F. (1952c). A quantitative description of membrane current and its application to conduction and excitation in nerve. *J. Physiol.* (in the press).

HODGKIN, A. L., HUXLEY, A. F. & KATZ, B. (1949). Ionic currents underlying activity in the giant axon of the squid. *Arch. Sci. physiol.* **3**, 129–150.

HODGKIN, A. L., HUXLEY, A. F. & KATZ, B. (1952). Measurement of current-voltage relations in the membrane of the giant axon of *Loligo. J. Physiol.* **116**, 424–448.

HODGKIN, A. L. & KATZ, B. (1949). The effect of sodium ions on the electrical activity of the giant axon of the squid. *J. Physiol.* **108**, 37–77.

KEYNES, R. D. & LEWIS, P. R. (1951). The sodium and potassium content of cephalopod nerve fibres. *J. Physiol.* **114**, 151–182.

MANERY, J. F. (1939). Electrolytes in squid blood and muscle. *J. cell. comp. Physiol.* **14**, 365–369.

STEINBACH, H. B. & SPIEGELMAN, S. (1943). The sodium and potassium balance in squid nerve axoplasm. *J. cell. comp. Physiol.* **22**, 187–196.

TEORELL, T. (1949a). *Annu. Rev. Physiol.* **11**, 545–564.

TEORELL, T. (1949b). Membrane electrophoresis in relation to bio-electrical polarization effects. *Arch. Sci. physiol.* **3**, 205–218.

USSING, H. H. (1949). The distinction by means of tracers between active transport and diffusion. *Acta physiol. scand.* **19**, 43–56.

WEBB, D. A. (1939). The sodium and potassium content of sea water. *J. exp. Biol.* **16**, 178–183.

WEBB, D. A. (1940). Ionic regulation in *Carcinus maenas. Proc. Roy. Soc.* B, **129**, 107–135.

Paper 3

Hodgkin, A. L. and Huxley, A. F. (1952b) The components of membrane conductance in the giant axon of *Loligo*. *Journal of Physiology* 116:473–496.

In the third paper, the tail-current paper, H&H directly investigate whether the permeability of the membrane truly acts like a conductance in the manner predicted by Ohm's law. For this hypothesis to be supported, the ionic current (sodium or potassium) must be linearly related to driving force for a fixed conducting state of the membrane. Because conductance changes as a function of both voltage and time, however, it is rarely constant. To address this issue, H&H activate a conductance by depolarizing the membrane to a given potential, and then measure the current instantaneously upon changing the driving force to different levels on different trials, before the conductance relaxes over time to a new value. The instantaneous measurement allows the assessment of current at different voltages but always with the membrane in the same state, ultimately providing evidence for the ohmic nature of the conductance and unequivocally identifying voltage as the underlying variable that controls ionic currents.

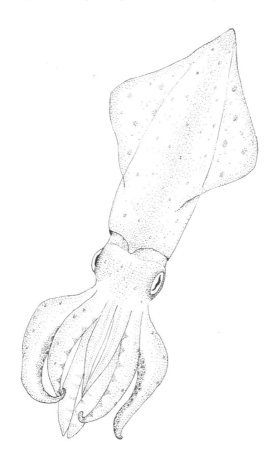

THE COMPONENTS OF MEMBRANE CONDUCTANCE IN THE GIANT AXON OF *LOLIGO*

By A. L. HODGKIN AND A. F. HUXLEY

*From the Laboratory of the Marine Biological Association, Plymouth,
and the Physiological Laboratory, University of Cambridge*

(*Received* 24 *October* 1951)

The flow of current associated with depolarizations of the giant axon of *Loligo* has been described in two previous papers (Hodgkin, Huxley & Katz, 1952; Hodgkin & Huxley, 1952). These experiments were concerned with the effect of sudden displacements of the membrane potential from its resting level ($V = 0$) to a new level ($V = V_1$). This paper describes the converse situation in which the membrane potential is suddenly restored from $V = V_1$ to $V = 0$. It also deals with certain aspects of the more general case in which V is changed suddenly from V_1 to a new value V_2. The experiments may be conveniently divided into those in which the period of depolarization is brief compared to the time scale of the nerve and those in which it is relatively long. The first group is largely concerned with movements of sodium ions and the second with movements of potassium ions.

METHODS

The apparatus and method were similar to those described by Hodgkin *et al.* (1952). The only new technique employed was that on some occasions two pulses, beginning at the same moment but lasting for different times, were applied to the feed-back amplifier in order to give a wave form of the type shown in Fig. 6. The amplitude of the shorter pulse was proportional to $V_1 - V_2$, while the amplitude of the longer pulse was proportional to V_2. The resulting changes in membrane potential consisted of a step of amplitude V_1, during the period when the two pulses overlap, followed by a second step of amplitude V_2.

RESULTS

Experiments with relatively brief depolarizations

Discontinuities in the sodium current

The effect of restoring the membrane potential after a brief period of depolarization is illustrated by Fig. 1. Record *A* gives the current associated with a maintained depolarization of 41 mV. As in previous experiments, this consisted of a wave of inward current followed by a maintained phase of

1. *'membrane potential is suddenly restored from $V = V_1$ to $V = 0$.'* This 'restoration' refers to the nearly instantaneous repolarization back to the resting membrane potential (H&H's 0 mV) after a period of depolarization.

2. *'brief compared to the time scale of the nerve'* The 'time scale of the nerve' refers to the duration of an action potential.

3. *'beginning at the same moment but lasting for different times'* The apparatus could not give two steps of different amplitudes in immediate succession. H&H therefore applied a command voltage that was the sum of two steps starting at the same time but with different durations: a longer one of the magnitude of the repolarization step (V_2) and a shorter one that provided the extra depolarization necessary to get to the desired initial voltage ($V_1 - V_2$). Because this short step sums with the long step, the result is a step to V_1 followed by a step to V_2.

4. *'Discontinuities in the sodium current.'* Note the emphasis on discontinuities in *current*, which will ultimately be contrasted with continuity of *conductance*.

outward current. Only the beginning of the second phase can be seen at the relatively high time base speed employed. At 0·85 msec. the ionic current 5 reached a value of 1·4 mA./cm.². Record B shows the effect of cutting short the period of depolarization at this time. The sudden change in potential was associated with a rapid surge of capacity current which is barely visible on the time scale employed. This was followed by a 'tail' of ionic current which 6

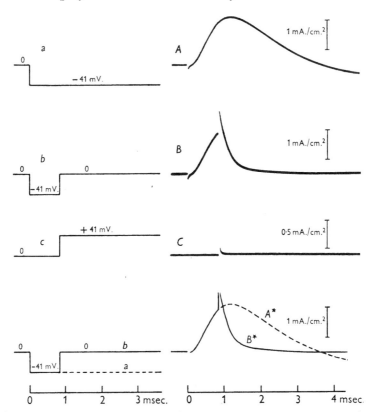

Fig. 1. Left-hand column: a, b, c, time course of potential difference between external and internal electrode. Right-hand column: A, B, C, records of membrane current associated with changes in membrane potential shown in left-hand column. (The amplification in C was 90% greater than that in A and B.) A*, B*, time course of ionic currents obtained by subtracting capacity current in C from A and B. Axon 25; temperature 5° C.; uncompensated feed-back. Inward current is shown upward in this and all other figures except Fig. 13.

started at about 2·2 mA./cm.² and declined to zero with a time constant of 0·27 msec. The residual effects of the capacitative surge were small and could be eliminated by subtracting the record obtained with a corresponding anodal displacement (C). Curves corrected by this method are shown in A* and B*.

The first point which emerges from this experiment is that the total period of inward current is greatly reduced by cutting short the period of depolarization. This suggests that the process underlying the increase in sodium permeability is reversible, and that repolarization causes the sodium current to

5. **'At 0.85 msec.'** Note that the kinetics of sodium current of the squid, which are adapted to living in cold water, are remarkably fast even at 5°C: In a mammalian neuron, sodium current evoked at nearly the same voltage (~-20 mV *abs.*) rises in ~0.7 msec at about 25°C. With a Q_{10} of 3, the rate of rise would be expected to be some nine-fold slower, or about 6 msec, at 5°C.

6. **'a "tail" of ionic current'** This is the first use of the term 'tail current.' The tail refers to the discontinuity at the onset of the repolarization on current trace *B* (*arrow* in the version of **Figure 1** below). At that point, the current increases as a result of the instantaneous increase in driving force, and then decays as the conductance deactivates. The gap in the record arises because the electron beam on the oscilloscope moves so rapidly during the step that it does not leave a detectable trace on the screen.

The identification of this surge as *ionic* current is important. The tail current flows in the same direction as that predicted for a capacity current associated with repolarization, raising the possibility that the tail is simply capacity current. To test this possibility, H&H apply a voltage step of equal magnitude from rest, without previously activating the current by depolarization. In trace *C* in **Figure 1**, the capacity current that flows with a 41-mV *hyperpolarization* appears much smaller than the tail current in trace *B*. The capacity current is in fact much larger than the ionic current, but it is so fast (subsiding in tens of microseconds) that its early phase is not recorded by the oscilloscope camera. Only the later, longer-lasting part of the decay is visible in the traces. The 'extra' current following the depolarization in trace *B* must therefore be an ionic current, the only other option.

Since the ionic and capacity currents sum linearly to give the total current ($I_{total} = I_c + I_i$), the ionic current can be isolated by subtracting the capacity current from the total current, as is done in the bottom record, *B**. This process of 'capacitance subtraction,' in which a (usually) hyperpolarizing step that evokes no ionic current is used to estimate the I_c associated with a (usually) depolarizing step that evokes both ionic and capacity current, has been in use ever since to isolate ionic current. Even after subtraction of the capacity current, *B** still has a tail, providing further evidence that it must be ionic current.

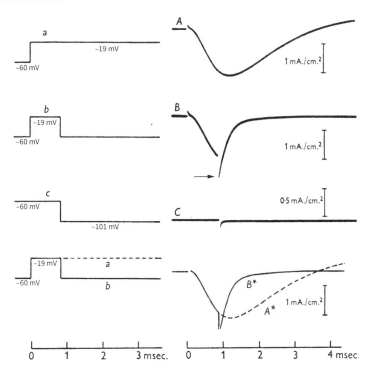

fall more rapidly than it would with a maintained depolarization. Further   7
experiments dealing with this phenomenon are described on p. 482. At present
our principal concern is with the discontinuity in ionic current associated with
a sudden change of membrane potential. Fig. 2 *D* illustrates the discontinuity
in a more striking manner. In this experiment the nerve was depolarized
nearly to the sodium potential, so that the ionic current was relatively small   8
during the pulse.

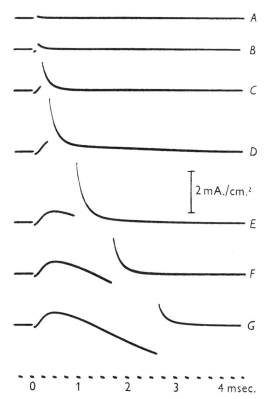

Fig. 2. Records of membrane current associated with depolarization of 97·5 mV. lasting, 0·05,
0·08, 0·19, 0·32, 0·91, 1·6 and 2·6 msec. The time and current calibration apply to all records.
Axon 41; temperature 3·5° C.; compensated feed-back.

The other records in Fig. 2 illustrate the effect of altering the duration of the
pulse. The surge of ionic current was small when the pulse was very short; it
reached a maximum at a duration of 0·5 msec. and then declined with a time
constant of about 1·4 msec. For durations less than 0·3 msec. the surge of
ionic current was roughly proportional to the inward current at the end of the
pulse. Since previous experiments suggest that this inward current is carried   9
by sodium ions (Hodgkin & Huxley, 1952), it seems likely that the tail of
inward current after the pulse also consists of sodium current. Fig. 3 illustrates
an experiment to test this point. In *A*, the membrane was initially depolarized
to the sodium potential. The ionic current was very small during the pulse but

7. **'reversible . . . maintained depolarization.'** H&H draw attention to the observations that a positive ΔV makes current flow and a negative ΔV makes that current stop flowing. Thus, the process elicited by depolarization is not ballistic, or irrevocable once initiated. Its magnitude and kinetics can be altered midcourse by subsequent changes in membrane potential, which is an indication that the current is controlled directly by voltage.

8. **'depolarized nearly to the sodium potential.'** The key inference here is that at E_{Na}, where the driving force is zero, little or no current is evident, but an underlying conductance change must nevertheless take place, since currents flow upon repolarization when the driving force on sodium ions is rapidly changed.

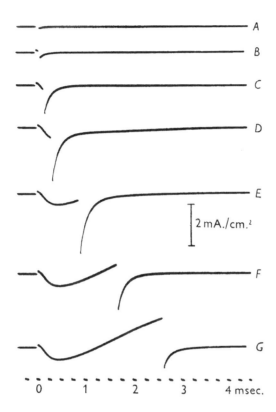

9. **'surge of ionic current was roughly proportional to the inward current at the end of the pulse.'** Here is an ideal example of the scientific method: *observations* giving rise to a *hypothesis*, yielding *predictions*, which are *experimentally tested*, producing results that permit *interpretations* that support or refute the hypothesis.

Specifically, the *observations* are as follows: With progressively longer depolarizing steps, the tail gets first larger and then smaller, just like the depolarization-evoked current that precedes it. This match, or proportionality, strengthens the idea that there is a link between the current just before and the current just after the repolarization; it thus generates the *hypothesis* that the inward tail current is probably carried by sodium, just as the depolarization-evoked current was shown to be. The hypothesis yields a *prediction*: if the hypothesis is correct, then removing external sodium ions should remove the inward tail current. The *test* of the prediction is to substitute external sodium ions with choline (continued in note 10).

the usual tail followed the restoration of the resting potential. The sequence of events was entirely different when choline was substituted for the sodium in the external fluid (Fig. 3 *B*). In this case there was a phase of outward current during the pulse but no tail of ionic current when the membrane potential was restored. The absence of ionic current after the pulse is proved by the fact that the capacitative surges obtained with anodal and cathodal displacements were almost perfectly symmetrical (records *B* and *C*). These effects are explained quite simply by supposing that sodium permeability rises when the membrane is depolarized and falls exponentially after it has been repolarized.

10

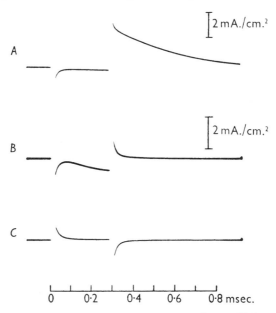

Fig. 3. *A*, membrane current associated with depolarization of 110 mV. lasting 0·28 msec.; nerve in sea water. *B*, same, but with nerve in choline sea water. *C*, membrane currents associated with an increase of 110 mV. in membrane potential; nerve in choline sea water. Axon 25; temperature 5° C.; uncompensated feed-back.

In record *A* the increase in permeability did not lead to any current during the pulse, since inward and outward movements of sodium are equal at the sodium potential. After the pulse the tendency of external sodium ions to enter the fibre is very much greater than that of internal sodium ions to leave. This means that there must be a large inward current after the pulse unless the sodium permeability reverts instantaneously to a low value. Record *B* is different because there were no external sodium ions to carry the current in an inward direction. The increase in sodium permeability therefore gave a substantial outward current during the period of depolarization but no inward current after the pulse. One might expect to see a 'tail' of outward current in *B* corresponding to the tail of inward current in *A*. However, the tendency of the internal sodium ions to leave the fibre against the resting

10. ***'absence of ionic currents after the pulse . . . explained quite simply'*** The visually unimpressive traces of ***Figure 3*** turn out to be remarkably important in testing the preceding hypothesis. The results of the experiment fulfill the prediction—the inward tail is indeed absent in choline. The tail current therefore must be carried by sodium, even though no (net) sodium current flows at the sodium potential. Note that the *interpretation* goes beyond the simple restatement of the hypothesis. H&H's 'simple explanation' is that sodium *permeability* rises upon depolarization and falls (exponentially) upon repolarization. The focus shifts here intentionally from current to permeability. The current is secondary to permeability (eventually equated functionally with conductance); the latter is therefore the primary variable.

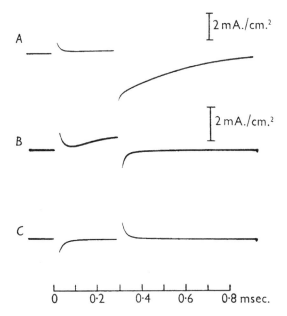

potential difference would be so small that the resulting outward current would be indistinguishable from the capacitative surge. According to the 'independence principle' (Hodgkin & Huxley, 1952, equation 12), the outward current in B should be only 1/97 of the inward current in A.   11

Continuity of sodium conductance   12

Discontinuities such as those in Figs. 1, 2 and 3 A disappear if the results are expressed in terms of the sodium conductance (g_{Na}). This quantity was defined 13 previously by the following equation:

$$g_{Na} = I_{Na}/(V - V_{Na}), \qquad (1)$$

where V is the displacement of the membrane potential from its resting value and V_{Na} is the difference between the equilibrium potential for sodium ions and the resting potential (Hodgkin & Huxley, 1952).

The records in Fig. 4 allow g_{Na} to be estimated as a function of time. Curves α and A give the total ionic current for a nerve in sea water. Curve α was obtained with a maintained depolarization of 51 mV. and A with the same depolarization cut short at 1·1 msec. Curves β and B are a similar pair with the nerve in choline sea water. Curves γ and C give the sodium current obtained from the two previous curves by essentially the method used in the preceding paper (see Hodgkin & Huxley, 1952). In this experiment the depolarization was 51 mV. and the sodium potential was found to be -112 mV. To convert sodium current into sodium conductance the former must therefore be divided by 61 mV. during the depolarization or by 112 mV. after the pulse. Curves δ and D were obtained by this procedure and show that the conductance reverts to its resting level without any appreciable discontinuity at the 14 end of the pulse. Fig. 5 illustrates the results of a similar analysis using the records shown in Fig. 2. In this experiment no tests were made in choline sea water, but the early part of the curve of sodium current was obtained by assuming that sodium current was zero initially and that the contribution of other ions remained at the level observed at the beginning of the pulse. Records made at the sodium potential (-117 mV.) indicated that the error introduced by this approximation should not exceed 5% for pulses shorter than 0·5 msec.

The instantaneous relation between ionic current and membrane potential

The results described in the preceding section suggest that the membrane obeys Ohm's law if the ionic current is measured immediately after a sudden change in membrane potential. In order to establish this point we carried 15 out the more complicated experiment illustrated by Fig. 6. Two rectangular pulses were fed into the feed-back amplifier in order to produce a double step of membrane potential of the type shown inset in Fig. 6. The first step had a duration of 1·53 msec. and an amplitude of -29 mV. The second step was relatively long and its amplitude was varied between -60 mV. and $+30$ mV.

11. *'should be only 1/97 of the inward current in A.'* A characteristic feature of the papers is that even when a qualitative result is obvious and provides a substantial discovery, H&H extract quantitative parameters that either consolidate the interpretation, offer a cross-check of previous conclusions, or lead the way to additional insights. Here, the calculation of the predicted magnitude of the inward current in *Figure 3B* supports the independence principle derived in Paper 2 ('the chance that any individual ion will cross the membrane in a specified interval of time is independent of the other ions which are present,' Paper 2, note 75).

12. *'Continuity of sodium conductance.'* By demonstrating the discontinuities in current upon changes in membrane potential, H&H have oriented the reader to appreciate the significance of the continuity of *conductance*. They will show that conductance rises (over time) to some value upon depolarization, and conductance falls from that same value upon repolarization. Hence, it is conductance that responds directly to voltage. Every step that establishes this point incontrovertibly helps lead H&H to a description that has its basis in the physics of classical circuit elements (e.g., Ohm's law), but with specific, quantifiable variations (voltage- and time-dependent conductances).

13. *'Discontinuities . . . disappear if the results are expressed in terms of the sodium conductance'* **Equation 1** is a rearrangement of Ohm's law, as an expression for conductance, g, instead of the usual resistance (its inverse). Calculations of conductance will show that the discontinuities in *current* at the start of the tail current can be accounted for by the discontinuity in *driving force* caused by the sudden repolarization.

14. *'conductance reverts to its resting level without any appreciable discontinuity'* In *Figure 4,* H&H illustrate that conductance is continuous even at the point of the instantaneous voltage change. The conductance values were obtained by (meticulous) division of the measured current values by the driving force values (V was applied; V_{Na} was observed experimentally).

15. *'the membrane obeys Ohm's law if the ionic current is measured immediately after a sudden change in membrane potential.'* The data in *Figure 6* constitute a quantitative proof of H&H's key proposal that ion permeability can be described as a conductance that follows Ohm's law. Plot *A*, which H&H refer to as an 'instantaneous' *I-V* curve, shows that current recorded immediately after a depolarization of fixed voltage and duration is linearly related to driving force. This protocol provides a measure of the chord conductance discussed in Paper 2, note 54.

The underlying idea is that the first depolarizing step activates the current, therefore bringing g to a nonzero value. When the driving force ($V - V_{Na}$) is instantaneously stepped to a new potential, the new magnitude of current can be measured *before the conductance begins to change with time*. Thus, by making g constant, the equation $I = g(V - V_{Na})$ describes a straight line of slope g, which intersects the x-axis at V_{Na}.

The first depolarizing step (–31 mV *abs*.) is only 29 mV more depolarized than the resting potential, and the second step either further depolarizes or repolarizes the membrane. One benefit of this intermediate value for the first voltage step is that the sodium current activates and decays relatively more slowly at –31 mV *abs*. than it would at greater depolarizations, reaching its peak in about 1.5 msec. The slower kinetics provide a broader time window during which to measure conductance and a lower chance of errors produced by small variations in the duration of the step.

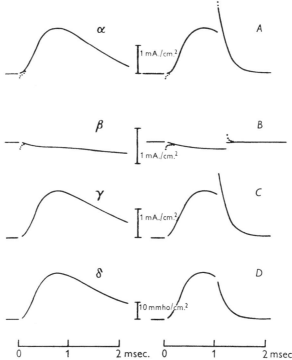

Fig. 4. α, ionic current in sea water associated with maintained depolarization of 51 mV. applied   16
at $t = 0$. (The dotted line shows the form of the original record before correcting for capacity
current.) β, same in choline sea water. γ, sodium current estimated as $(\alpha - \beta) \times 0.92$.
δ, sodium conductance estimated as $\gamma/61$ mV. A, B, same as α and β respectively, but with
depolarization lasting about 1·1 msec. C, sodium current estimated as $(A - B) \times 0.92$ during
pulse or $(A - B) \times 0.99$ after pulse. D, sodium conductance estimated as $C/61$ mV. during
pulse or $C/112$ mV. after pulse. The factors 0·92 and 0·99 allow for the outward sodium
current in choline sea water and were obtained from the 'independence principle'. Axon 17;
temperature 6° C.; V_{Na} in sea water $= -112$ mV.; uncompensated feed-back.

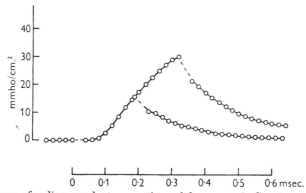

Fig. 5. Time course of sodium conductance estimated from records C and D (Fig. 2) by method   17
described in text. At zero time the membrane potential was reduced by 97·5 mV. and was
restored to its resting value at 0·19 msec. (lower curve) or 0·32 msec. (upper curve). The
broken part of the curve has been interpolated in the region occupied by the capacitative
surge. Axon 41; temperature 3·5° C.; compensated feed-back; $V_{Na} = -117$ mV.

16. **Figure 4.** This figure compares responses for long (left) and short (right) voltage steps. The raw currents in control and choline seawater are shown in the top two rows, with the isolated sodium current in the third row, illustrating the current discontinuities at the end of brief steps. When the current is transformed to conductance in the bottom row, the discontinuity at the end of the brief step disappears; it is also evident that the conductance decays more rapidly upon termination of depolarization than during sustained depolarization.

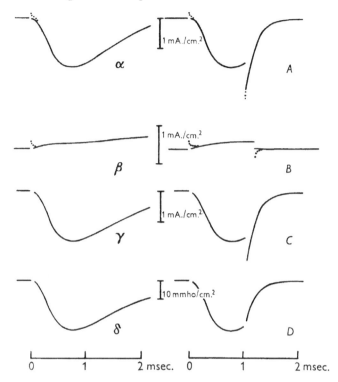

17. **Figure 5.** An illustration of sodium conductance similar to trace D of **Figure 4** is shown for traces C and D of **Figure 2**. Note the very cold temperature: recording at 3.5°C slows the kinetics enough to record and calculate the changes with high resolution, and the dashed lines emphasize the continuity of the conductance upon repolarization with the conductance during depolarization.

The ordinate (I_2) is the ionic current at the beginning of the second step and the abscissa (V_2) is the potential during the second step. Measurement of I_2 depends on the extrapolation shown in Fig. 6A. This should introduce little error over most of the range but is uncertain for $V_2 > 0$, since the ionic current then declined so rapidly that it was initially obscured by capacity current. There was some variation in the magnitude of the current observed during the first pulse. This arose partly from progressive changes in the condition of the

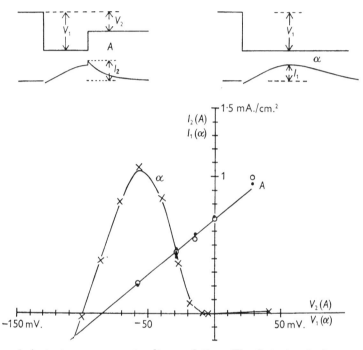

Fig. 6. Line A, instantaneous current-voltage relation. The first step had an amplitude of -29 mV. and a duration of 1·53 msec. The abscissa (V_2) gives the amplitude of the second step. The ordinate (I_2) is the ionic current at the beginning of the second step. The dots are observed currents. Hollow circles are these currents multiplied by factors which equalize the currents at the end of the first step. Inset A, method of measuring V_2 and I_2. Curve α and crosses, relation between maximum inward current (I_1) and membrane potential using single pulse of amplitude V_1. Inset α, method of measuring V_1 and I_1. Axon 31; temperature 4° C.; uncompensated feed-back.

nerve and partly from small changes in V_1 which cause large variations in current in the region of $V = -29$ mV. Both effects were allowed for by scaling all records so that the current had the same amplitude at the end of the first step. This procedure is justified by the fact that records made with $V_2 = 0$ show that the amplitude of the current immediately after the step was directly proportional to the current immediately before it.

The results are plotted in curve A and show that the relation between I_2 and V_2 is approximately linear. This is in striking contrast to the extremely non-

18. ***'scaling all records so that the current had the same amplitude at the end of the first step.'*** In this 'more complicated' experiment, H&H first apply a fixed depolarization and then step the membrane potential back to different voltages on different trials. They recognize that even small drifts in resting potential over successive trials can shift the amount of inactivation of the sodium conductance (described more fully in Paper 4). Since the resting potential can vary over time, repeated 29-mV depolarizing steps end up stepping the potential to slightly different values, thereby evoking slightly different currents. Restating with modern conventions, sometimes the step might be from −60 mV *abs.* to −31 mV *abs.*, and other times from −55 mV *abs.* to −26 mV *abs.* Therefore, in ***Figure 6***, H&H first scale all the measurements to the size of the sodium current at the *end* of the first step, V_1, and then illustrate the proportionate change evoked when the driving force changes by a fixed amount. This is another example of H&H coping with imperfections of preparation and technique; instead of discarding the data, they use their understanding of the error to extract the desired information. It is interesting to note that they plot both the measured currents (*dots*) and the corrected currents (*hollow circles*), so that the raw (primary) as well as the extracted data are shown. The difference is not large, but all the information is available for inspection.

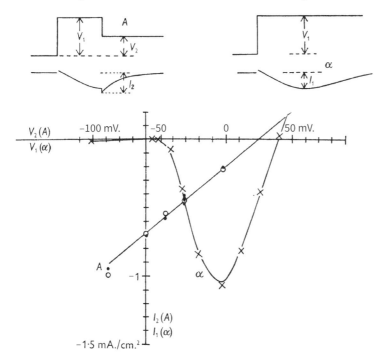

linear relation obtained when the current is measured at longer intervals. An 19
example of the second type is provided by curve α which shows how the
maximum inward current varied with membrane potential in the same axon.
In this case only a single pulse of variable amplitude was employed and current
was measured at times of 0·5–2·0 msec. Under these conditions the sodium
conductance had time to reach the value appropriate to each depolarization
and the current-voltage relation is therefore far from linear.

The line A and the curve α intersect at -29 mV. since the two methods of 20
measurement are identical if $V_2 = V_1$. A second intersection occurs at
-106 mV. which is close to the sodium potential in this fibre.   21

A similar pair of curves obtained with a larger initial depolarization is shown
by A and α in Fig. 7. In this case the nerve was depolarized to the sodium
potential so that one would expect the line A to be tangential to the curve α. 22
This is approximately true, although any exact comparison is invalidated by
the fact that the two curves could not be obtained at exactly the same time.

The instantaneous current-voltage relation in sodium-free solution

The measurements described in the preceding section indicate that the
instantaneous behaviour of the membrane is linear when the nerve is in sea
water. The conclusion cannot be expected to apply for all sodium concentrations. The method of defining a chord conductance breaks down altogether if
there is no sodium in the external medium. In this case $V_{Na} = \infty$ and g_{Na}
must be zero if the sodium current is to be finite. This condition could not be
realized in practice but the theoretical possibility of its existence indicates that
the concept of sodium conductance must be used with caution.   23

The lower part of Fig. 7 illustrates an attempt to determine the instantaneous current-voltage relation in a sodium-free solution. The upper curves
(A and α) were measured in sea water and have already been described. The
crosses in the lower part of the figure give the instantaneous currents in choline
sea water, determined in the same way as the circles which give the corresponding relation in sea water. The effect of the change in resting potential has
been allowed for by shifting the origin to the right by 4 mV. (see Hodgkin & 24
Huxley, 1952). The series of records from which these measurements were
made was started shortly after replacing normal sea water by choline sea
water and was continued, in the order shown, with an interval of about 40 sec.
between records. On analysis it was found that the earliest records (e.g. 1)
showed a small inward current, whereas records taken later (e.g. 11 or 15) gave
no such effect. It is evident that the series was started before all the sodium
had diffused away from the nerve and that only the later records (e.g. 6–15)
can be regarded as representative of a nerve in a sodium-free solution. Never- 25
theless, it is clear that the instantaneous current-voltage relation shows
a marked curvature and is quite different from the linear relation in sea water.

19. *'in striking contrast to the extremely nonlinear relation obtained when the current is measured at longer intervals.'* The point made from **Figure 6** is that the instantaneous relationship (curve *A*) is linear: *I* is indeed proportional to $(V_2 - E_{Na})$ in accordance with Ohm's law. The slope of this straight line is the conductance at the onset of the V_2 step; owing to continuity, this value of *g* is the same as it was at the end of the V_1 step. Because a 29-mV depolarization brings the voltage to about −30 mV *abs.*, which activates about half the conductance, the slope of the line is about ½ g_{max}.

20. *'intersect at –29 mV.'* For the nonlinear plot in **Figure 6**, which is similar to the sodium *I-V* relations illustrated in the earlier papers, the peak current is plotted against depolarizations to different values of V_1. For the instantaneous (linear) plot, the *x*-value indicates the repolarization voltage (V_2) after the initial step to 29 mV depolarized from rest (−31 mV *abs.*). Therefore, −29 mV indicates the point on the instantaneous curve at which the membrane potential stayed constant ($V_2 = V_1$). Because the repolarization step was applied near the peak of the current (at 1.53 msec), the point on the instantaneous curve has the same value as the −29-mV point on the curve obtained with depolarizations.

21. *'close to the sodium potential in this fibre.'* Remember that H&H are exploring the idea that Ohm's law holds: $I = g(V - V_{Na})$. The *y*-value (current) should go to zero where $V = V_{Na}$, as in the *I-V* curve for depolarization-evoked currents. Since 'tail currents' were measured only by repolarization (more hyperpolarized than −30 mV *abs.*), no data were obtained near V_{Na}, but the straight line determined by the points can be extrapolated to see whether it crosses zero and intersects with the curved line near V_{Na}. The extrapolation is imperfect but close.

22. *'In this case the nerve was depolarized to the sodium potential'* In **Figure 7**, the initial depolarization that defines line *A* is to the sodium potential, or approximately +50 mV *abs.*, which is much greater than in **Figure 6**. Such a large depolarization is likely to activate the sodium conductance maximally ($g = g_{max}$). Therefore, the instantaneous current flowing on the repolarizing step is $I = g_{max}(V - V_{Na})$, which is a straight line with a slope of g_{max}. This straight line converges with the nonlinear relation defined by depolarization-evoked currents (line α), which straightens out where $g = g_{max}$. This graph illustrates that the nonlinear part of this curve is where *g* is not maximal but ranges from zero at the most hyperpolarized potentials (no conductance activated) up to g_{max} at the most depolarized potentials.

In fact, current-voltage relations that are linear through the reversal potential are rarely seen with different internal and external concentrations of permeant ions (as predicted by the Goldman-Hodgkin-Katz equation); later work indicated that the inward sodium current in the squid giant axon is reduced by the external calcium ions in seawater, making the curve more linear at hyperpolarized potentials than it would be in divalent-free solutions (Yamamoto et al. 1985). The assumption of linearity, which implies a constant g_{max}, is therefore an adequate approximation for the physiological ionic conditions of H&H's experiments.

23. *'the concept of sodium conductance must be used with caution.'* This caution is validated by **Figure 7**, which shows recordings made in sodium-free choline seawater (x's). The instantaneous currents define a curved line, rather than obeying Ohm's law, as they do in physiological solution. This deviation makes physical sense: the linearity of the instantaneous *I-V* relation *depends on the presence of ions* to carry the current; without ions, nothing can be conducted, even through an open ion conducting pathway. This idea is consistent with the Goldman-Hodgkin-Katz current equation (see Paper 2, note 76).

24. *'shifting the origin to the right by 4 mV.'* Recall that the resting potential hyperpolarizes in choline. The shift of the origin by 4 mV corrects for that change.

25. *'later records . . . can be regarded as representative of a nerve in sodium-free solution.'* The points are numbered in the sequence in which they were recorded. H&H recorded the control currents, and then applied the sodium-free solution, but started recording before the sodium was fully washed away. For the latest currents, when the sodium was definitely removed, no inward current is measured, and the line defined by the points is strongly curved. Here again, H&H acknowledge the error of incomplete exchange of choline for sodium in the early records. Because replication trials were performed within the experiment and all the data are presented, it is possible to observe the deviation from expectation in the early records, as well as to understand the explanation for the deviation.

The results are, in fact, reasonably close to those predicted by the 'independence principle'. This is illustrated by a comparison of the crosses in Fig. 7 with the theoretical curves B and C which were calculated from A on the assumption that the independence principle holds and that the sodium

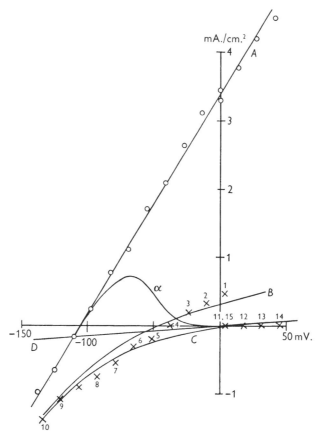

Fig. 7. Current-voltage relations in sea water and choline sea water. Ordinate: current density. 26 Abscissa: displacement of membrane potential from resting potential in sea water. Line A and curve α were obtained in sea water in the same way as in Fig. 6, except that the current for α was not measured at the maximum but at a fixed time (0·28 msec.) after application of a single step. The initial depolarization for A was 110 mV. and the duration of the first step was 0·28 msec. The crosses give the instantaneous currents in choline sea water, determined in the same way as the circles in A. The numbers show the order in which the measurements were made. B and C, instantaneous current in 10 % sodium sea water and in choline sea water respectively, derived from A by means of the 'independence principle' using the equations

$$\frac{(I_{\mathrm{Na}})_B}{(I_{\mathrm{Na}})_A} = \frac{0 \cdot 1 \exp{(V - V_{\mathrm{Na}})/24 - 1}}{\exp{(V - V_{\mathrm{Na}})/24 - 1}}$$

and

$$\frac{(I_{\mathrm{Na}})_C}{(I_{\mathrm{Na}})_A} = \frac{-1}{\exp{(V - V_{\mathrm{Na}})/24 - 1}},$$

V_{Na} = sodium potential in sea water = − 110 mV. Sodium currents measured from the line D which passes through the origin and the point for the small current observed at the sodium potential in sea water. Axon 25; temperature 5° C.; uncompensated feed-back.

26. *Figure 7.* The linearization of the sodium *I-V* curve by tail current measurement is dramatically illustrated by curve *A*. H&H also take advantage of the recordings made before complete solution exchange to compute two curves, *B* and *C*, assuming 10% and 0% sodium, respectively (see note 27), providing further support for their developing theory.

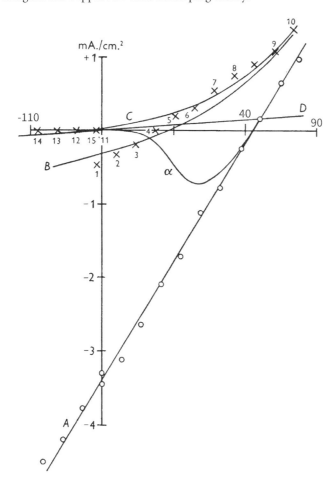

concentrations in the external solution were 10% (B) and 0% (C) of that in A. It will be seen that there is general agreement between calculated and observed results, although the change in external sodium and the possibility of progressive changes invalidates any exact comparison. In the preceding paper it 27 was shown that the observed sodium currents in a choline solution were usually larger than those calculated from the independence principle. This deviation is not seen here, probably because the measurements in choline were made later than those in sea water and no attempt was made to correct for deterioration, which is likely to have reduced the currents by 30% between the sea water and choline runs.

The experiment described in the preceding paragraph indicates that the linear relation between current and voltage observed in sea water is not a general property of the membrane since it fails in sodium-free solutions. This does not greatly detract from the usefulness of the result, for the primary 28 concern of this paper is to determine the laws governing ionic movements under conditions which allow a normal action potential to be propagated.

The reversible nature of the change in sodium conductance

The results described in the first part of this paper show that the sodium conductance reverts rapidly to a low value when the membrane potential is restored to its resting value. Figs. 2 and 5 suggest that this is true at all stages of the response and that the rate at which the conductance declines is roughly proportional to the value of the conductance. A rate constant (b_{Na}) can be defined by fitting a curve of the form $\exp(-b_{Na}t)$ to the experimental results. 29 Values obtained by this method are given in Table 1.

In order to investigate the effect of repolarizing the membrane to different levels on the rate of decline of sodium conductance we carried out the experiment illustrated by Fig. 8. The curves in the left-hand column are tracings of the membrane current while those on the right give the sodium conductance, calculated on the assumption that the contribution of ions other than sodium is negligible (records made in a solution containing 10% of the normal sodium concentration show that the error introduced by this approximation should not exceed 5% of the maximum current). The initial depolarization was 29 mV. and the sodium conductance reached its maximum value in 1·53 msec. When the membrane potential was restored to its resting level the conductance fell towards zero with a rate constant of about 4·3 msec.$^{-1}$ (curve γ). If V_2 was made $+28$ mV. the rate constant increased to about 10 msec.$^{-1}$ and a further increase to 15 msec.$^{-1}$ occurred with $V_2 = +57$ mV. On the other hand, if V_2 was reduced to -14 mV. the conductance returned with a rate constant of only 1·6 msec.$^{-1}$. When $V_2 = -57$ mV. the conductance no longer fell but increased towards an 'equilibrium' value which was greater than that attained at -29 mV. (The curve cannot be followed beyond about 2 msec.

27. ***'there is general agreement between calculated and observed results, although the . . . possibility of progressive changes invalidates any exact comparison.'*** The basic method of hypothesis testing used here is to gather quantitative data and then to check how well those values can be predicted by equations that represent the hypothesis: not curve *fitting* but curve *matching*. Here, the equation is based on the 'independence principle' (Goldman-Hodgkin-Katz current equation; Paper 2, Equation 12). It states that the magnitude of the current varies with the concentration of ions and that, far from the reversal potential, current and concentration are approximately proportional. H&H show that the early data points are reasonably well described by the equation, assuming that 90% of the seawater had been replaced by the choline solution (curve *B*), and the later points are well described by assuming that all the seawater had been replaced by the choline solution (curve *C*). Taken at face value, the data support the hypothesis. H&H acknowledge, however, a difference with the previous paper, in which sodium currents actually became larger than predicted in choline solution (which may have had to do with hyperpolarization, addressed later), so they leave open the possibility that other variables might have influenced the goodness of fit (choline may have made currents larger, or deterioration of the axon may have made them smaller).

28. ***'This does not greatly detract from the usefulness of the result'*** The result that the concentration of ions influences conductance shows that biological systems (in which ions flow across cell membranes) do not always display the properties of nonbiological circuits (in which electrons flow through electronic devices). This deviation is *biophysically* true, but H&H point out that it is most pronounced in the extreme experimental case of no external sodium ions. In the *physiological* condition, which is 'the primary concern of this paper,' ions are present. Thus, the approximation of conductances that obey Ohm's law effectively describes the membrane when the axon is functioning normally.

29. ***'A rate constant . . . can be defined'*** These are the first measurements of *deactivation* kinetics, which describe the time course of the offset of the conductance, now known to reflect the closing rate of ion channels. The comment that 'the rate at which the conductance declines is roughly proportional to the value of the conductance' simply states that $dy/dt \propto y$ throughout the decay phase. Because of this proportionality, the equation takes the form of an exponential decay, which can be written in terms of a time constant, τ, as $y = \exp(-t/\tau)$, or in terms of a rate constant, b_{Na}, as $y = \exp(-b_{Na}t)$, where $b_{Na} = 1/\tau$ (see Appendix 1.3).

Table 1. Apparent values of rate constant determining decline of sodium conductance   30
following repolarization to resting potential

Axon	Membrane potential during pulse (mV.)	Duration of pulse (msec.)	Temperature (°C.)	Average rate constant (msec.⁻¹)	Rate constant at 6° C. (msec.⁻¹)
15	− 32	0·4–1·1	11	9·4	5·4
15	− 91	0·1–0·5	11	9·0	5·2
17	− 32	0·7–1·6	6	5·9	5·8
17	− 51	0·2–1·0	6	6·7	6·6
24	− 42	0·2	20	18·5	4·1
24	− 84	0·1	20	17·2	3·9
25	− 41	1·0	4	3·8	4·8
25	− 110	0·3	4	3·3	4·0
31	− 100	0·3	4	3·0	3·8
31	− 29	1·5	4	4·2	5·3
32	− 116	0·2	5	6·3*	6·9*
32	− 67	0·7	5	6·3*	6·8*
41	− 98	0·1–4	3	7·1*	9·6*
41	− 117	0·1–3	3	7·7*	10·5*

Results marked with an asterisk were obtained with compensated feed-back. The last column is calculated on the assumption that the temperature coefficient (Q_{10}) of the rate constant is 3.

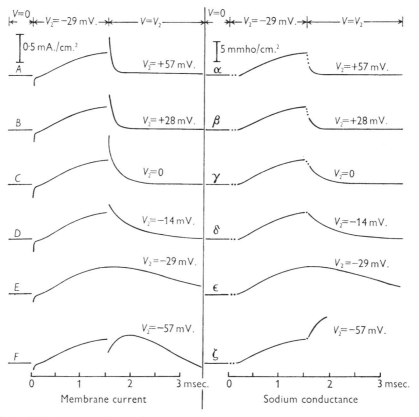

Fig. 8. *A–F*, time course of membrane current associated with change in membrane potential   31
shown at top of figure. α–ζ; time course of sodium conductance obtained by dividing *A–F*
by $V + 100$ mV. Axon 31; temperature 4° C.; uncompensated feed-back.

30. **Table 1.** Note that the measured rate constants are considerably faster with series resistance compensation (marked with asterisks in the table) than without it. The series resistance slows the time course of the voltage change across the membrane by adding to the membrane resistance and prolonging the time constant of the system. Because the deactivation rate for sodium currents is so fast, even a small slowing has a proportionately large effect on the time course of the measured currents.

31. **Figure 8.** This figure is the first time that a family of curves resulting from a pair of consecutive voltage steps is shown. Sodium current was activated with a depolarization to −31 mV *abs.*, and after about 1.5 msec the membrane was stepped to a range of (mostly) more hyperpolarized voltages. The currents (left) were then converted to conductance by dividing by the driving force (right). From the decay of the conductance, H&H measure the deactivation rate (rate of decline of sodium conductance) by fitting to an exponential decay and estimating the value of b_{Na}. The deactivation rate gets faster at more hyperpolarized potentials, indicating that it is voltage dependent. In contrast to the decrease in conductance seen with repolarizing steps, further *depolarizing* the membrane, from −31 mV *abs.* to −3 mV *abs.*, generates an *increase* in conductance (bottom right record in original figure; top right record in redrawn figure).

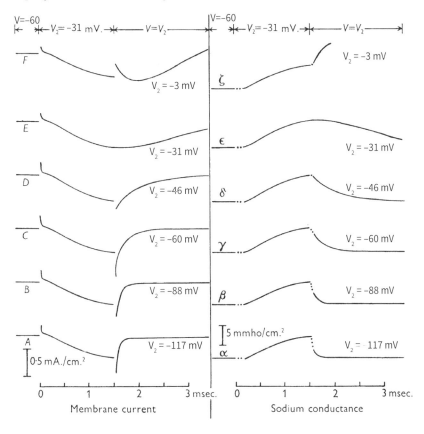

because the contribution of potassium ions soon becomes important at large depolarizations.) The whole family of curves suggests that the conductance reached at any depolarization depends on the balance of two processes occurring at rates which vary in opposite directions with membrane potential.

32

The observation that the rate constant increases with membrane potential does not depend on the details of the method used to estimate sodium conductance, for the tracings of current in the left-hand column of Fig. 8 show exactly the same phenomenon. Similar results were obtained in all the experiments of this type, and are plotted against V_2 in Fig. 9. It will be seen that there is good agreement between different experiments, and that a tenfold increase of rate constant occurs between $V_2 = -20$ and $+50$ mV.

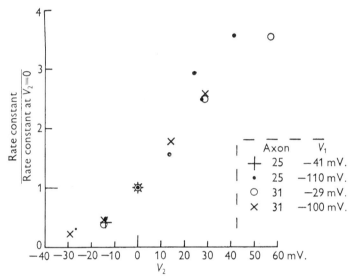

Fig. 9. Relation between rate constant determining decline of sodium conductance and potential to which membrane is repolarized. Abscissa: membrane potential during second step (V_2). Ordinate: relative value of rate constant.

Errors due to the series resistance

Most of the experiments in this paper were obtained with uncompensated feed-back and must therefore have been somewhat affected by the small resistance in series with the membrane (Hodgkin *et al.* 1952). The linearity of the relation between current and voltage illustrated by Figs. 6 and 7 can clearly stand, for the effect of a series resistance would simply be to change the slope of the straight line and not to introduce any curvature. From our estimates of the value of the series resistance it can be shown that the true slopes in these two figures should be 7 and 30% greater than those shown. A more serious error is introduced in the measurement of the rate constant. In Fig. 8 the total current at the beginning of the record was about 0·5 mA./cm.². This means that the true membrane potential was not zero but about -4 mV. At this potential the rate of return of membrane conductance would be slowed by about 8%. In some of the experiments used in compiling Table 1 this error may be as great as 50%. However, it should be small in axons 32 and 41 which were examined with compensated feed-back. We are also uncertain about the extent to which the rate of return of sodium conductance can be regarded as exponential. Axon 41, which was in excellent condition and was examined with compensated feed-back, showed clear depar-

32. *'the balance of two processes occurring at rates which vary in opposite directions with membrane potential.'* The data from the previous paper supported the idea that the conductance *magnitude* depends on membrane potential; the present data suggest that the *rates* governing the time course of conductance change (the approach either to a lesser or to a greater conductance) also depend on membrane potential. The activation rate constants become faster at more depolarized voltages; the deactivation rate constants become faster at more hyperpolarized voltages. The voltage dependence of the deactivation rate constant is shown in *Figure 9*.

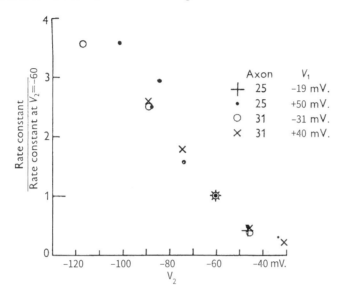

tures from exponential behaviour in that the initial fall of conductance was too rapid (see Figs. 2 and 5). In all other experiments the curves of current against time were reasonably close to exponentials, but in many cases this may have been due to an error introduced by the series resistance.

33

Errors due to polarization effects

If the membrane was maintained in the depolarized condition for long periods of time the outward current declined as a result of a 'polarization effect' (Hodgkin *et al.* 1952). At the end of such a pulse a phase of inward current was observed and was found to be roughly proportional to the amount of 'polarization'. This was quite distinct from the inward current described in the preceding sections, since it only appeared with long pulses and was unaffected by removing external sodium. With the possible exception of Fig. 2*G*, the results described in the preceding sections are unlikely to have been affected by polarization since the duration of the pulse was always kept short.

The time course of the sodium conductance during a maintained depolarization

In a previous paper we showed that the time course of the sodium conductance could be obtained from records of membrane current in solutions of different sodium concentration (Hodgkin & Huxley, 1952). An alternative method of calculating these curves is illustrated by Fig. 10. The method

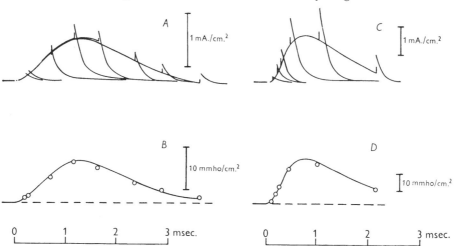

Fig. 10. *A*, time course of ionic currents associated with depolarizations of 32 mV. lasting from $t=0$ to times indicated by vertical strokes on tracings. Nerve in sea water. *B*, time course of sodium conductance. The circles were obtained by dividing the peak currents in *A* by 112 mV. and the continuous curve from the difference between the continuous curve in *A* and a similar curve in choline sea water (see text and legend to Fig. 4). *C*, same as *A* but employing depolarization of 51 mV. *D*, sodium conductance obtained from *C* by similar methods to those used for *B*. (The smooth curve is the same as that shown in Fig. 4.) Axon 17; temperature 6° C.; V_{Na} in sea water = − 112 mV.; uncompensated feed-back.

depends on the fact that the inward current immediately after a pulse is proportional to the sodium conductance at the end of the preceding depolarization. The variation of inward current with pulse duration is illustrated by the   34 tracings in Fig. 10*A*. The time course of the sodium conductance can be measured by determining the maximum ionic current associated with repolar-

33. **'may have been due to an error introduced by the series resistance.'** As elsewhere, the expression of a result and its interpretation is immediately followed by the consideration of possible errors, so that all sides of the argument are expressed and available to the reader. Here, the issue at stake is whether the fall of conductance is actually exponential. H&H are trying to achieve a description of biological phenomena based on physical principles. It is necessary to determine the extent to which the biology can be mimicked by mathematical equations known to describe specific physical processes. The kinetics of deactivation matter, because if the decay phase is accurately expressed by a single exponential, then it can be represented as a physical system in which a single 'gate' closes (governed by a single rate constant) to turn off the conductance.

34. **'inward current immediately after a pulse is proportional to the sodium conductance at the end of the preceding depolarization.'** This is a restatement of the idea underlying tail currents: the proportionality constant between current and conductance is the driving force, $V - V_{Na}$. When the conductance, g_{Na}, is activated by depolarization, and a repolarization step is applied, then the driving force will change nearly instantaneously, but the conductance will change only slowly. So, under the hypothesis that Ohm's law holds and $I_{Na} = g_{Na}(V - V_{Na})$, the magnitude of the conductance at the end of the depolarizing step can be obtained by dividing the magnitude of the tail current upon repolarization by the driving force, $g_{Na} = I_{Na} / (V - V_{Na})$. In **Figure 10**, H&H apply repolarizing steps at different times while current is flowing to observe the time course of the conductance. Notably, when both the depolarization-evoked currents and the tail currents (top records) are converted to conductances (bottom records), the conductance values at both V_1 and the onset of V_2 fall along a single continuous curve.

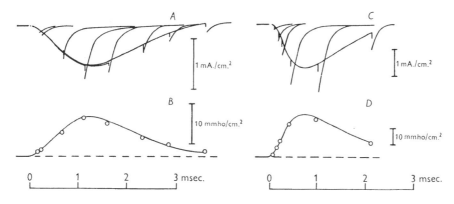

ization and dividing this quantity by the difference between the resting potential and the sodium potential (about −112 mV. in this experiment). A series of points obtained by this method is shown in Fig. 10 B. These may be compared with the smooth curve, which represents sodium conductance obtained by the method described in the preceding paper (subtraction of ionic current in choline from that in sea water). Good agreement was obtained, and also when other depolarizations were employed, for example C and D at −51 mV. The only occasions on which the two methods did not agree were those in which the sodium conductance was measured at long times, with a large depolarization. In these experiments the subtraction method sometimes gave an apparent negative conductance which we regarded as an error due to slight differences between the potassium currents in the two records. This conclusion was confirmed by the fact that the alternative method never showed a 'negative conductance' but only a residual positive conductance.

Experiments with relatively long depolarizations

The instantaneous relation between potassium current and membrane potential

In a previous paper (Hodgkin & Huxley, 1952) we gave reasons for thinking that potassium ions were largely responsible for carrying the maintained outward current associated with prolonged depolarization of the membrane. In order to investigate the instantaneous relation between potassium current and membrane potential it is necessary to employ depolarizations lasting for much longer periods than those used to study sodium current. Polarization effects made such experiments difficult at 5° C., but errors from this cause could be greatly reduced by working at 20° C. In this case the polarization effect occurred at the same rate but the potassium conductance rose in about one-fifth of the time required at 5° C.

A typical experiment with a nerve in choline sea water is illustrated by Fig. 11. Its general object is to measure the ionic currents associated with repolarization of the membrane when the potassium conductance is much greater than the sodium conductance. The amplitude of the first step was −84 mV. and its duration 0·63 msec., which is equivalent to about 4 msec. at 5° C. Under these conditions 90–95 % of the outward current should be potassium current and only 5–10 % should be sodium current (see Fig. 10 for an indication of the rate of fall of sodium conductance from its initial maximum). After the pulse, sodium current should be negligible since the nerve was in choline sea water (see Fig. 3).

The simplest record in Fig. 11 is E, in which the membrane potential was restored to its resting value at the end of the first step. The sequence of events was as follows. At $t=0$ the membrane was depolarized by 84 mV. and was held at this level until $t=0·63$ msec. The current was outward during the whole period and consisted of a hump of sodium current followed by a rise of

35

35. **'to investigate the instantaneous relation between potassium current and membrane potential'**
This section begins the analysis of tail currents after long pulses, to test explicitly the assumption made in Paper 2 that the delayed outward current is carried by potassium. For these experiments, the depolarizing steps must be long enough for the transient inward current (the sodium current) to decay nearly completely and for the sustained outward current to activate. H&H note that long depolarizations tended to polarize the electrode (the same problems noted in Paper 2). Many of the sodium tail current experiments were done at cold temperature (5°C), which slows the kinetics; here they worked at 20°C to speed the kinetics so the potassium current would activate with shorter steps. The polarization effect itself was not temperature dependent, consistent with the later findings that it resulted from the accumulation of extracellular potassium (Frankenhaeuser & Hodgkin 1956), rather than from protein conformational changes per se.

potassium current which reached 1·83 mA./cm.² at $t = 0.63$ msec. At this 36
moment the membrane potential was restored to its resting value. The sudden
increase in potential was associated with a brief capacity current in an inward
direction. This was followed by an outward current which declined exponentially to zero. A record at higher amplification (*e*) shows this 'tail' of outward current more clearly. The dots give the ionic current extrapolated to

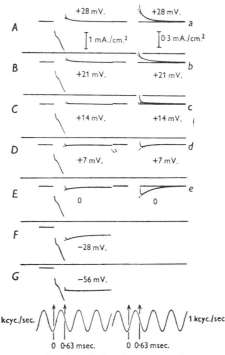

Fig. 11. Membrane currents associated with depolarization of 84 mV. followed by repolarization
to value shown on each record. The duration of the first step was 0·63 msec. The second step
lasted longer than these records. *A* to *G*, records at low amplification showing current during
both steps. *a–e*, records at higher amplification showing only the current during the second
step. The dots give the ionic current after correcting for capacity current. Axon 26 in choline
sea water; temperature 20° C.; uncompensated feed-back.

$t = 0.63$ msec. after correcting for the residual effect of the capacitative surge. 37
The 'tail' of outward current can be explained by supposing that the equilibrium potential for potassium is about 12 mV. greater than the resting
potential in choline sea water, and that the instantaneous value of the potassium conductance (g_K) is independent of V. At $t = 0.63$ msec. the current is
1·83 mA./cm.² when $V = -84$ mV., or 0·22 mA./cm.² when $V = 0$. Taking the
potassium potential (V_K) as $+12$ mV. and neglecting the contribution of
chloride and other ions it follows that the potassium conductance (g_K) was
approximately the same in the two cases. Thus

$$g_K = I_K / (V - V_K), \tag{2}$$

36. *'potassium current which reached 1.83 mA./cm.^{2} at $t = 0.63$ msec.'* Note again how rapidly the squid giant axon potassium current activates: in much less than a millisecond, the conductance is already substantial at the room temperature of 20°C.

37. *'correcting for the residual effect of the capacitative surge.'* In *Figure 11*, the initial brief depolarizing step is to about +24 mV *abs*. Since the recording is made in choline seawater, the sodium current is outward (the first hump). The second phase of outward current is the potassium current. With repolarization to voltages near or more depolarized than the resting potential (*D, E, F, G*), the ionic current is outward; with progressively greater hyperpolarizations (*C, B, A*), the ionic current is *inward*. Because the earliest part of the current recording during V_2 includes the capacity current that repolarizes the membrane, H&H fit the decay phases of the tail currents with exponentials (dotted curves immediately after the repolarization steps, expanded in the column of currents at right, *a–e*) and extrapolate back to the onset of the repolarizing voltage steps to find the instantaneous amplitudes. The results also show that the sustained current, which was always outward with previous voltage steps, can indeed be made to reverse direction. Therefore, nothing intrinsic to the sustained current makes it *necessarily* flow only outward. Because this current is *usually* outward, however, H&H later refer to it as a 'rectifier,' which refers to a circuit element that passes current only in one direction.

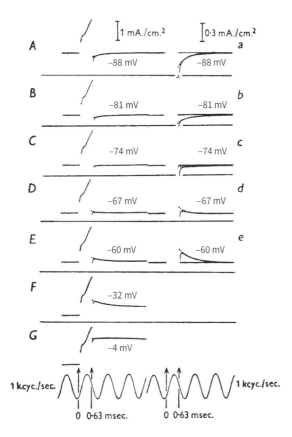

so that

$$g_K = \frac{-1\cdot83 \text{ mA./cm.}^2}{-96 \text{ mV.}} = 19 \text{ m.mho/cm.}^2 \quad \text{when } V = -84 \text{ mV.,}$$

and

$$g_K = \frac{-0\cdot22 \text{ mA./cm.}^2}{-12 \text{ mV.}} = 18 \text{ m.mho/cm.}^2 \quad \text{when } V = 0. \qquad 38$$

As soon as the membrane potential is restored to its resting value the potassium conductance reverts exponentially to its resting level and therefore gives the tail of current seen in E and e. If this explanation is right, the exponential tail of current should disappear at $V = V_K$ and should be reversed in sign when $V > V_K$. Records a to d show that the current is inward for $V > 21$ mV. and is outward for $V < 7$ mV. There is practically no inward current at $V = 14$ mV. and 13 ± 1 mV. would seem to be a reasonable estimate of V_K.   39

This method of measuring the potassium potential depends on the assumption that potassium ions are the only charged particles responsible for the component of the current which varies with time after the end of the pulse. It is not affected by the fact that chloride and other ions may carry appreciable quantities of current, provided that the resistance to the motion of these ions is constant at any given value of membrane potential. The magnitude of the   40 'leak' due to chloride and other ions may be estimated from the current needed to maintain the membrane at the potassium potential. In the experiment illustrated by Fig. 11 this current was about $0\cdot008$ mA./cm.^{2} which is small   41 compared with the maximum potassium current at $V = 0$ or $V = +28$ mV.

Fig. 12 was prepared from the records in Fig. 11 by essentially the same method as that used in studying the instantaneous relation between sodium current and membrane potential. Curve α gives the relation between current and voltage $0\cdot63$ msec. after the application of a single step of amplitude V_1. In curve A, V_1 was fixed at -84 mV. and the potential was changed suddenly to a new level V_2. The abscissa is V_2, while the ordinate is the ionic current immediately after the sudden change. The experimental points in A are seen to fall very close to a straight line which crosses curve α at the potassium potential ($+13$ mV.). In this experiment no measurements were made with   42 $V_2 < V_1$ but records obtained with other fibres showed that the instantaneous current-voltage relation was linear for $V_2 < V_1$ as well as for $V_2 > V_1$.

In the experiment considered in the previous paragraphs the initial rise of potassium current was obscured by sodium current and the plateau was not reached because the pulse was kept short in order to reduce possible errors from the 'polarization effect'. A clearer picture of the sequence of events is provided by Fig. 13. In this experiment the amplitude of the pulse was -25 mV. and its duration nearly 5 msec. The polarization effect was not appreciable since the current density was relatively small. Sodium current was also small since the nerve was in choline sea water, and the depolarization

38. **'when V = 0.'** The calculation is based on the hypothesis that the currents obey Ohm's law, $I_K = g_K(V - V_K)$, so $g_K = I_K/(V - V_K)$. The result supports the idea that the conductance is the same at the end of the depolarizing step and at the onset of the repolarizing step.

39. **'would seem to be a reasonable estimate of V_K.'** This paragraph provides another vivid illustration of the scientific method in miniature. Building upon the *observation* just made in the preceding calculation, namely that conductance is continuous, even as the voltage changes by 84 mV, the four sentences state the following: (1) *Hypothesis:* immediately upon repolarization, the potassium conductance remains briefly at the value attained during the previous step, and the driving force dictates the new magnitude and direction of the current. (2) *Prediction:* if this explanation is correct, then the current should be zero (the tail should disappear) at the potassium equilibrium potential and should reverse direction at more hyperpolarized membrane potentials. (3) *Test and result:* the current indeed flows inward at the predicted voltages. (4) *Discovery and interpretation:* the membrane potential at which the tail current is near zero turns out to be about 13 mV more hyperpolarized than rest, so this value is likely close to V_K. Recalling that the membrane tends to hyperpolarize somewhat in choline seawater, the reversal might indicate an absolute E_K near −80 mV.

40. **'provided that the resistance to the motion of these ions is constant at any given value of membrane potential.'** This statement refers to the idea that currents under voltage clamp sum linearly, such that the time-varying potassium conductance will be unaffected by the leak. The leak conductance (here 'resistance') may be 'appreciable,' but as long as it is constant, that unchanging conductance value will simply superimpose upon the dynamic (voltage-gated, time-varying) currents that are measured.

41. **'this current was about 0.008 mA./cm.²'** This value provides a (first) direct measurement of the 'leak' conductance, which is not voltage-gated but steady at all potentials. The leak was later found to be carried mostly by potassium ions and to a lesser extent by chloride ions. Note that the leak *current* will change with driving force, but the conductance can be taken as relatively constant. In many neurons, the leak current is not quite linear with voltage but instead shows mild rectification or curvature, although to a much lower degree than voltage-gated channels.

42. **'experimental points in A are seen to fall very close to a straight line which crosses curve α at the potassium potential'** *Figure 12* further supports the hypothesis by showing that, when g_K is (instantaneously) identical across potentials, at the magnitude achieved by the preceding step depolarization, the equation $I_K = g_K(V - V_K)$ is a straight line with slope g_K.

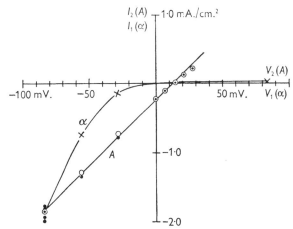

Fig. 12. Current-voltage relations during period of high potassium permeability. Line *A*, 43 instantaneous current-voltage relation determined by changing membrane potential in two steps. The first step had a constant amplitude of −84 mV. and a constant duration of 0·63 msec. The abscissa (V_2) gives the amplitude of the second step in millivolts. The ordinate (I_2) is the ionic current density at the beginning of the second step. The dots are observed currents. Hollow circles are these currents multiplied by factors which equalize the currents at the end of the first step. Curve α and crosses, relation between current and membrane potential at 0·63 msec. after beginning of single step of amplitude V_1. Experimental details are as in Fig. 11.

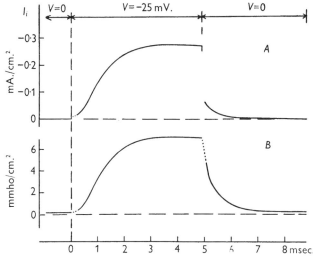

Fig. 13. *A*, ionic current associated with depolarization of 25 mV. lasting 4·9 msec. Axon 18 in 44 choline sea water at a temperature of 21° C. The curve is a direct replot of the original current record except in the regions 0–0·3 msec. and 4·9–5·2 msec., where corrections for capacity current were made by the usual method. Outward current shown upward. *B*, potassium conductance estimated from *A* by the equation $g_K = I_K/(V − V_K)$, where V_K is 12 mV. and I_K is taken as the ionic current (I_i) minus a leakage current of 0·5 m.mho/cm.² × ($V + 4$ mV.).

43. ***Figure 12***. The hollow circles and dots represent data corrections for drift as in ***Figure 6*** (see note 18). The two curves meet at the voltage 84 mV more depolarized than rest (+24 mV *abs.*) because that potential is used for the initial depolarization; hence the current does not change during the 'repolarization' step. If the initial depolarization were sufficient to activate the potassium conductance maximally, the straight line defined by A would have a steeper slope equal to the maximal g_K. If the measurement of potassium current during step depolarizations (curve α) were to be continued to higher values, the curve would eventually linearize and converge with a straight line with a slope of the maximal g_K, which would be about 20% steeper than A.

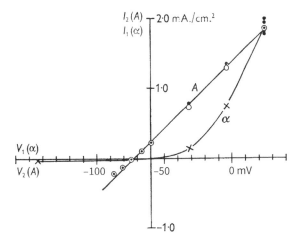

44. ***Figure 13.*** As in the case of sodium, conversion of potassium current (A) to conductance (B) reveals the continuity of the latter. H&H deviate from their general convention and depict outward current as *upward* in this figure to emphasize the difference between the curves at the time of the repolarization. With a repolarizing step toward the reversal potential of the permeating ion, the instantaneous current is *reduced* as the driving force decreases, and thus creates a gap rather than a 'tail,' although the term is applied nonetheless.

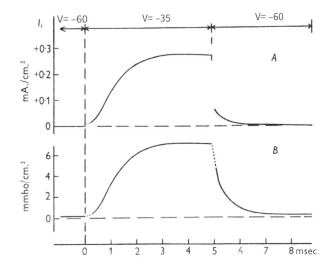

was less than that at which a hump of outward sodium current first became appreciable. On the other hand, it was desirable to make a small correction for the leakage current due to ions other than sodium and potassium. The method of estimating this current at different voltages is indicated on p. 494. The experiment shows that whereas the potassium conductance rises with a marked delay it falls along an exponential type of curve which has no inflexion corresponding to that on the rising phase. This difference was present   45 in all records except, possibly, those with very small depolarizations. It was also present in the curves calculated for the rise and fall of sodium conductance (e.g. Figs. 4, 5 and 8).

The rate constant determining the decline of potassium conductance

The experiments described in the preceding sections indicate that the potassium conductance returns to a low level when the membrane is repolarized to its resting value. The restoration of the condition of low conductance leads to a 'tail' of potassium current which can be fitted with reasonable accuracy by a curve of the form $\exp(-b_K t)$. Table 2 gives the values of the rate constant (b_K) determined by this method. It shows that b_K varies markedly with temperature and also to some extent with the amplitude of the step used   46 to depolarize the axon. The second effect was particularly noticeable in axon 1 which was in poor condition and had a high potassium conductance in the resting state.

The effect of repolarizing the membrane to different levels is shown in Fig. 14. For $V_2 > -20$ mV. the rate constant increased with membrane potential but the relation is less steep than in the corresponding curve for sodium conductance (Fig. 9). Thus, changing V_2 from 0 to $+40$ mV. increases b_{Na} about 3·2-fold and b_K about 1·6-fold. Another important difference between the two processes is that b_{Na} is about 30 times greater than b_K at the resting potential.   47

The potassium potential

Table 3 summarizes a number of measurements of the potential at which potassium current reverses its direction. At 22° C. the apparent potassium potential is about 19 mV. higher than the resting potential if the axon is in sea water and about 13 mV. higher if it is in choline sea water. Corresponding figures at 6–11° C. are 13 mV. in sea water and 8 mV. in choline sea water. Since the resting potential is about 4 mV. higher in choline sea water (Hodgkin & Huxley, 1952) it seems likely that the absolute value of the potassium potential is unaffected by substituting choline for sodium ions. At 20° C. the absolute value of the potassium potential would be 80–85 mV. if the resting potential is taken as 60–65 mV. (Hodgkin & Huxley, 1952). This is nearly equal to the potential of 91 mV. estimated from chemical analyses

45. *'whereas the potassium conductance rises with a marked delay it falls along an exponential type of curve which has no inflexion corresponding to that on the rising phase.'* This analysis of *Figure 13* emphasizes that the potassium conductance has an S-shaped rise but an exponential fall. When a single exponential decay describes a change from one state to another, it suggests that only one transition is necessary to induce that change (see Appendix 1.3). This idea will be expanded upon in Paper 5. H&H note that these attributes are shared by the sodium conductance. (For H&H's p. 494, see note 51 and associated text.)

46. *'can be fitted . . . by a curve of the form exp(−$b_K t$) . . . varies markedly with temperature'* Here, as for the sodium conductance, actual values for the rate constant of deactivation are obtained from the data. The strong temperature dependence is consistent with an underlying biological or biochemical mechanism, comparable to protein-dependent processes such as enzyme-substrate reactions, rather than a nonbiological mechanism such as free diffusion of ions in solution.

47. *'b_{Na} is about 30 times greater than b_K at the resting potential.'* The sodium conductance deactivates much more rapidly than the potassium conductance across a range of voltages, and the change in deactivation rate with voltage—the voltage sensitivity—is more extreme for the sodium conductance. This difference can be seen by comparing the graph of *Figure 14* (the potassium deactivation rate constants) with *Figure 9* (the sodium deactivation rate constants). The observation that the potassium conductance decays relatively slowly at voltages near the resting potential will ultimately offer an explanation for the afterhyperpolarization of the squid action potential. In contrast, neurons with rapid potassium current deactivation kinetics may not have an afterhyperpolarization and may even have afterdepolarizations.

(Hodgkin, 1951). A similar conclusion applies to the results at 6–11° C. 48 In squid fibres, cooling from 20 to 8° C. either has no effect or increases the resting potential by 1 or 2 mV. (Hodgkin & Katz, 1949). The observed

TABLE 2. Rate constant determining decline of potassium conductance following repolarization to resting potential

Axon	Depolarization (mV.)	Temperature (° C.)	Rate constant (msec.$^{-1}$)	Rate constant at 6° C. (msec.$^{-1}$)	Average rate constant at 6° C. (msec.$^{-1}$)
1	6	23	1·2	0·14	
1	13	23	1·3	0·15	
1	21	23	1·3	0·16	
15	13	11	0·36	0·19	
15	20	11	0·35	0·19	
17	10	6	0·20	0·20	
18	6	22	1·5	0·20	
18	13	22	1·6	0·22	
A 20	21	6	0·17	0·17	0·17
21	14	7	0·19	0·16	
38*	10	5	0·12	0·13	
39*	20	19	1·0	0·20	
39*	10	19	0·83	0·16	
39*	10	3	0·10	0·15	
41*	20	4	0·10	0·12	
41*	10	4	0·11	0·14	
1	36	23	1·5	0·18	
1	54	23	1·8	0·22	
18	50	22	2·0	0·27	
B 18	63	22	1·7	0·23	0·22
18	112	22	1·8	0·24	
27	28	21	1·4	0·22	
28	28	21	1·3	0·20	
15	13	11	0·49	0·26	
17	10	6	0·21	0·21	
C 18	6	22	1·6	0·21	0·23
18	13	22	1·7	0·23	
18	19	22	1·9	0·25	
18	25	22	2·0	0·27	
18	50	22	2·1	0·28	
18	63	22	2·1	0·28	
D 23	84	21	1·8	0·28	0·28
24	84	20	2·0	0·35	
26	84	20	1·8	0·30	
27	28	21	1·3	0·19	

Groups A and B in sea water; C and D in choline sea water. Groups A and C: depolarization less than 25 mV.; B and D: depolarization greater than 25 mV. An asterisk denotes the use of compensated feed-back. Rate constants at 6° C. are calculated for a Q_{10} of 3·5, which was found suitable for groups A and C.

potassium potential should therefore be taken as 75–80 mV. while the theoretical potassium potential would be reduced from 91 to 87 mV.

The effect of changing the external concentration of potassium on the potassium potential

Experimental determinations of V_K such as that illustrated by Fig. 11 were made in choline solutions containing different concentrations of potassium.

48. ***'nearly equal to the potential of 91 mV. estimated from chemical analyses (Hodgkin 1951).'*** The first part of the preceding paragraph discusses membrane potentials *relative* to the resting potential, as they have been measured throughout the study, with 'higher' meaning more hyperpolarized. Later in the paragraph comes an important transformation, as each value is translated into *absolute* transmembrane potential, which ultimately is determined to be the independent variable of most importance. Conductance depends on absolute voltages. Additionally, the observation that the delayed current reverses near the potassium equilibrium potential provides evidence that potassium is the charge carrier. Interestingly, H&H make use of the membrane hyperpolarization by choline, with the associated shift of reversal potential, which could have posed a problem, to guide their reasoning and lead them to this fundamental discovery.

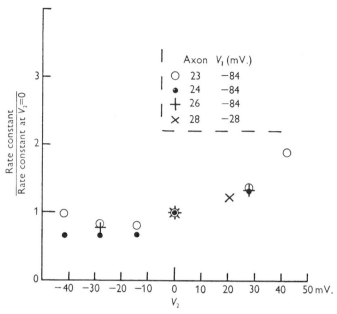

Fig. 14. Effect of membrane potential on the rate constant determining decline of potassium   49
conductance. Abscissa (V_2): membrane potential during second step. Ordinate: relative
value of rate constant.

TABLE 3. Apparent values of potassium potential

Axon	Medium	Temperature (°C.)	Potassium potential minus resting potential (mV.)	Average (mV.)
1	S	23	18	19
28	S	21	19	
15	S	11	14	
20	S	6	10	11
21	S	8	9	
18	C	22	12	
23	C	21	14	
24	C	20	7	13
26	C	20	13	
27	C	21	15	
28	C	21	16	
15	C	11	8	
17	C	6	7	8
20	0·7C:0·3S	6	8	
21	0·9C:0·1S	8	10	

S denotes sea water; C choline sea water. In axons 23–28 the potassium potential was found by
the method illustrated by Fig. 12. In other cases it was taken as the potential at which the steady
state current-voltage curve intersects the line joining the potassium currents before and after
repolarization to the resting potential (e.g. line A and curve α intersect at 13 mV. in Fig. 12).

49. *Figure 14*. Compare to rate constant ratios for sodium, in *Figure 9*, which are plotted with similar *y*-axis limits.

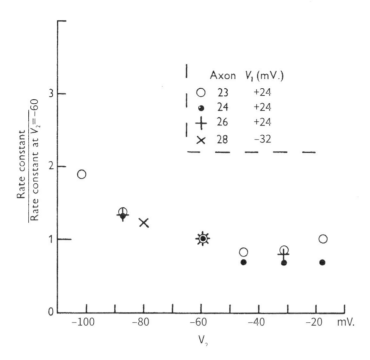

TABLE 4. Effect of potassium concentration on the apparent value of the potassium potential

Axon	Medium	Change in resting potential (ΔE_r) (mV.)	Change in applied potential at which I_K is zero (ΔV_K) (mV.)	Change in absolute membrane potential at which I_K is zero ($\Delta E_K = \Delta E_r + \Delta V_K$) (mV.)
27	A ($\frac{1}{2}$K)	$+2$	$+6$	$+8$
27	B (2K)	-3	-6	-9
28	B (2K)	-3	-7	-10
28	C (5K)	-13	-9	-22

Changes are given relative to the mean potentials observed in a 1K choline sea water before and after application of the test solution. The 1K choline sea water was identical with that described by Hodgkin & Huxley (1952) and contained choline at a concentration of 484 and potassium at 10 g.ions/kg. H$_2$O. The test solutions A, B, C and D were similar but contained potassium at concentrations of 5, 10, 20, 50 g.ions/kg. H$_2$O and correspondingly reduced concentrations of choline. Potentials are given as 'outside potential' minus 'inside potential'.

It was not possible to use a wide range of potassium concentrations, since squid axons tend to undergo irreversible changes if left in solutions containing high concentrations of potassium for any length of time. The results obtained with the two axons studied by this method are given in Table 4. They show that the potassium potential (E_K) is sensitive to the external concentration of potassium but that it changes by only about half the amount calculated   50 for a concentration cell. Thus solutions A, B, C should give changes of $+17$, -17 and -41 mV. if E_K obeyed the ordinary equation for a concentration cell.

One possible explanation of this result is that potassium ions are not the only charged particles responsible for the delayed rise in conductance associated with depolarization (delayed rectification). Thus the discrepancy would be explained if choline or sodium, which are present in relatively high concentrations in the external solution, take part in the process with an affinity only 5 % of that of potassium. This explanation might be consistent with the evidence which suggests that potassium ions are responsible for carrying most of the outward current through a depolarized membrane. For the concentration of potassium inside a fibre is about 10 times greater than that of sodium and the internal concentration of choline is almost certainly negligible. The participation of chloride ions in the process responsible for delayed rectification can probably be eliminated since one experiment showed that replacing all the choline chloride and two-thirds of the magnesium chloride in the choline sea water by dextrose gave an apparent *increase* of 3 mV. in the 'potassium potential'. The magnitude of the change was less certain in this experiment, since the solution employed gave a junction potential of 5–7 mV. which had to be allowed for in estimating the shift in resting potential. But it was clear that any change in E_K was small compared with the *reduction* of 45 mV. expected on the hypothesis that delayed rectification is entirely due to chloride ions.

Another way of accounting for the relatively small changes seen in Table 4 is to suppose that the potassium concentration in the immediate vicinity of the surface membrane is not the same as that in the external solution. Isolated cephalopod axons leak potassium ions at a fairly high rate and these must diffuse through layers of connective tissue and other structures between the excitable membrane and the external solution. This leakage is likely to increase in potassium-deficient solutions and to decrease in potassium-rich solutions. Hence the changes in effective potassium concentration might be less than those in Table 4. A related possibility is that the potassium concentration may be raised locally by the large outward currents used in these

50. **'changes by only about half the amount calculated'** The changes in external potassium should shift the reversal potential by an amount that can be calculated from the Nernst equation. The measured shifts, however, match the prediction of the Nernst equation qualitatively but not quantitatively. This outcome, which could be taken as a refutation of the hypothesis, is considered in detail in the section in small print, and again at the very end of the series of studies (see Paper 5, note 100). The alternative idea would be that ions other than potassium, such as chloride, contribute to the late conductance. H&H tested this alternative by replacing chloride with sucrose, which will depolarize E_{Cl} greatly, but the reversal potential for the sustained voltage-gated current slightly *hyperpolarizes*, ruling out chloride as a likely charge carrier. A different explanation for the discrepancy between prediction and outcome may be that another previously unknown variable may have interfered with the quantitative measurement. H&H consider the possibility that the concentrations right along the membrane may differ from those in the bulk solution: if the axons leak potassium, as cephalopod axons were known to do, the higher *local* concentration of potassium may account for the discrepancy, especially if residual connective tissue impeded the flow of extruded potassium away from the surface of the membrane. Even without the supposition of leak, the current itself includes an efflux of potassium, which may shift the local reversal potential, as later demonstrated by Frankenhaeuser and Hodgkin (1956). Given that the concentration of extracellular potassium is low compared to sodium, the equilibrium potential is likely to be more sensitive even to a small addition of potassium into the local extracellular space.

experiments. This hypothesis is of interest since it might account for the slow polarization effect which is not otherwise explained except in terms of a complicated polarization process at the internal electrode. In a former paper (Hodgkin *et al.* 1952) we obtained evidence of an external layer with a resistance of about 3 Ω.cm.^{2}. The transient change in potassium concentration due to current cannot be calculated without knowing the thickness of this layer. The steady change due to leakage might be large enough to explain the deviations in Table 4, if the leakage of potassium had been several times greater than that found by Steinbach & Spiegelman (1943).

It may be asked why effects similar to those discussed in the preceding paragraph do not upset the relation between the external sodium concentration and V_{Na}. The answer, probably, is that similar effects are present but that they are small because the sodium concentration in sea water is 45 times greater than that of potassium. Changes in concentration due to current would also be smaller in the case of sodium because the sodium currents are of relatively short duration.

The contribution of ions other than sodium and potassium

The experimental results described in this series of papers point to the existence of special mechanisms which allow first sodium and then potassium to cross the membrane at a high rate when it is depolarized. In addition it is likely that charge can be carried through the membrane by other means. Steinbach's (1941) experiments suggest that chloride ions can cross the membrane and there is probably a small leakage of sodium, potassium and choline through cut branches

TABLE 5. Tentative values of leakage conductance and 'equilibrium' potential for leakage current. Five nerves in choline sea water at 6–22° C.

	Average	Range
Leakage conductance (g_l) (m.mho/cm.^{2})	0·26	0·13 to 0·50
Equilibrium potential for leakage current (V_l) (mV.)	−11	−4 to −22
Resting potassium conductance (g_K)$_r$ (m.mho/cm.^{2})	0·23	0·12 to 0·39
Equilibrium potential for potassium (V_K)	+10	+7 to +13

or through parts of the membrane other than those concerned with the selective system. All these minor currents may be thought of as contributing towards a leakage current (I_l) which has a conductance (g_l) and an apparent equilibrium potential (V_l) at which I_l is zero. In this leakage current we should probably also include ions transferred by metabolism against concentration gradients. So many processes may contribute towards a leakage current that measurement of its properties is unlikely to give useful information about the nature of the charged particles on which it depends. Nevertheless, a knowledge of the approximate magnitude of g_l and V_l is important since it is needed for any calculation of threshold or electrical stability. Various methods of measurement were tried but only the simplest will be considered since the orders of magnitude of g_l and V_l were unaffected by the precise method employed. In the experiment of Fig. 11 the steady current needed to maintain the membrane at the potassium potential (+13 mV.) was 8 μA./cm.^{2}. According to our definitions this inward current must have been almost entirely leakage current, for the nerve was in choline sea water and $I_K = 0$ when $V = V_K$. Hence

$$(13 \text{ mV.} - V_l)\, g_l = 8 \ \mu\text{A./cm.}^2.$$

In order to estimate g_l we make use of the fact that the inward current associated with $V = +84$ mV. was not appreciably affected by a fourfold change of potassium concentration (from 5 to 20 mM.). We therefore assume that the potassium conductance was reduced to a negligible value at this membrane potential and that the inward current of 24 μA./cm.^{2} was entirely leakage current. Hence

$$(84 \text{ mV.} - V_l)\, g_l = 24 \ \mu\text{A./cm.}^2.$$

From these two equations we find a value of −22 mV. for V_l and one of 0·23 m.mho/cm.^{2} for g_l. An estimate of the resting value of g_K may also be obtained by this method. At the resting potential in choline sea water

$$V_l g_l + V_K (g_K)_r = 0.$$

Hence

$$(g_K)_r = 0·39 \text{ m.mho/cm.}^2.$$

Tentative values obtained by this type of method are given in Table 5.

51. **'Tentative values obtained by this type of method are given in Table 5.'** At this point, H&H have gathered values for E_{Na} and E_K, as well as g as a function of voltage for both ionic currents. Here, they obtain corresponding values for the leak conductance. They know that the current required to voltage clamp the cell at E_K must all be leak current, because the potassium current is net zero. This observation gives them one point on the I-V relation. They also measure the current required to voltage-clamp the cell at a potential 84 mV more hyperpolarized than rest (–150 mV *abs.*)! They confirm that the magnitude of this current is relatively unaffected by the large changes in the external potassium concentration, making it reasonable to attribute it to non-potassium leak. This measurement gives them a second point on the current-voltage relation. The slope of the line defined by these two points gives g_{leak} and the x-intercept gives the reversal potential, E_{leak}, given in **Table 5**.

DISCUSSION

At this stage all that will be attempted by way of a discussion is a brief comparison of the processes underlying the changes in sodium and potassium conductance. The main points of resemblance are: (1) both sodium and potassium conductances rise along an inflected curve when the membrane is depolarized and fall without any appreciable inflexion when the membrane is repolarized; (2) the rate of rise of conductance increases continuously as the membrane potential is reduced whereas the rate of fall associated with repolarization increases continuously as the membrane potential is raised; (3) the rates at which the conductances rise or fall have high temperature coefficients whereas the absolute values attained depend only slightly on temperature; (4) the instantaneous relation between sodium or potassium current and membrane potential normally consists of a straight line with zero current at the sodium or potassium potential.

The main differences are: (1) the rise and fall of sodium conductance occurs 10–30 times faster than the corresponding rates for potassium; (2) the variation of peak conductance with membrane potential is greater for sodium than for potassium; (3) if the axon is held in the depolarized condition the potassium conductance is maintained but the sodium conductance declines to a low level after reaching its peak.

SUMMARY   52

1. Repolarization of the giant axon of *Loligo* during the period of high sodium permeability is associated with a large inward current which declines rapidly along an approximately exponential curve.

2. The 'tail' of inward current disappears if sodium ions are removed from the external medium.

3. These results are explained quantitatively by supposing that the sodium conductance is a continuous function of time which rises when the membrane is depolarized and falls when it is repolarized.

4. For nerves in sea water the instantaneous relation between sodium current and membrane potential is a straight line passing through zero current about 110 mV. below the resting potential.

5. The rate at which sodium conductance is reduced when the fibre is repolarized increases markedly with membrane potential.

6. The time course of the sodium conductance during a voltage clamp can be calculated from the variation of the 'tail' of inward current with the duration of depolarization. The curves obtained by this method agree with those described in previous paper.

7. Repolarization of the membrane during the period of high potassium permeability is associated with a 'tail' of current which is outward at the

52. ***'Summary'*** The main accomplishment of Paper 3 is to demonstrate that ion-selective permeabilities of the axonal membrane behave like conductances. Both the sodium and potassium conductances vary directly with (absolute) membrane potential (g is a function of V). In physiological saline, the ionic conductances are ohmic when their time dependence is removed by instantaneously changing the voltage and measuring the current. In response to changes in membrane voltage, conductance values relax over time to a new equilibrium value, without discontinuities (g is a function of t). The time course of this change is faster for the sodium than the potassium conductance; in both cases the rates but not magnitudes are strongly temperature dependent. All these observations set up for the reconstruction of the action potential in the final paper.

resting potential and inward above a critical potential about 10–20 mV. above the resting potential.

8. The instantaneous relation between potassium current and membrane potential is a straight line passing through zero at 10–20 mV. above the resting potential.

9. These results suggest that the potassium conductance is a continuous function of time which rises when the nerve is depolarized and falls when it is repolarized.

10. The rate at which the potassium conductance is reduced on repolarization increases with membrane potential.

11. The critical potential at which the 'potassium current' appears to reverse in sign varies with external potassium concentration but less steeply than the theoretical potential of a potassium electrode.

REFERENCES

HODGKIN, A. L. (1951). The ionic basis of electrical activity in nerve and muscle. *Biol. Rev.* **26**, 339–409.

HODGKIN, A. L. & HUXLEY, A. F. (1952). Currents carried by sodium and potassium ions through the membrane of the giant axon of *Loligo*. *J. Physiol.* **116**, 449–272.

HODGKIN, A. L., HUXLEY, A. F. & KATZ, B. (1952). Measurement of current-voltage relations in the membrane of the giant axon of *Loligo*. *J. Physiol.* **116**, 424–448.

HODGKIN, A. L. & KATZ, B. (1949). The effect of temperature on the electrical activity of the giant axon of the squid. *J. Physiol.* **109**, 240–249.

STEINBACH, H. B. (1941). Chloride in the giant axons of the squid. *J. cell. comp. Physiol.* **17**, 57–64.

STEINBACH, H. B. & SPIEGELMAN, S. (1943). The sodium and potassium balance in squid nerve axoplasm. *J. cell. comp. Physiol.* **22**, 187–196.

Paper 4

Hodgkin, A. L. and Huxley, A. F. (1952c) The dual effect of membrane potential on sodium conductance in the giant axon of *Loligo*. *Journal of Physiology* 116:497–506.

In the fourth paper, H&H examine the time dependence of the sodium conductance and provide evidence for the phenomenon they call 'inactivation.' In the previous papers, they observed that the early current activates upon depolarization, but then seems to decay as the depolarization is sustained. Although they were initially uncertain about the validity of this observation, they concluded that inactivation was a real phenomenon that can account for experimental results in all three preceding papers: In Paper 1, when the early sodium current was outward, it decayed before the relatively slow potassium conductance activated. In Paper 2, even small prolonged hyperpolarizations increased the magnitude of the sodium current evoked by subsequent depolarization, and when this increase was accounted for, the rest of the data were quantitatively predictable by the independence principle. In Paper 3, the sodium conductance evoked by depolarizations of constant magnitude changed as the resting potential progressively depolarized over the course of the experiment. Here, H&H directly measure the voltage dependence and time course of inactivation of the early current, justifying the conclusion that the membrane potential has a 'dual effect,' to activate and inactivate the sodium conductance.

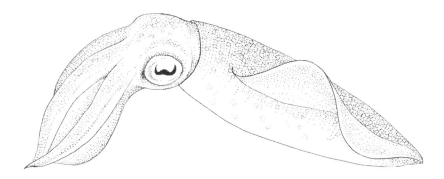

THE DUAL EFFECT OF MEMBRANE POTENTIAL ON SODIUM CONDUCTANCE IN THE GIANT AXON OF *LOLIGO*

By A. L. HODGKIN and A. F. HUXLEY

From the Laboratory of the Marine Biological Association, Plymouth, and the Physiological Laboratory, University of Cambridge

(*Received 24 October* 1951)

This paper contains a further account of the electrical properties of the giant axon of *Loligo*. It deals with the 'inactivation' process which gradually reduces sodium permeability after it has undergone the initial rise associated with depolarization. Experiments described previously (Hodgkin & Huxley, 1952 *a*, *b*) show that the sodium conductance always declines from its initial maximum, but they leave a number of important points unresolved. Thus they give no information about the rate at which repolarization restores the ability of the membrane to respond with its characteristic increase of sodium conductance. Nor do they provide much quantitative evidence about the influence of membrane potential on the process responsible for inactivation. These are the main problems with which this paper is concerned. The experimental method needs no special description, since it was essentially the same as that used previously (Hodgkin, Huxley & Katz, 1952; Hodgkin & Huxley, 1952 *b*).

RESULTS

The influence of a small change in membrane potential on the ability of the membrane to undergo its increase in sodium permeability is illustrated by Fig. 1. In this experiment the membrane potential was changed in two steps. The amplitude of the first step was -8 mV. and its duration varied between 0 and 50 msec. This step will be called the conditioning voltage (V_1). It was followed by a second step called the test voltage (V_2) which was kept at a constant amplitude of -44 mV.

Record A gives the current observed with the test voltage alone. B–F show the effect of preceding this by a conditioning pulse of varying duration. Although the depolarization of 8 mV. was not associated with any appreciable inward current it greatly altered the subsequent response of the nerve. Thus, if the conditioning voltage lasted longer than 20 msec., it reduced the inward

1. *'the "inactivation" process which gradually reduces sodium permeability'* The term 'inactivation' is formally introduced here. Inactivation is the decrease of conductance when the stimulus (depolarization) is still present; it is distinct from 'deactivation,' which is the decrease of conductance when the stimulus is removed (repolarization).

2. *'These are the main problems with which this paper is concerned.'* The main issues to be addressed are the usual two dimensions of magnitude and kinetics: the extent and time course of inactivation and recovery.

3. *'the conditioning voltage . . . the test voltage'* H&H introduce the concept of a conditioning step, which sets the conductance (in modern terminology, the channels) in a particular state or condition. An attribute of the conditioning step—its duration or its magnitude—is the independent variable in the experiments. The test step, which is the same on every trial, assays the consequence of that conditioning step on the current of interest, which is the dependent variable. Here, the conditioning steps consist of depolarizations of constant magnitude (to −52 mV *abs.*) but different durations; the test step is a depolarization to −16 mV *abs.*

4. *'the depolarization . . . greatly altered the subsequent response of the nerve.'* The only parameter that can be measured with the voltage-clamp technique is current. When current flows, the membrane is conducting. When no current flows, assuming a nonzero driving force, the membrane can be said to be nonconducting. In modern terms, current signifies open channels, and no current signifies nonconducting channels. If multiple nonconducting or silent states exist, however, they cannot be distinguished. Experimentally, therefore, it is necessary to make the conducting state into a *reporter* for those silent states. Here, H&H point out that the conditioning step was too small to evoke appreciable current, and yet it is clear that the current evoked by the test depolarization has changed. Therefore, the membrane must have gone from one kind of silent state (from which the test step *does* evoke a transition to a conducting state) into a distinct silent state (from which the test step *does not* evoke a transition to a conducting state). This is the evidence for a second nonconducting state, which H&H term 'inactivation.'

current during the test pulse by about 40%. At intermediate durations the inward current decreased along a smooth exponential curve with a time constant of about 7 msec. The outward current, on the other hand, evidently behaved in a different manner; for it may be seen to approach a final level 5 which was independent of the duration of the conditioning step. This is consistent with the observation that depolarization is associated with a maintained increase in potassium conductance (Hodgkin & Huxley, 1952a).

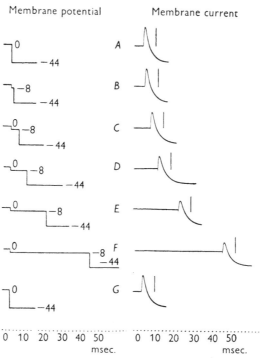

Fig. 1. Development of 'inactivation' during constant depolarization of 8 mV. Left-hand 6 column: time course of membrane potential (the numbers show the displacement of the membrane potential from its resting value in mV.). Right-hand column: time course of membrane current density. Inward current is shown as an upward deflexion. (The vertical lines show the 'sodium current' expected in the absence of a conditioning step; they vary between 1·03 mA./cm.² in A and 0·87 mA./cm.² in G). Axon 38; compensated feed-back; temperature 5° C.

Fig. 2 illustrates the converse process of raising the membrane potential before applying the test pulse. In this case the conditioning voltage improved the state of the nerve for the inward current increased by about 70% if the 7 first step lasted longer than 15 msec. This finding is not altogether surprising, for the resting potential of isolated squid axons is less than that of other excitable cells (Hodgkin, 1951) and is probably lower than that in the living animal.

A convenient way of expressing these results is to plot the amplitude of the sodium current during the test pulse against the duration of the conditioning

5. *'The outward current . . . behaved in a different manner'* This observation provides another piece of evidence that the sodium and potassium currents are separate entities with distinct voltage dependences. The outward (potassium) current does not inactivate, whereas the inward sodium current shows a dependence of current magnitude on the duration of the conditioning step (illustrated in *Figure 3*).

6. *Figure 1.* The longer the depolarizing conditioning step, the smaller the inward (sodium) current evoked by the test step. Note that the vertical bars are not calibration bars, but the magnitude of the sodium current with no conditioning step.

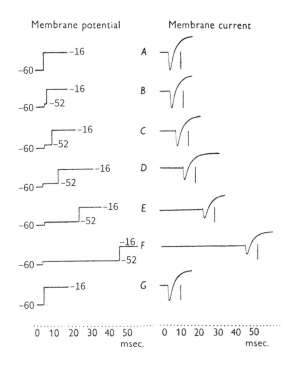

7. *'raising the membrane potential . . . improved the state of the nerve'* Hyperpolarization during conditioning led to larger sodium currents evoked by the test step. In other words, the sodium conductance is partly inactivated at rest. H&H refer to the larger currents as an 'improvement' for the inward current, possibly indicating their suspicion that hyperpolarization brings the membrane toward a more physiologically natural condition. They cite Hodgkin's 1951 review article, in which he had summarized absolute resting membrane potential measurements and estimates from squid (*Loligo*), cuttlefish (*Sepia*), and crustacean (*Homarus, Carcinus*) giant axons (about −50 to −60 mV) and contrasted them with comparable measurements from striated muscle and cardiac Purkinje fibers from frog, dog, and goat (about −80 to −90 mV). At the time, H&H had no means to assess the extent to which the more depolarized resting potentials of axons relative to muscle were indicative of genuine (biological) differences versus artifactual (technical) consequences of deterioration of the axon. In fact, muscles do have resting potentials much closer to E_K than do most neurons.

pulse. For this purpose we used the simple method of measurement illustrated   8
by Fig. 3 (inset). This procedure avoids the error introduced by variations of
potassium conductance during the first step and should give reasonable results
for $V > -15$ mV. With larger depolarizations both the method of measurement and the interpretation of the results become somewhat doubtful, since
there may be appreciable sodium current during the conditioning period. Two

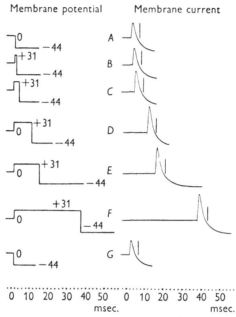

Fig. 2. Removal of 'inactivation' at membrane potential of $+31$ mV. Experimental details are   9
    as in Fig. 1. The vertical lines show the 'sodium current' with no conditioning step; they
    vary between 0·82 mA./cm.² in A and 0·75 mA./cm.² in G.

of the curves in Fig. 3 were obtained from the families of records illustrated in
Figs. 1 and 2. The other two were determined from similar families obtained
on the same axon. All four curves show that inactivation developed or was
removed in an approximately exponential manner with a time constant which
varied with membrane potential and had a maximum near $V = 0$. They also
indicate that inactivation tended to a definite steady level at any particular
membrane potential. Values of the exponential time-constant (τ_h) of the   10
inactivation process are given in Table 1.   11

The influence of membrane potential on the steady level of inactivation is
illustrated by the records in Fig. 4. In this experiment the conditioning step
lasted long enough to allow inactivation to attain its final level at all voltages.   12
Its amplitude was varied between $+46$ and -29 mV., while that of the test
step was again kept constant at -44 mV. The effect of a small progressive
change was allowed for in calculating the vertical lines on each record. These
give the inward current expected in the absence of a conditioning step and

8. *'plot the amplitude . . . during the test pulse against the duration of the conditioning pulse'* This plot (in *Figure 3*) illustrates the *time course* of inactivation for different conditioning potentials.

9. *Figure 2.* This figure complements *Figure 1*: the longer the hyperpolarizing conditioning step, the larger the sodium current evoked by the test step. As in *Figure 1*, the vertical lines show the magnitude of the sodium current in the absence of a conditioning step.

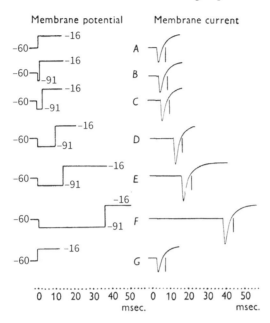

10. *'inactivation tended to a definite steady level at any particular membrane potential.'* This observation emphasizes the important point that, with voltage steps away from rest, inactivation changes over time (kinetics) but reaches a final constant or steady-state level at each voltage (magnitude). The fact that each voltage can be associated with a steady-state level of inactivation makes the inactivation process quantitatively tractable and physically interpretable.

11. *Table 1.* See note 22.

12. *'the conditioning step lasted long enough to allow inactivation to attain its final level at all voltages.'* This experiment, shown in *Figure 4* and analyzed in *Figure 5*, leads to the first steady-state inactivation curve (see note 15).

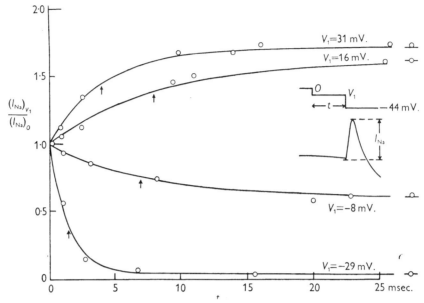

Fig. 3. Time course of inactivation at four different membrane potentials. Abscissa: duration of   13
conditioning step. Ordinate: circles, sodium current (measured as inset) relative to normal
sodium current; smooth curve, $y = y_\infty - (y_\infty - 1) \exp(-t/\tau_h)$, where y_∞ is the ordinate at $t = \infty$
and τ_h is the time constant (shown by arrows). Experimental details as in Figs. 1 and 2.

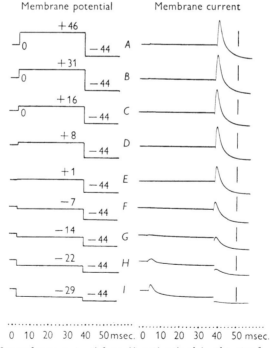

Fig. 4. Influence of membrane potential on 'inactivation' in the steady state. Experimental   14
details are as in Fig. 1. The vertical lines show the sodium current with no conditioning step;
they vary between 0·74 mA./cm.² in A and 0·70 mA./cm.² in I.

13. ***Figure 3.*** This figure quantifies the data from ***Figure 1*** and ***Figure 2***, as well as from two comparable experiments with a greater depolarizing or hyperpolarizing conditioning step. The amplitude of current evoked by a test step to −16 mV *abs.* after conditioning is normalized to the current evoked without conditioning, so all curves start at 1. The normalized test-step response is plotted against the duration of the conditioning step. For each conditioning step, the distribution of the points follows a single exponential relaxation, which is defined by the time constant, τ. The time constant (arrow on each trace) defines the time required for the function to decay from its initial value (1 at time zero) to 63% of the way to its final value. The final value, indicated by the point at the far right of each curve, is the normalized current evoked by stepping to −16 mV *abs.* after holding at the conditioning voltage for longer than 25 msec. Note that larger conditioning steps give shorter time constants.

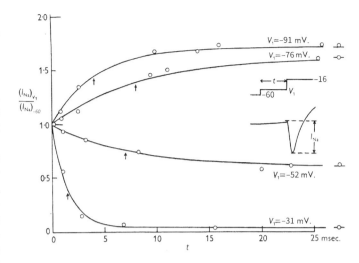

H&H measured the peak inward current relative to the baseline value from which the current rose, thereby compensating for the tiny increase in *outward* current that could be evoked during depolarizing conditioning steps. This outward current can be seen in the inset as the small, slow deflection in the current trace just before the sodium current onset.

This figure forms the basis for the discovery that the rate of the onset of inactivation is faster at more depolarized potentials, and the rate of recovery from inactivation is faster at more hyperpolarized potentials. This form of inactivation is now referred to as 'fast inactivation' to distinguish it from 'slow inactivation' (Chandler & Meves 1970b), from which recovery occurs more slowly, and which becomes appreciable only with longer conditioning depolarizations than those used here.

14. ***Figure 4.*** These recordings provide the data for the steady state inactivation curve of ***Figure 5*** (see note 15). As before, the vertical lines show the magnitude of the sodium current with no conditioning step. With hyperpolarizations as well as with small depolarizations, the amplitude of the sodium current evoked by the test step varies even when no detectable sodium current is evoked by the conditioning step itself, demonstrating that inactivation (or the recovery from inactivation) can proceed even without activation. H&H later assumed that activation and inactivation are independent processes (see Paper 5, note 23). Although later experiments provided evidence for coupling between activation and inactivation (Armstrong & Bezanilla 1977; Bezanilla & Armstrong 1977), H&H's assumption of independence was a sufficient first approximation for the reconstruction of the action potential (see Paper 5).

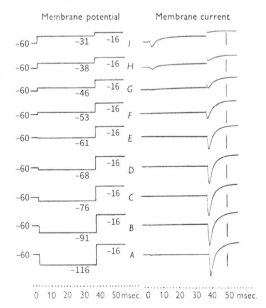

were obtained by interpolating between records made with the test step alone. The conditioning voltage clearly had a marked influence on the inward current during the second step, for the amplitude of the sodium current varied between $1\cdot3$ mA./cm.^{2} with $V_1 = +46$ mV. and about $0\cdot03$ mA./cm.^{2} with $V_1 = -29$ mV.

The quantitative relation between the sodium current during the test pulse and the membrane potential during the conditioning period is given in Fig. 5.

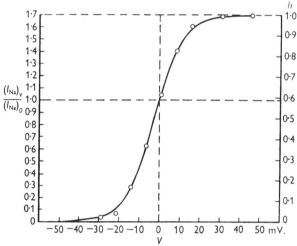

Fig. 5. Influence of membrane potential on 'inactivation' in the steady state. Abscissa: displace-   15
ment of membrane potential from its resting value during conditioning step. Ordinate:
circles, sodium current during test step relative to sodium current in unconditioned test step
(left-hand scale) or relative to maximum sodium current (right-hand scale). The smooth
curve was drawn according to equation 1 with a value of $-2\cdot5$ mV. for V_h. This graph is
based on the records illustrated in Fig. 4. Sodium currents were determined in the manner
shown in Fig. 3.

This shows that the two variables are related by a smooth symmetrical curve which has a definite limiting value at large membrane potentials. In discussing   16
this curve it is convenient to adopt the following nomenclature. We shall denote the ability of the nerve to undergo a change in sodium permeability by a variable, h, which covers a range from 0 to 1 and is proportional to the ordinate in Fig. 5. In these terms $(1-h)$ is a measure of inactivation, while h is the fraction of the sodium-carrying system which is not inactivated and is therefore rapidly available for carrying sodium ions when the membrane is depolarized. If these definitions are adopted we may say that inactivation is almost complete when $V < -20$ mV. and is almost absent when $V > 30$ mV. At the resting potential h is about $0\cdot6$ which implies that inactivation is 40% complete.

The smooth curve in Fig. 5 was calculated from the equation

$$(h)_{\text{steady state}} = \frac{1}{1 + \exp(V_h - V)/7}, \tag{1}$$

15. **Figure 5.** This key figure is the first illustration of an inactivation versus voltage curve, also called an h_∞ curve or, later, an 'availability' curve. It illustrates the data from **Figure 4** as the extent of inactivation (peak amplitude of sodium current during the test step relative to that with no conditioning step) plotted as a function of the voltage during the conditioning step. The conditioning step was of sufficient duration (30 msec) for inactivation to reach a steady-state value; consequently, this type of plot has no information about kinetics. The left-hand y-axis shows the peak inward currents evoked by the test step, normalized to the current evoked without a conditioning step. The plot reveals that the sodium conductance is nearly half-inactivated at the resting potential. A comparable degree of inactivation at the resting potential is also characteristic of most mammalian neurons. The right-hand y-axis provides an alternative scale, with the current normalized to the maximum sodium current, measured as the current evoked after equilibration at the most hyperpolarized conditioning potential. This scaling describes inactivation with a factor that H&H call h, which can vary from 0 (complete inactivation) to 1 (no inactivation). To make the meaning of the value more intuitive, it has become conventional to refer to h as a measure of *availability* of the conductance to be activated upon depolarization. Thus, 0 indicates no availability, and 1 indicates complete availability. As stated in the text, $1-h$ is therefore a measure of the degree of inactivation.

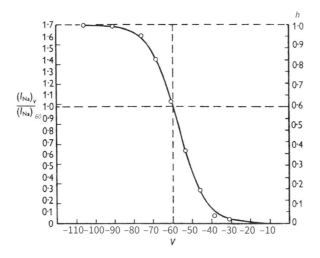

16. **'the two variables are related by a smooth, symmetrical curve which has a definite limiting value at large membrane potentials.'** Two points are made here: The smooth symmetry indicates that a simple mathematical function may describe the dependence of inactivation on voltage. The limiting values are 'definite' (defined), indicating that the equation relating inactivation and voltage will be a saturating function.

where V is expressed in millivolts and V_h is the value of V at which $h = \frac{1}{2}$ in the steady state. The same equation gave a satisfactory fit in all experiments but there was some variation in the value of V_h. Five experiments with three fresh fibres gave resting values of h between 0·55 and 0·62. In these cases V_h varied between $-1·5$ and $-3·5$ mV. On the other hand, two experiments with a fibre which had been used for some time gave a resting h of only about 0·25; V_h was then $+7·5$ mV. Since the resting potential was found to decline by 10–15 mV. during the course of a long experiment it is reasonable to suppose that the change in V_h arose solely from this cause and that the relation between h and the absolute membrane potential was independent of the condition of the fibre.

In a former paper we examined the relation between the concentration of sodium ions in the external medium and the sodium current through the membrane (Hodgkin & Huxley, 1952a). The results were reasonably close to those predicted by the 'independence principle' except that the currents were 20–60 % too large in the sodium-deficient solutions. This effect was attributed to the small increase in resting potential associated with the substitution of choline ions for sodium ions. This explanation now seems very reasonable. The resting potential probably increased by about 4 mV. in choline sea water and this would raise h from 0·6 to 0·73 in a fresh fibre and from 0·25 to 0·37 in a fibre which had been used for some time.

The quantitative results obtained in this series of experiments are summarized in Table 1. Most of the experiments were made at 3–7° C. but a temperature of 19° C. was used on one occasion. The results suggest that temperature has little effect on the equilibrium relation between h and V, but greatly alters the rate at which this equilibrium is attained. The Q_{10} of the rate constants cannot be stated with certainty but is clearly of the order of 3.

Two-pulse experiment

This section deals with a single experiment which gave an independent measurement of the time constant of inactivation.

Two pulses of amplitude -44 mV. and duration 1·8 msec. were applied to the membrane. Fig. 6A is a record obtained with the second pulse alone. The ionic current was inward and reached a maximum of about 0·25 mA./cm.². As in all other records, the inward current was not maintained but declined as a result of inactivation. Restoration of the normal membrane potential was associated with a tail of inward current due to the rapid fall of sodium conductance (see Hodgkin & Huxley, 1952b). When two pulses were applied in quick succession the effect of the first was similar to that in A, but the inward current during the second was reduced to about one half (record B). A gradual recovery to the normal level is shown in records C–G.

The curve in Fig. 7 was obtained by estimating sodium current in the manner

17. **'The same equation gave a satisfactory fit in all experiments'** *Equation 1* is a saturating function that describes a two-state model based on the Boltzmann principle (see Appendix 1.4). As described in Paper 2, note 64, this equation has the form

$$y = \frac{1}{1 + e^{-x}}$$

(see Appendix 1.4). The curve can be shifted on the *x*-axis by subtracting a constant term $x_{1/2}$ from the *x*-value, and the slope can be made shallower by dividing the *x*-value by a constant term *k*. Here the *x*-values are voltages and the *y*-values are the steady-state availabilities, so the equation becomes

$$h_{steady\text{-}state} = \frac{1}{1 + e^{-(V - V_h)/k}},$$

which, with H&H's conventions for the sign on *V*, simplifies to

$$h_{steady\text{-}state} = \frac{1}{1 + e^{(V_h - V)/k}}.$$

(H&H use V_h, where the '*h*' refers to the inactivation parameter *h;* in modern conventions the term $V_{1/2}$ is also used.) The value $V_{1/2}$ indicates the voltage at which the current is half inactivated ($h_{steady\text{-}state} = 0.5$). Note that H&H are careful to distinguish the measured steady-state value from the theoretical true equilibrium reached at infinite time, which is the formal definition of h_∞.

In physics, this model becomes applicable at equilibrium when a particle can exist in either one of two distinct states, and the probability of occupancy of either state depends on another variable, in this case *V*. The good fit of this curve suggests that it may be physically correct to think of the conductance underlying the current as existing in two states: inactivated and noninactivated (available).

18. **'the relation between h and the absolute membrane potential was independent of the condition of the fibre.'** Recall that H&H generally estimate all voltages, including V_h, relative to the resting potential, which depolarizes over the course of recording. Here they note that the shifts in V_h correlate with the depolarization as the axon deteriorates, which supports the idea that inactivation depends on *absolute* rather than relative membrane potential.

19. **'This explanation now seems very reasonable.'** H&H point out that the observation that inactivation depends directly on absolute membrane potential can account for why sodium currents become smaller over time: they become progressively more inactivated as the resting potential depolarizes because of slow deterioration of the axon. Inactivation also provides an explanation for why the current-voltage relations predicted in Paper 2 *underestimated* the current measured in choline-rich solutions. The resting potential is slightly more hyperpolarized in choline-rich solutions, which reduces the extent of inactivation (see Paper 2, note 84).

20. **'temperature has little effect on the equilibrium relation . . . but greatly alters the rate at which this equilibrium is attained.'** As in the case of the activation of conductance, the *kinetics* of inactivation are temperature-sensitive, but the *magnitude* (voltage dependence) is not.

21. **'When two pulses were applied in quick succession . . . inward current during the second was reduced to about one half'** This two-pulse voltage protocol is used to quantify 'recovery from inactivation.' Once again, only current, which provides a measure of the activated state of the membrane, can be recorded directly; a voltage protocol must therefore be designed with a test current as the reporter for distinct nonconducting states. Here, the first (conditioning) depolarization activates a sodium current that decays slightly during the 1.8-msec step and is followed by a tail current upon repolarization. The second (test) depolarization also activates a sodium current, but it is smaller, from which H&H can infer that the conductance is at least partly inactivated. The longer the interval between pulses, the smaller the degree of inactivation.

TABLE 1. Experiments with conditioning voltage   22

Axon	Temperature (°C.)	Variable	Displacement of membrane potential (mV.)									
			−29	−22	−14	−8	−7	0	9	16	31	46
38	5	h^* (steady	0·02	0·04	0·17	—	0·37	0·59	0·82	0·94	0·99	1·00
39	19	state)	0·02	0·04	0·09	—	0·28	0·55	0·83	0·94	0·98	0·99
39†	3		0·01	0·03	0·04	—	0·11	0·26	0·50	0·69	0·93	0·99
38	5	h‡ (steady	0·02	—	—	0·43	—	0·58	—	0·92	0·99	—
39	19	state)	0·03	—	—	0·40	—	0·61	—	0·94	—	—
39†	3		—	—	—	—	—	0·22	—	0·75	0·93	—
37	3		—	0·04	—	0·34	—	0·55	0·81	0·96	—	—
38	5	τ_h‡ (msec.)	1·5	—	—	7	—	[8–10]	—	8	4	—
39	19		0·35	—	—	1·5	—	[1·7–2·1]	—	1·8	—	—
39†	3		—	—	—	—	—	—	—	13	7	—
37	3		—	3	—	6	—	[8–10]	9	7	—	—
38	6	τ_h§ (msec.)	1·3	—	—	6	—	[7–9]	—	7	3·6	—
39	6		1·5	—	—	6	—	[7–9]	—	8	—	—
39†	6		—	—	—	—	—	—	—	9	5	—
37	6		—	2·2	—	4	—	[6–7]	7	5	—	—

Two-pulse experiment

Axon 31 at 4·5° C. $\tau_h = 1·8$ msec. at $V = -44$ mV. $\tau_h = 12$ msec. at $V = 0$
Axon 31 at 6° C. τ_h§ $= 1·5$ msec. at $V = -44$ mV. τ_h§ $= 10$ msec. at $V = 0$

* Measurements made by methods illustrated in Figs. 4 and 5.

† The axon had been used for some time and was in poor condition when these measurements were made.

‡ Methods illustrated in Figs. 1–3.

§ Calculated from above assuming Q_{10} of 3.

[ ] Interpolated.

h is the fraction of the sodium system which is rapidly available, $(1 - h)$ is the fraction inactivated.

τ_h determines the rate at which h approaches its steady state.

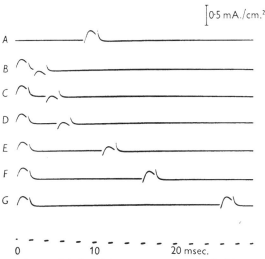

Fig. 6. Membrane currents associated with two square waves applied in succession. The amplitude   23
of each square wave was −44 mV. and the duration 1·8 msec. Record *A* shows the second
square wave alone, *B–G* both square waves at various intervals. Axon 31; uncompensated
feed-back; temperature 4·5° C.

22. **Table 1.** The $h_{steady\text{-}state}$ values represent magnitude; the τ values (time constants) represent kinetics. The τ's for depolarization represent the time course of onset of inactivation and those for hyperpolarization represent the time course of recovery from inactivation. Axon 39—reported to be in poor condition for the second measurement, from which H&H infer that its resting potential has become depolarized—shows only ~22% availability (78% inactivation) at rest, whereas all the other experiments give resting availabilities between 55% and 61%.

23. **Figure 6.** These currents are evoked by the two-pulse protocol. Note that the peak of the second current in B is smaller than the amplitude of the first current at the end of the step, but its amplitude is measured relative to the preceding baseline (as in **Figure 3**), reducing the contribution of leak and capacity current. Measured in this way, the second current in B is just under half the amplitude of the first current.

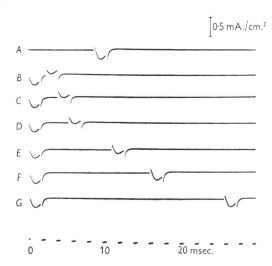

shown in Fig. 3 and plotting this against the interval between the two pulses. It will be seen that recovery from inactivation took place in an approximately exponential manner with a time constant of about 12 msec. A similar curve   24 and a similar time constant were obtained by plotting   25

$$\left(\frac{dI_{Na}}{dt}\right)_{max.} \quad \text{instead of} \quad (I_{Na})_{max.}.$$

This time constant is clearly of the same order as that given by the method using weak conditioning voltages (see Table 1). An estimate of the inactivation time constant at -44 mV. may be obtained by extrapolating the curve in Fig. 7 to zero time. This indicates that the available fraction of the sodium-carrying system was reduced to 0.37 at the end of a pulse of amplitude -44 mV. and duration 1.8 msec. Hence the inactivation time constant at -44 mV. is about 1.8 msec., which is of the same order as the values obtained with large   26 depolarizations by the first method (Table 1). It is also in satisfactory agreement with the time constant obtained by fitting a curve to the variation of sodium conductance during a maintained depolarization of 40–50 mV. (Hodgkin & Huxley, 1952c).

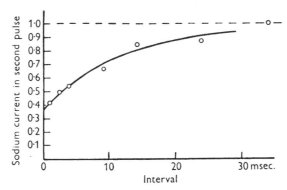

Fig. 7. Recovery from inactivation. Abscissa: interval between end of first pulse and beginning of second pulse. Ordinate: sodium current in second pulse measured as shown in Fig. 3 and expressed as a fraction of the sodium current in an unconditioned pulse. The circles are experimental points derived from the records in Fig. 6. The smooth curve is drawn according to the expression $1-0.63 \exp(-t/\tau_h)$, where $\tau_h = 12$ msec.

The two-pulse experiment is interesting because it emphasizes the difference between the rapid fall of sodium conductance associated with repolarization and the slower decline during a maintained depolarization. Both events lead   27 to a decrease in sodium current, but the underlying mechanisms are clearly different. In the first case it must be supposed that repolarization converts active membrane into resting membrane; in the second that prolonged depolarization turns it into a refractory or inactivated condition from which it recovers at a relatively slow rate when the fibre is repolarized. It cannot be   28 argued that repolarization reduces sodium conductance by making the active

24. *'recovery from inactivation . . . in an approximately exponential manner with a time constant of about 12 msec.'* The ratio of the second to the first current (from **Figure 3**) gives the fractional availability (or 1 minus the fractional inactivation), which is plotted against the recovery interval in **Figure 7**. The observation that the time course of recovery from inactivation can be described well by a single exponential function suggests that only one process—such as a single gate opening—is necessary for the conductance to go from an inactivated to an available state (see Paper 3, note 45). Also, this time constant of about 12 msec for recovery of sodium current after intervals at the resting potential fits into the series of time constants for the growth of sodium current amplitude after conditioning at hyperpolarized potentials shown in **Figure 3**. The time constants become briefer (the recovery rate gets faster) with greater hyperpolarization.

25. *'A similar curve and a similar time constant were obtained'* These data are not shown, but the equation indicates that the plot of the maximum rate of rise of sodium currents was equivalent to the plot of the sodium current against the recovery interval.

26. *'the inactivation time constant . . . is about 1.8 msec.'* H&H extrapolate the exponential fit in **Figure 7** to time 0. This intercept corresponds to the extent of inactivation at the end of the first depolarizing pulse, which is 1.8 msec in duration. The value (intentionally or fortuitously) is 0.37, meaning that the current has decayed to 37% of its initial value after 1.8 msec. Since 37% is $1/e$, and the time for the current to decay to $1/e$ of its initial value is the definition of τ, the time constant of inactivation at this voltage (−26 mV *abs.*) must be 1.8 msec. This value corroborates the measurements in **Table 1**, which indicate that a depolarization to −31 mV *abs.* gives a τ_h of about 1.5 msec.

27. *'the difference between the rapid fall of sodium conductance associated with repolarization and the slower decline during a maintained depolarization.'* This sentence emphasizes the contrast between what are now called *deactivation* and *inactivation*. The distinction is important because it provides evidence that the 'membrane' can exist in at least two separate nonconducting states, as well as in a third, conducting (open) state of *activation*. The ability to define such states, as well as the mono- or multiexponential transition rates between them, begins to validate the idea that the conducting elements of the membrane, which are now known to be ion channels, may be governed by laws of chemical kinetics. Because these laws are known and tractable, it may be possible to describe quantitatively the processes underlying the changes in conductance, and thereby obtain a model that can predict the action potential.

28. *'repolarization converts active membrane into resting membrane . . . prolonged depolarization turns it into a refractory or inactivated condition from which it recovers at a relatively slow rate when the fibre is repolarized.'* In modern terms, ion channel proteins can assume distinct conformations, described as closed, open, and inactivated states (symbolized as C, O, and I). The transitions between these states can be voltage dependent. H&H's first phrase translates into the idea that channels in open states deactivate into closed states upon hyperpolarization. (Their work also anticipates the converse, that channels in closed states activate into open states upon depolarization.) In other words, with hyperpolarization, deactivation is *favored* as the rate of deactivation becomes progressively faster (kinetics) and a greater proportion of channels end up in closed but available states (magnitude). Written symbolically with modern conventions for voltage polarity,

$$C \underset{-V}{\overset{+V}{\rightleftharpoons}} O$$

H&H's second phrase translates into the idea that longer periods at depolarized potentials ultimately favor a stable nonconducting 'inactivated' state, I, which must be different from C, since C is favored at hyperpolarized voltages. The exit from the inactivated state (recovery) is favored at hyperpolarized potentials. Thus,

$$C \underset{-V}{\overset{+V}{\rightleftharpoons}} I$$

Also, the recovery rate from I to C is slower than the deactivation rates from O to C, when compared at the same voltages.

membrane refractory. If this were so, one would expect that the inward current during the second pulse would be reduced to zero at short intervals, instead of to 37 % as in Fig. 7. The reduction to 37 % is clearly associated with the incomplete decline of sodium conductance during the first pulse and not with the rapid and complete decline due to repolarization at the end of the first pulse.

29

DISCUSSION

The experimental evidence in this paper and in those which precede it (Hodgkin & Huxley, 1952a, b) suggests that the membrane potential has two distinct influences on the system which allows sodium ions to flow through the membrane. The early effects of changes in membrane potential are a rapid increase in sodium conductance when the fibre is depolarized and a rapid decrease when it is repolarized. The late effects are a slow onset of a refractory or inactive condition during a maintained depolarization and a slow recovery following repolarization. A membrane in the refractory or inactive condition resembles one in the resting state in having a low sodium conductance. It differs in that it cannot undergo an increase in sodium conductance if the fibre is depolarized. The difference allows inactivation to be measured by methods such as those described in this paper. The results show that both the final level of inactivation and the rate at which this level is approached are greatly influenced by membrane potential. At high membrane potentials inactivation appears to be absent, at low membrane potentials it approaches completion with a time constant of about 1·5 msec. at 6° C. This conclusion is clearly consistent with former evidence which suggests that the sodium conductance declines to a low level with a time constant of 1–2 msec. during a large and maintained depolarization (Hodgkin & Huxley, 1952a). Both sets of experiments may be summarized by stating that changes in sodium conductance are transient over a wide range of membrane potentials.

30

The persistence of inactivation after a depolarization is clearly connected with the existence of a refractory state and with accommodation. It is not the only factor concerned, since the persistence of the raised potassium conductance will also help to hold the membrane potential at a positive value and will therefore tend to make the fibre inexcitable. The relative importance of the two processes can only be judged by numerical analysis of the type described in the final paper of this series (Hodgkin & Huxley, 1952c).

SUMMARY   32

1. Small changes in the membrane potential of the giant axon of *Loligo* are associated with large alterations in the ability of the surface membrane to undergo its characteristic increase in sodium conductance.

2. A steady depolarization of 10 mV. reduces the sodium current associated with a sudden depolarization of 45 mV. by about 60 %. A steady rise of 10 mV.

29. *'It cannot be argued that repolarization reduces sodium conductance by making the active membrane refractory. . . . inward current . . . would be reduced to zero'* Here, the alternative hypothesis is stated and rejected. H&H note that when a repolarizing step is applied while sodium current is still flowing (e.g., after 1.8 msec), the membrane changes from conducting to nonconducting (the current deactivates). Nevertheless, the conductance is still *available* to be opened with a second depolarization. Therefore, the transition to a nonconducting state with repolarization must be different from the transition to a nonconducting state with prolonged depolarization.

30. *'changes in sodium conductance are transient over a wide range of membrane potentials.'* The word 'transient' appears here for the first time as a descriptor of sodium current. It refers to the fact that the conducting (open) state exists only briefly as a short-lived transition state between the nonconducting (closed) state that is favored at hyperpolarized voltages and the nonconducting (inactivated) state that is favored at depolarized voltages. The inactivating component of sodium current is today still referred to as 'the transient sodium current.' Note that in many voltage-gated sodium channels the transition to inactivated states remains incomplete by a few percent, resulting in a small 'persistent' component of sodium current (Chandler & Meves 1970a).

31. *'connected with the existence of a refractory state and with accommodation.'* The refractory state is the brief period after an action potential during which generation of a second action potential is either impossible or requires a larger stimulus than the first action potential. The accommodated state arises when the frequency of action potential firing slows over time, during a prolonged depolarization. H&H draw attention to the fact that the inactivation of sodium channels may well account for both the refractory period and accommodation, but they defer further testing of this hypothesis to the quantitative modeling of Paper 5.

32. *'Summary'* The primary contribution of Paper 4 is the discovery and description of the phenomenon of inactivation of the sodium conductance. Originally, the semipermeable membrane was thought either to be permeable or impermeable to any given ion. Here, however, H&H demonstrate that 'impermeability' is not a single variable. The membrane can be nonconducting in a condition—closed—that readily switches to conducting upon application of the appropriate stimulus (depolarization), but it can also be nonconducting in a different condition—inactivation—which is unresponsive to the stimulus. With prolonged periods at depolarized potentials, the membrane gradually converts to the second nonconducting state; at hyperpolarized potentials, it converts to the first nonconducting state. (In later years, a correlate of inactivation in ligand-gated channels would be called 'desensitization.') Along with the ion-selective, depolarization-dependent conductance increases studied in Papers 2 and 3, the quantitative descriptions of inactivation from this paper provide H&H with the means with which to reconstruct the action potential in Paper 5.

increases the sodium current associated with subsequent depolarization by about 50%.

3. These effects are described by stating that depolarization gradually inactivates the system which enables sodium ions to cross the membrane.

4. In the steady state, inactivation appears to be almost complete if the membrane potential is reduced by 30 mV. and is almost absent if it is increased by 30 mV. Between these limits the amount of inactivation is determined by a smooth symmetrical curve and is about 40% complete in a resting fibre at the beginning of an experiment.

5. At 6° C. the time constant of the inactivation process is about 10 msec. with $V = 0$, about 1·5 msec. with $V = -30$ mV. and about 5 msec. at $V = +30$ mV.

REFERENCES

HODGKIN, A. L. (1951). The ionic basis of electrical activity in nerve and muscle. *Biol. Rev.* **26**, 339–409.

HODGKIN, A. L. & HUXLEY, A. F. (1952a). Currents carried by sodium and potassium ions through the membrane of the giant axon of *Loligo*. *J. Physiol.* **116**, 449–472.

HODGKIN, A. L. & HUXLEY, A. F. (1952b). The components of membrane conductance in the giant axon of *Loligo*. *J. Physiol.* **116**, 473–496.

HODGKIN, A. L. & HUXLEY, A. F. (1952c). A quantitative description of membrane current and its application to conduction and excitation in nerve. *J. Physiol.* (in the press).

HODGKIN, A. L., HUXLEY, A. F. & KATZ, B. (1952). Measurement of current-voltage relations in the membrane of the giant axon of *Loligo*. *J. Physiol.* **116**, 424–448.

Paper 5

Hodgkin, A. L. and Huxley, A. F. (1952d) A quantitative description of membrane current and its application to conduction and excitation in nerve. *Journal of Physiology* 117:500–544.

In the final paper, H&H collect their experimental data and formulate it into a comprehensive, quantitative model that describes the voltage-dependent conductances, replicates the action potential, and predicts the responses of the squid axonal membrane to a variety of stimuli.

At the outset of this paper, H&H have gathered substantial evidence that voltage is the variable that controls the conductances, but the question remains of how the voltage might link to the conductance and thence the currents. Ideally, any model would be grounded in knowledge about physical elements—here, the molecules in the membrane that might facilitate the flow of current. In the absence of such specific molecular information, H&H base their reasoning on general physical models of how charges—which likely are involved in 'voltage-dependent gating'—behave according to known electrochemical principles. With this foundation, H&H infer that voltage-dependent rate constants control the transitions of hypothetical particles in the membrane that gate ion-specific conductances.

It is worth noting that the process of developing the model has two distinct phases. First, H&H use their voltage-clamped current data to deduce a plausible physical mechanism based on the Boltzmann principle, formulate equations consistent with this theory, and calculate the voltage-dependent rate constants required by the theory. They are then able to demonstrate that the equations incorporating the calculated rate constants are sufficient to *replicate* the magnitude and kinetics of the currents at all potentials. Second, H&H use the model to *predict* electrical behavior of the axon, including the stationary and propagated action potentials, as well as absolute and relative refractory periods, responses to relief of hyperpolarization and to prolonged depolarization, and subthreshold responses. Importantly, these data were recorded from non-voltage-clamped axons, and as such were not part of the datasets that were used to constrain the model. The demonstration that the solutions to the equations can re-create the behavior of the axon under a range of conditions illustrates the extensive and remarkable predictive power of the model, thereby giving credibility to the proposed underlying mechanisms.

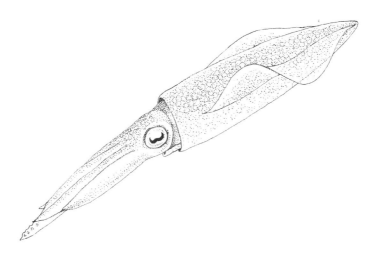

A QUANTITATIVE DESCRIPTION OF MEMBRANE CURRENT AND ITS APPLICATION TO CONDUCTION AND EXCITATION IN NERVE

By A. L. HODGKIN and A. F. HUXLEY

From the Physiological Laboratory, University of Cambridge

(*Received* 10 *March* 1952)

This article concludes a series of papers concerned with the flow of electric   1
current through the surface membrane of a giant nerve fibre (Hodgkin, Huxley & Katz, 1952; Hodgkin & Huxley, 1952 *a–c*). Its general object is to discuss the results of the preceding papers (Part I), to put them into mathematical form (Part II) and to show that they will account for conduction and excitation in quantitative terms (Part III).

PART I. DISCUSSION OF EXPERIMENTAL RESULTS

The results described in the preceding papers suggest that the electrical behaviour of the membrane may be represented by the network shown in Fig. 1. Current can be carried through the membrane either by charging the   2
membrane capacity or by movement of ions through the resistances in parallel with the capacity. The ionic current is divided into components carried by sodium and potassium ions (I_{Na} and I_K), and a small 'leakage current' (I_l) made up by chloride and other ions. Each component of the ionic current is determined by a driving force which may conveniently be measured as an electrical potential difference and a permeability coefficient which has the dimensions of a conductance. Thus the sodium current (I_{Na}) is equal to the sodium conductance (g_{Na}) multiplied by the difference between the membrane potential (E) and the equilibrium potential for the sodium ion (E_{Na}). Similar   3
equations apply to I_K and I_l and are collected on p. 505.

Our experiments suggest that g_{Na} and g_K are functions of time and membrane potential, but that E_{Na}, E_K, E_l, C_M and $\bar{g}_l$ may be taken as constant. The influence of membrane potential on permeability can be summarized by stating: first, that depolarization causes a transient increase in sodium conductance and a slower but maintained increase in potassium conductance; secondly, that these changes are graded and that they can be reversed by repolarizing the membrane. In order to decide whether these effects are sufficient to account for complicated phenomena such as the action potential and refractory period, it is necessary to obtain expressions relating

1. **'This article concludes a series of papers'** Hodgkin recounts in his autobiography that he and Huxley had hoped to use the Cambridge University computer to calculate action potentials in about March 1951, but it went down for six months of repairs before they could begin. Therefore, Huxley performed the computations by hand on a Brunsviga mechanical calculator (see Appendix 5). The first four experimental papers were submitted in October 1951 and appeared consecutively in a single issue of the *Journal of Physiology* the following April; the present paper—the 'modeling paper'—was submitted in March 1952 and appeared in August.

2. **'electrical behaviour of the membrane may be represented by the network shown in Fig. 1.'** The circuit diagram in ***Figure 1*** is a landmark in the history of biophysics, illustrating how physical laws that govern nonliving material are also manifested in, and indeed give rise to, life processes. Many different circuits had been proposed earlier by Cole and others (see Historical Background); the circuit in ***Figure 1*** has yet to be superseded. The four parallel elements of the circuit (see Appendix 2.1) all arise from the extensive characterization of current flow from the previous papers. The membrane capacitance, C_M, which was shown in Paper 1 to obey, within experimental error, the relationship, $Q = CV$, which defines a classical capacitor. The three resistors represent the sodium, potassium, and leak conductances. Paper 2 demonstrated that the conductances behave as independent mechanisms and so can be represented separately in the circuit; Paper 3 demonstrated that each acts as a classical conductance, such that $I = gV$, or $I = V/R$. The arrows through the resistors associated with sodium and potassium flux indicate that these conductances can change, in this case as functions of voltage and time, as quantified in Papers 2, 3, and 4. Papers 2 and 3 also demonstrated that each of the three currents has a different electrochemical driving force, arising from the concentration gradient of the corresponding ion, and here represented by batteries the values of which can be calculated from the Nernst equation. The sodium battery faces inward such that at the resting potential positive ions will tend to flow inward; the potassium battery has the opposite polarity. Given that H&H have identified and quantitatively characterized all components of the circuit, particularly the dependence of R_{Na} and R_K (or their inverses, g_{Na} and g_K) on V and t, they can now make quantitative predictions regarding the electrical behavior of the axon. The rest of paper describes those predictions, and evaluates how well they match the measured behavior of the axon.

3. **'the sodium current (I_{Na}) is equal to the sodium conductance (g_{Na}) multiplied by the difference between the membrane potential (E) and the equilibrium potential for the sodium ion (E_{Na}).'** This sentence is a verbal rendering of Ohm's law: $I_{Na} = g_{Na}(E - E_{Na})$. It is unusual that an equation would be given exclusively in words, but here the idea is not only the validity of the mathematical relationship but also the equivalence between the variables of interest in the nerve and the quantities in the familiar equation.

the sodium and potassium conductances to time and membrane potential.   4
Before attempting this we shall consider briefly what types of physical system
are likely to be consistent with the observed changes in permeability.

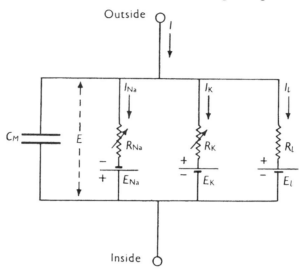

Fig. 1. Electrical circuit representing membrane. $R_{\mathrm{Na}} = 1/g_{\mathrm{Na}}$; $R_{\mathrm{K}} = 1/g_{\mathrm{K}}$; $R_l = 1/\bar{g}_l$. R_{Na} and
R_{K} vary with time and membrane potential; the other components are constant.

The nature of the permeability changes

At present the thickness and composition of the excitable membrane are
unknown. Our experiments are therefore unlikely to give any certain information about the nature of the molecular events underlying changes in permeability. The object of this section is to show that certain types of theory are
excluded by our experiments and that others are consistent with them.   5

The first point which emerges is that the changes in permeability appear to
depend on membrane potential and not on membrane current. At a fixed   6
depolarization the sodium current follows a time course whose form is independent of the current through the membrane. If the sodium concentration
is such that $E_{\mathrm{Na}} < E$, the sodium current is inward; if it is reduced until
$E_{\mathrm{Na}} > E$ the current changes in sign but still appears to follow the same time
course. Further support for the view that membrane potential is the variable   7
controlling permeability is provided by the observation that restoration of the
normal membrane potential causes the sodium or potassium conductance to
decline to a low value at any stage of the response.

The dependence of g_{Na} and g_{K} on membrane potential suggests that the
permeability changes arise from the effect of the electric field on the distribution or orientation of molecules with a charge or dipole moment. By this we   8
do not mean to exclude chemical reactions, for the rate at which these occur
might depend on the position of a charged substrate or catalyst. All that is
intended is that small changes in membrane potential would be most unlikely

4. *'In order to decide whether these effects are sufficient to account for . . . necessary to obtain expressions relating the sodium and potassium conductances to time and membrane potential.'* Note the logical formulation. The discovery as stated is of a fixed (invariant) capacitance and voltage- and time- dependent (variable) conductances: g is a function of V and t, but H&H still must investigate whether they have captured *all* the variables that operate to produce an action potential. This idea is not a 'straw man' or pro forma control experiment. The notion of biochemical reactions generating electrical changes was still an actively competing hypothesis at the time. Also, the time dependence of g presents a significant deviation from classic analog electronics. The proof (here meaning 'test') will be in determining, first, whether the raw measurements made from the voltage-clamp studies, when the voltage was artificially *fixed,* are able to provide mathematical descriptions of how conductances increase or decrease as the voltage *changes;* second, whether these equations will allow H&H to calculate each ionic current and the voltage changes they induce; and third, whether doing so can recreate the action potential de novo. If so, the description of voltage-gated conductances will be 'sufficient to account for'—in the scientific sense, meaning no other variables are needed to fully explain the mechanism of—the action potential.

5. *'Our experiments are . . . unlikely to give any certain information about the nature of the molecular events. . . . The object . . . is to show that certain types of theory are excluded by our experiments and that others are consistent with them.'* These sentences present two key points. First, H&H (and their contemporaries) do not know what materials compose the membrane: the identification of conductances as ion channel proteins is still about two decades in the future. Nevertheless, they can describe how the materials work in precise detail, and by doing so, place constraints on the attributes of physical substance. Second, H&H are fully transparent, to use a modern term, about what they know and what they do not know, and they explicitly define the scientific scope and limitations of their analysis from the outset.

6. *'changes in permeability appear to depend on membrane potential and not on membrane current.'* Membrane potential, V, is the independent variable; permeability, here characterized as conductance, g, is the dependent variable. This discovery was fundamental, as competing hypotheses had suggested that changes in permeability depended instead on current (see Historical Background).

7. *'If . . . $E_{Na} < E$, the sodium current is inward . . . same time course.'* In modern conventions, the greater and less than signs would be flipped; generalizing, positive ions always flow inward at voltages more hyperpolarized than the equilibrium potential of that ion and outward at voltages more depolarized than the equilibrium potential. Kinetics, however, are generally not sensitive to the polarity of the current.

8. *'changes arise from the effect of the electric field on the distribution or orientation of molecules with a charge or dipole moment.'* H&H offer a key inference about the molecular mechanism of voltage sensing. Molecular biological studies later demonstrated that voltage-gated channels indeed contain a voltage sensor that includes charged amino acids in membrane-spanning domains of the protein. The voltage sensor is poised to detect and respond to the electric field within the membrane by moving and changing the conformation of the protein, thereby opening and closing the structures that form the 'gate' of the ion channel.

to cause large alterations in the state of a membrane which was composed entirely of electrically neutral molecules.

The next question to consider is how changes in the distribution of a charged particle might affect the ease with which sodium ions cross the membrane. Here we can do little more than reject a suggestion which formed the original basis of our experiments (Hodgkin, Huxley & Katz, 1949). According to this view, sodium ions do not cross the membrane in ionic form but in combination with a lipoid soluble carrier which bears a large negative charge and which can   9 combine with one sodium ion but no more. Since both combined and un-combined carrier molecules bear a negative charge they are attracted to the outside of the membrane in the resting state. Depolarization allows the carrier molecules to move, so that sodium current increases as the membrane potential is reduced. The steady state relation between sodium current and voltage could be calculated for this system and was found to agree reasonably with the observed curve at 0·2 msec after the onset of a sudden depolarization. This was encouraging, but the analogy breaks down if it is pursued further. In the model the first effect of depolarization is a movement of negatively charged molecules from the outside to the inside of the membrane. This gives an initial outward current, and an inward current does not occur until combined carriers lose sodium to the internal solution and return to the outside of the membrane. In our original treatment the initial outward current was reduced to vanishingly small proportions by assuming a low density of carriers and a high rate of movement and combination. Since we now know that the sodium current takes an appreciable time to reach its maximum, it is necessary to suppose that there are more carriers and that they react or move more slowly. This means that any inward current should be preceded by a large outward current. Our experiments show no sign of a component large enough to be consistent with the model. This invalidates the detailed mechanism assumed for the permeability change but it does not exclude the more general possibility that   10 sodium ions cross the membrane in combination with a lipoid soluble carrier.

A different form of hypothesis is to suppose that sodium movement depends on the distribution of charged particles which do not act as carriers in the usual sense, but which allow sodium to pass through the membrane when they   11 occupy particular sites in the membrane. On this view the rate of movement of the activating particles determines the rate at which the sodium con-ductance approaches its maximum but has little effect on the magnitude of the conductance. It is therefore reasonable to find that temperature has a large effect on the rate of rise of sodium conductance but a relatively small effect on its maximum value. In terms of this hypothesis one might explain the transient nature of the rise in sodium conductance by supposing that the activating particles undergo a chemical change after moving from the position which they occupy when the membrane potential is high. An alternative is to

9. *'According to this view, sodium ions . . . cross the membrane . . . in combination with a lipoid soluble carrier which bears a large negative charge'* H&H point out that they have refuted their original hypothesis of a carrier mechanism, which had directed the design of experiments in the first paper. This rejection or ruling out of hypotheses is a particularly clear example of the scientific method. A hypothesis cannot formally be proven true, since regardless of the quantity of evidence amassed in its favor, new data to the contrary may still emerge in the future, leaving even a well-supported hypothesis still vulnerable. Refutation is more powerful: once a logical fallacy is shown, the hypothesis must be rejected, as has been done here.

10. *'Our experiments show no sign. . . . This invalidates the detailed mechanism assumed for the permeability change'* H&H explain that the negatively charged hypothetical carrier would bind a single sodium ion, but owing to its 'large' charge would remain negative. Upon depolarization, the concerted inward movement of negatively charged carrier molecules across the membrane, equivalent to the outward movement of positive charges, would be measured as an outward current. With sustained depolarization, the outward current would be largely canceled by inward current from carriers asynchronously returning to the extracellular face of the membrane to bind more sodium, and the carrier current would subside from its initial high value. The predicted early outward current is not observed, and hence, the hypothesis is rejected. An alternative hypothesis is developed in the next sentence.

11. *'A different form of hypothesis . . . particles which do not act as carriers . . . but which allow sodium to pass through the membrane'* The alternative presented by H&H is the theoretical articulation of ion channels, which are now known to be proteins that span the membrane. They state two properties of ionic currents that can be accounted for by the proposal: first, the high temperature-dependence of current kinetics ($Q_{10} \approx 3$) is consistent with a biochemical mechanism controlling the *gating* of the current, and second, the relative temperature-insensitivity of the current magnitude can be explained by the unimpeded *permeation* of ions through the membrane once the particles have moved.

attribute the decline of sodium conductance to the relatively slow movement   12
of another particle which blocks the flow of sodium ions when it reaches a
certain position in the membrane.

Much of what has been said about the changes in sodium permeability
applies equally to the mechanism underlying the change in potassium permeability. In this case one might suppose that there is a completely separate
system which differs from the sodium system in the following respects: (1) the
activating molecules have an affinity for potassium but not for sodium;
(2) they move more slowly; (3) they are not blocked or inactivated. An
alternative hypothesis is that only one system is present but that its selectivity
changes soon after the membrane is depolarized. A situation of this kind would   13
arise if inactivation of the particles selective for sodium converted them into
particles selective for potassium. However, this hypothesis cannot be applied
in a simple form since the potassium conductance rises too slowly for a direct
conversion from a state of sodium permeability to one of potassium
permeability.

One of the most striking properties of the membrane is the extreme steepness
of the relation between ionic conductance and membrane potential. Thus g_{Na}
may be increased e-fold by a reduction of only 4 mV, while the corresponding
figure for g_K is 5–6 mV (Hodgkin & Huxley, 1952a, figs. 9, 10). In order to
illustrate the possible meaning of this result we shall suppose that a charged
molecule which has some special affinity for sodium may rest either on the
inside or the outside of the membrane but is present in negligible concentrations elsewhere. We shall also suppose that the sodium conductance is proportional to the number of such molecules on the inside of the membrane but
is independent of the number on the outside. From Boltzmann's principle the
proportion P_i of the molecules on the inside of the membrane is related to the
proportion on the outside, P_o, by   14

$$\frac{P_i}{P_o} = \exp[(w + zeE)/kT],$$

where E is the potential difference between the outside and the inside of the
membrane, w is the work required to move the molecule from the inside to the
outside of the membrane when $E = 0$, e is the absolute value of the electronic
charge, z is the valency of the molecule (i.e. the number of positive electronic
charges on it), k is Boltzmann's constant and T is the absolute temperature.
Since we have assumed that $P_i + P_o = 1$ the expression for P_i is

$$P_i = 1 \bigg/ \left[1 + \exp - \left(\frac{w + zeE}{kT} \right) \right].$$

For negative values of z and with E sufficiently large and positive this gives

$$P_i = \text{constant} \times \exp[zeE/kT].$$

12. **'one might explain the transient nature of the rise. . . . An alternative is to attribute the decline of sodium conductance'** By spelling out the conditions that must be met to account for the observations, H&H set up the plausibility of the upcoming model: a potential physical referent is identified for each element of the equations, namely particles whose movement in the membrane permits a conductance to be activated and inactivated. Even though the specific identities of the molecules are unknown, the actions of those molecules can be inferred (see note 35).

13. **'An alternative hypothesis is that only one system is present but that its selectivity changes'** The consideration of this alternative underscores how novel the notion of an ion-selective channel was; here H&H offer the supposition that one conductance mechanism (channel) might change form to become alternately permeable to sodium and then potassium. This idea, which came to be referred to as 'ion channel conversion,' persisted for more than a decade until rendered implausible by pharmacological and enzymatic manipulation of individual sodium channels (Narahashi et al. 1964; Armstrong et al. 1973). H&H note, however, that the sodium conductance does not always inactivate to zero before the potassium conductance activates, so if this idea is valid, it is not straightforward.

14. **'From Boltzmann's principle the proportion P_i of the molecules on the inside of the membrane is related to the proportion on the outside'** Here begins the quantitative elaboration of H&H's hypothesis, that the movement of charged 'particles' might serve to activate the sodium conductance. H&H did not know what these particles were, but they treat them as molecules that can be either inside or outside the membrane. These positions correspond to two distinct states, one of which is permissive and the other nonpermissive of current flow. In this way, they resemble 'gates' that can be open or closed; although never used by H&H, the term 'gates' has become the standard terminology and is used here. It is now evident that H&H's quantitative description of particles pertains most closely to the voltage-sensing domains of ion channel proteins. H&H, however, also had to account for the selectivity of the conductance, which they proposed might occur if the particles had an 'affinity' for the ion of interest. It is now known that the affinity arises not from the voltage sensors, but from the selectivity filters in the ion channel pores. Also, the voltage sensors of ion channel proteins have turned out to be distinct from the domain that physically occludes the permeation pathway; that domain—called the 'main structural gate'—is moved as *a result of* voltage sensor movement (see Appendix 4).

For simplicity, below we use the terms 'particles,' 'gates,' and 'gating particles' interchangeably to refer to H&H's 'molecules,' with the molecular referent of the voltage-sensing domain. To be consistent with H&H's formulation, an 'inward' negatively charged particle corresponds to an 'open' gate. (Modern structural data indicates that voltage sensing domains contain positively charged amino acids, which reverses this relationship.)

On the assumption of a two-state model, H&H develop equations by drawing upon the well-established physics of how particles distribute among states within an electric field at equilibrium, namely, Boltzmann's principle (see Paper 2, note 64; Paper 4, note 17). The equation for P_i expresses the *proportion* of total charged particles in the inward position—equivalently, the *probability* that any one charged particle will be in the inward position—in terms of two physical quantities: (1) temperature, which represents the thermal kinetic energy that keeps the particles in constant motion and causes them to change state randomly, but does not favor one state over the other; and (2) the membrane electric field, proportional to membrane voltage, which applies a force to the charged particles, and thereby increases the probability of finding particles in one state rather than the other.

In the full expression (second equation on the facing page), the proportion of particles in the inward position, P_i, is a saturating function of voltage,

$$P_i = \frac{1}{1 + \exp(-x)}$$

(see Appendix 1.4), where $x = (w + zeE)/kT$. Here, the effect of the membrane field is expressed as zeE, the work (in joules) required to move a charged gating particle through the membrane, where

(continued on page 193)

In order to explain our results z must be about -6 since $\dfrac{kT}{e}\left(=\dfrac{RT}{F}\right)$ is 25 mV at room temperature and $g_{Na} \propto \exp - E/4$ for E large. This suggests that the particle whose distribution changes must bear six negative electronic charges, or, if a similar theory is developed in terms of the orientation of a long molecule with a dipole moment, it must have at least three negative charges on one end and three positive charges on the other. A different but related approach is to suppose that sodium movement depends on the presence of six singly charged molecules at a particular site near the inside of the membrane. The proportion of the time that each of the charged molecules spends at the inside is determined by $\exp - E/25$ so that the proportion of sites at which all six are at the inside is $\exp - E/4\cdot17$. This suggestion may be given plausibility but not mathematical simplicity by imagining that a number of charges form a bridge or chain which allows sodium ions to flow through the membrane when it is depolarized. Details of the mechanism will probably not be settled for some time, but it seems difficult to escape the conclusion that the changes in ionic permeability depend on the movement of some component of the membrane which behaves as though it had a large charge or dipole moment. 15 If such components exist it is necessary to suppose that their density is relatively low and that a number of sodium ions cross the membrane at a single active patch. Unless this were true one would expect the increase in sodium permeability to be accompanied by an outward current comparable in magnitude to the current carried by sodium ions. For movement of any charged particle in the membrane should contribute to the total current and 16 the effect would be particularly marked with a molecule, or aggregate, bearing a large charge. As was mentioned earlier, there is no evidence from our experiments of any current associated with the change in sodium permeability, apart from the contribution of the sodium ion itself. We cannot set a definite upper limit to this hypothetical current, but it could hardly have been more than a few per cent of the maximum sodium current without producing a conspicuous effect at the sodium potential.

PART II. MATHEMATICAL DESCRIPTION OF MEMBRANE CURRENT DURING A VOLTAGE CLAMP

Total membrane current

The first step in our analysis is to divide the total membrane current into a capacity current and an ionic current. Thus

$$I = C_M \frac{dV}{dt} + I_i, \tag{1}$$

(continued from page 191)

E is the membrane voltage in joules/coulomb, e is the absolute value of the charge on the electron in coulombs, and z is the valence (number of charges) on the negatively charged particle. The term w, also in joules, describes any bias or tendency for the particles to occupy one or the other state in the absence of a voltage difference, and therefore shifts the curve on the x-axis. In modern terms, w represents molecular interactions or other structural constraints that predispose the ion channel protein to occupy an open state when the membrane potential is 0 mV *abs*. The effect of temperature is expressed as T, in degrees kelvin, scaled by Boltzmann's constant, k, in joules/degree kelvin, which makes the denominator of the exponent also in joules. The ratio x is thus dimensionless.

The equation illustrates that if the temperature, T, were to rise to an extremely high value, then the exponent would become close to 0, $\exp(-x)$ would approach 1, and P_i would approach ½. Thus, when the thermal energy dominates, the particle has a 50% chance of being in either state. Conversely, with negative z, as the membrane potential hyperpolarizes and E becomes more positive in H&H's conventions, the term zeE becomes large and negative and the exponential term becomes large and positive. As a result, P_i becomes small, meaning that the particle has a low probability of occupying the open state. Under these conditions, the 1 in the denominator of the expression for P_i becomes negligible. Thus, the equation reduces to

$$P_i = \frac{1}{\exp\left[\dfrac{-(w + zeE)}{kT}\right]}$$

$$= \exp\left[\frac{w + zeE}{kT}\right].$$

This expression can be further simplified into H&H's next equation, in which 'constant' is $\exp(w)$, and which states that the probability of finding the gate open at the most hyperpolarized potentials is related exponentially to voltage. The theory, therefore, accounts for H&H's observation that at the foot of the activation curve, conductance increases exponentially with voltage—e-fold for every 4-mV depolarization.

15. **'Details of the mechanism will probably not be settled for some time, but it seems difficult to escape the conclusion that the changes in ionic permeability depend on the movement of . . . a large charge or dipole moment.'** This statement is strikingly accurate in its predictions: it anticipates the charged voltage sensor on voltage-gated ion channels. H&H reason that the movement of these hypothetical charges (z) to activate the conductance must make a current—now called *gating current*. Two decades later, with advances in technologies, the tiny, brief gating currents were successfully recorded by Armstrong and Bezanilla (1973). Their paper begins with the above quotation from H&H, and the study of these currents provided substantial information about the opening and closing of ion channels. H&H were also correct in assuming that the specific movements of the voltage-sensing mechanism would be debated for the next seven decades.

16. **'it is necessary to suppose that their density is relatively low and that a number of sodium ions cross the membrane at a single active patch. . . . For movement of any charged particle . . . should contribute to the total current'** The expression 'active patch' refers to an area of membrane where current flows under the control of a single set of gating particles, or what would now be called an ion channel. H&H acknowledge that not enough of the hypothetical gating current exists for the carrier mechanism to be plausible, which leads them to conclude that the movement of even a few gating charges must give rise to a great deal of ionic current through a given 'patch.'

where

I is the total membrane current density (inward current positive);

I_i is the ionic current density (inward current positive);

V is the displacement of the membrane potential from its resting value (depolarization negative);

C_M is the membrane capacity per unit area (assumed constant);

t is time.

The justification for this equation is that it is the simplest which can be used and that it gives values for the membrane capacity which are independent of the magnitude or sign of V and are little affected by the time course of V (see, for example, table 1 of Hodgkin *et al.* 1952). Evidence that the capacity current and ionic current are in parallel (as suggested by eqn. (1)) is provided by the similarity between ionic currents measured with $\dfrac{\mathrm{d}V}{\mathrm{d}t}=0$ and those calculated from $-C_M\dfrac{\mathrm{d}V}{\mathrm{d}t}$ with $I=0$ (Hodgkin *et al.* 1952).   17

The only major reservation which must be made about eqn. (1) is that it takes no account of dielectric loss in the membrane. There is no simple way of estimating the error introduced by this approximation, but it is not thought to be large since the time course of the capacitative surge was reasonably close to that calculated for a perfect condenser (Hodgkin *et al.* 1952).   18

The ionic current

A further subdivision of the membrane current can be made by splitting the ionic current into components carried by sodium ions (I_{Na}), potassium ions (I_K) and other ions (I_l):

$$I_i = I_{Na} + I_K + I_l. \tag{2}$$

19

The individual ionic currents

In the third paper of this series (Hodgkin & Huxley, 1952*b*), we showed that the ionic permeability of the membrane could be satisfactorily expressed in terms of ionic conductances (g_{Na}, g_K and $\bar{g}_l$). The individual ionic currents are obtained from these by the relations

$$I_{Na} = g_{Na}\,(E - E_{Na}),$$
$$I_K = g_K\,(E - E_K),$$
$$I_l = \bar{g}_l\,(E - E_l),$$

where E_{Na} and E_K are the equilibrium potentials for the sodium and potassium ions. E_l is the potential at which the 'leakage current' due to chloride and other ions is zero. For practical application it is convenient to write these equations in the form

$$I_{Na} = g_{Na}\,(V - V_{Na}), \tag{3}$$
$$I_K = g_K\,(V - V_K), \tag{4}$$
$$I_l = \bar{g}_l\,(V - V_l), \tag{5}$$

17. **'similarity between ionic currents measured with dV/dt =0 and those calculated from −C_M(dV/dt) with I =0'** H&H point out that the voltage-clamped ionic current (with $dV/dt = 0$) is approximately equal and opposite to the capacity current estimated from the action potential, $-C_M(dV/dt)$, in Paper 1 (Figures 10 and 13). Thus, for the stationary (membrane) action potential the sum of these currents is zero in the absence of any externally applied current. All charges that flow into the cell as ionic current must therefore be deposited on the membrane capacitance, consistent with the membrane as a parallel RC circuit. The match between the currents measured directly and those estimated from the membrane action potential provides further evidence that the membrane was not significantly disrupted either by the space clamp imposed by the axial wire or the feedback associated with voltage clamping.

18. **'takes no account of dielectric loss . . . reasonably close to that calculated for a perfect condenser.'** By 'dielectric loss' H&H refer to the possibility that the membrane does not act like an ideal capacitor, in which $Q = CV$. In real capacitors, the material between the plates of the capacitor (the dielectric material) absorbs or dissipates a small but finite amount of energy during each cycle of charging and discharging. As a result, the charge recovered upon discharging is slightly less than the charge supplied upon charging. Although such loss likely occurs in the membrane as well, Figure 16 in Paper 1 demonstrates that the loss is small enough for the membrane to be treated as an ideal capacitor.

19. **'splitting the ionic current into components carried by sodium ions (I_{Na}), potassium ions (I_K) and other ions (I_l)'** This split is justified because H&H demonstrated that the sodium, potassium, and leak currents were linearly separable in Papers 2 and 3. When the voltage is clamped, the total ionic current at any potential is composed of the sodium, potassium, and leak current, such that changing one—for example, by replacing sodium with choline in the external solution—has no effect on the others. Because currents sum linearly under voltage clamp, individual currents can be subtracted from the total to get an accurate measure of the remainder. In contrast to currents, *voltage* changes (across a membrane that is not voltage clamped) do *not* sum linearly: replacing external sodium to eliminate sodium current would alter the voltage trajectory of the membrane and thereby greatly affect potassium currents. Therefore, changes in voltage measured with and without an ionic current blocked cannot be subtracted to estimate the voltage change induced by that ionic current.

where
$$V = E - E_r,$$
$$V_{Na} = E_{Na} - E_r,$$
$$V_K = E_K - E_r,$$
$$V_l = E_l - E_r,$$

and E_r is the absolute value of the resting potential. V, V_{Na}, V_K and V_l can then be measured directly as displacements from the resting potential.   20

The ionic conductances

The discussion in Part I shows that there is little hope of calculating the time course of the sodium and potassium conductances from first principles. Our object here is to find equations which describe the conductances with reasonable accuracy and are sufficiently simple for theoretical calculation of   21 the action potential and refractory period. For the sake of illustration we shall try to provide a physical basis for the equations, but must emphasize that the interpretation given is unlikely to provide a correct picture of the membrane.

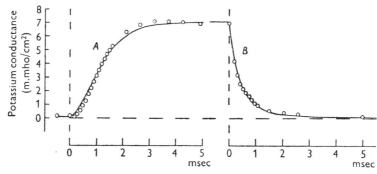

Fig. 2. *A*, rise of potassium conductance associated with depolarization of 25 mV; *B*, fall of potassium conductance associated with repolarization to the resting potential. Circles: experimental points replotted from Hodgkin & Huxley (1952*b*, Fig. 13). The last point of *A* is the same as the first point in *B*. Axon 18, 21° C in choline sea water. The smooth curve is drawn according to eqn. (11) with the following parameters:

	Curve *A* ($V = -25$ mV)	Curve *B* ($V = 0$)
g_{K0}	0·09 m.mho/cm²	7·06 m.mho/cm²
$g_{K\infty}$	7·06 m.mho/cm²	0·09 m.mho/cm²
τ_n	0·75 msec	1·1 msec

At the outset there is the difficulty that both sodium and potassium conductances increase with a delay when the axon is depolarized but fall with no appreciable inflexion when it is repolarized. This is illustrated by the circles in   22 Fig. 2, which shows the change in potassium conductance associated with a depolarization of 25 mV lasting 4·9 msec. If g_K is used as a variable the end of the record can be fitted by a first-order equation but a third- or fourth-order equation is needed to describe the beginning. A useful simplification is

20. *'displacements from the resting potential.'* The equations above restate the expressions for current and conductance according to Ohm's law, as justified by the experiments of Paper 3. Once again, instead of using absolute voltages, *E*, H&H use *V*, the voltage measured relative to the resting potential, as recorded in their experiments.

21. *'little hope of calculating . . . from first principles. Our object here is to find equations which describe the conductances with reasonable accuracy and are sufficiently simple'* In retrospect, the first sentence is amusingly forthright. In the absence of molecular knowledge that might provide a clear physical referent for the hypothetical gating particles, H&H could not derive the relationships from what they considered to be first principles, which they report with seeming regret. Their 'reasonably accurate' substitutes, however, which they inferred by drawing parallels with physical systems, have turned out to be remarkably similar to molecular realities and have driven the field to the present day. Nevertheless, the statements indicate that H&H were aware that their equations were an approximation that *described* the data, but that the variables did not necessarily have known physical referents, restricting the scope of the model's predictive value. They fully expected that their approximations would be supplanted as molecules and their motions were identified.

22. *'conductances increase with a delay when the axon is depolarized but fall with no appreciable inflexion when it is repolarized.'* In the earlier analysis (see note 14), H&H applied the two-state equation to describe the steady-state proportion of open gates (inward particles), P_i, which in turn determines the steady-state conductance *amplitude* at any voltage. Here, they consider the *kinetics* of particle movement, which determines the time course with which the new steady-state conductance is approached upon a voltage change. As shown in ***Figure 2*** for the potassium conductance, the falling phase (deactivation) upon repolarization has no 'inflexion,' but instead follows a single exponential decay (see Paper 3, note 45). An exponential time course is predicted by kinetic theory for the redistribution of particles from one state to another upon a change in voltage (known as a first-order process; see Appendix 1.3). The match of observation with theory sets the stage for H&H to conclude in the next sentence that the conductance offset can be modeled quantitatively as the closure of a single gate. In contrast, the rising phase (activation) describes a curve with an inflexion; the concave-up region introduces the delay. The 'difficulty' that H&H mention is how to account for this S-shaped rise, both quantitatively and mechanistically. This observation is the introduction to H&H's idea, elaborated quantitatively below, that *multiple* gating particles must move for the conductance to switch from rest to activated.

achieved by supposing that g_K is proportional to the fourth power of a variable which obeys a first-order equation. In this case the rise of potassium con- 23 ductance from zero to a finite value is described by $(1 - \exp(-t))^4$, while the fall is given by $\exp(-4t)$. The rise in conductance therefore shows a marked inflexion, while the fall is a simple exponential. A similar assumption using a cube instead of a fourth power describes the initial rise of sodium con- ductance, but a term representing inactivation must be included to cover the 24 behaviour at long times.

The potassium conductance

The formal assumptions used to describe the potassium conductance are:

$$g_K = \bar{g}_K n^4, \tag{6}$$

$$\frac{dn}{dt} = \alpha_n (1-n) - \beta_n n, \tag{7}$$

where $\bar{g}_K$ is a constant with the dimensions of conductance/cm^2, α_n and β_n are rate constants which vary with voltage but not with time and have dimensions of [time]$^{-1}$, n is a dimensionless variable which can vary between 0 and 1. 25

These equations may be given a physical basis if we assume that potassium ions can only cross the membrane when four similar particles occupy a certain region of the membrane. n represents the proportion of the particles in a certain position (for example at the inside of the membrane) and $1-n$ represents the proportion that are somewhere else (for example at the outside of the membrane). α_n determines the rate of transfer from outside to inside, while β_n determines the transfer in the opposite direction. If the particle has a negative charge α_n should increase and β_n should decrease when the membrane is depolarized.

Application of these equations will be discussed in terms of the family of curves in Fig. 3. Here the circles are experimental observations of the rise of potassium conductance associated with depolarization, while the smooth curves are theoretical solutions of eqns. (6) and (7).

In the resting state, defined by $V=0$, n has a resting value given by 26

$$n_0 = \frac{\alpha_{n0}}{\alpha_{n0} + \beta_{n0}}.$$

If V is changed suddenly α_n and β_n instantly take up values appropriate to the new voltage. The solution of (7) which satisfies the boundary condition that 27 $n = n_0$ when $t=0$ is

$$n = n_\infty - (n_\infty - n_0) \exp(-t/\tau_n), \tag{8}$$

where

$$n_\infty = \alpha_n/(\alpha_n + \beta_n), \tag{9}$$

and

$$\tau_n = 1/(\alpha_n + \beta_n). \tag{10}$$

23. *'A useful simplification is achieved by supposing that g_K is proportional to the fourth power of a variable which obeys a first-order equation.'* As stated in note 22, the proportion of open gates evolves over time according to a single exponential function, $1-e^{-t/\tau}$. This first-order equation relaxes from 0 at time 0 to 1 in the steady state (see Appendix 1.3). In contrast, the S-shaped rise in conductance indicates that a higher-order process is involved in activation. Although H&H do not know the underlying mechanism, they can replicate the S-shape by multiplying an exponential function by itself four times, raising it to the fourth power. The reasonable fits (solid lines in ***Figure 2*** and ***Figure 3***) hint that whatever activates the conductance *behaves as though* a single transition occurs four times over for the conductance to be activated. Activation can therefore be mimicked by the 'useful simplification' of assuming that four independent gates must open for current to flow. Thus, a simple route to a fourth-order equation—raising a first-order equation to the fourth power—can be interpreted physically in the context of the proposed gating mechanism. This idea was later substantiated when voltage-gated potassium channels were found to have four subunits (MacKinnon 1991), each with its own voltage sensor.

24. *'a cube instead of a fourth power describes the initial rise of sodium conductance, but a term representing inactivation must be included'* Likewise, the activation of sodium conductance can be represented as the cube of an exponential relaxation, suggesting that the sodium channel may have three gates, all of which must be open for the channel to conduct. Additionally, a transition to a nonconducting state at depolarized potentials—inactivation—must be incorporated, which H&H represent by another process going in the opposite direction (see note 35).

25. *'n is a dimensionless variable which can vary between 0 and 1.'* In this section, the key variables are formally presented in equation form with a theoretical physical basis. The equations express the two attributes of conductance that always require defining: magnitude (***Equation 6***) and kinetics (***Equation 7***). The activation of the potassium conductance seems to require four transitions (see note 23), which H&H conceptualize as gating particles moving in the membrane. Each gate can either be open or closed—there are no half-open gates. In H&H's formulation, many routes exist for potassium to flow (in modern terms, many channels exist), each controlled by four independently operating gating particles. The probability of finding a gating particle in the inward position is given by the equation for P_i (see note 14). This 'dimensionless variable,' renamed n in the context of the potassium conductance, represents the proportion of gates open across all gates in all patches; $1-n$ is the proportion of closed gates. Since n is a proportion, it is dimensionless. Since it varies from 0 to 1, it can equivalently be thought of as the probability of finding any single gate open. If every gate were open, the membrane would attain its maximal potassium conductance, $\bar{g}_K$ ($\bar{g}$ is pronounced 'g-bar'). With fewer gates open, however, the conductance will be lower. ***Equation 6*** expresses these ideas by saying that the magnitude of the potassium conductance, g_K, is the maximal conductance, $\bar{g}_K$, scaled by the proportion of gates open, raised to the fourth power. Note that this equation incorporates the idea that all four gates must open for the conductance to be activated; if even a single gate in a cluster remains closed, current cannot flow. In modern terminology, the maximal conductance, $\bar{g}_K$, would correspond to the single-channel conductance multiplied by the number of channels; and the scaling term n^4 would correspond to the fraction of the maximal conductance activated at any voltage.

H&H assert that n can be well approximated by a first-order differential equation, which means that n is proportional to its first derivative, dn/dt, the net rate of change of open gates (***Equation 7***). Only closed gates can open, which they do with some probability per unit time, expressed as the forward rate constant, α_n. Thus, the transition rate into the open state becomes $\alpha_n(1-n)$, the proportion of closed gates scaled by the forward rate constant. Conversely, only an open gate can close, with a closing probability expressed by the backward rate constant, β_n. Hence, the transition rate out of the open state is $\beta_n n$, the proportion of open gates scaled by the backward rate constant. Therefore, dn/dt, the net rate of change of n, is the difference between these two terms.

(continued on page 201)

From eqn. (6) this may be transformed into a form suitable for comparison with the experimental results, i.e.

$$g_{K} = \{(g_{K\infty})^{\frac{1}{4}} - [(g_{K\infty})^{\frac{1}{4}} - (g_{K0})^{\frac{1}{4}}] \exp(-t/\tau_{n})\}^{4}, \tag{11}$$

where $g_{K\infty}$ is the value which the conductance finally attains and g_{K0} is the conductance at $t=0$. The smooth curves in Fig. 3 were calculated from 28

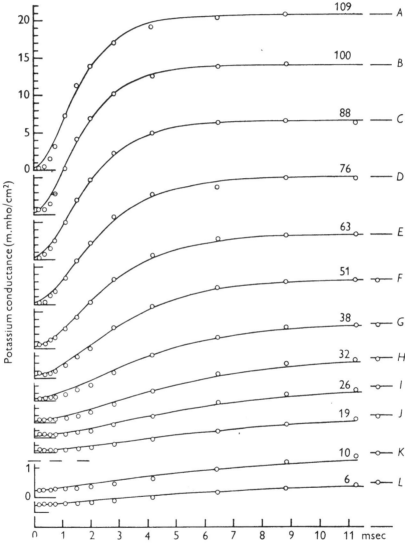

Fig. 3. Rise of potassium conductance associated with different depolarizations. The circles are experimental points obtained on axon 17, temperature 6–7° C, using observations in sea water and choline sea water (see Hodgkin & Huxley, 1952a). The smooth curves were drawn from eqn. (11) with $g_{K0} = 0{\cdot}24$ m.mho/cm² and other parameters as shown in Table 1. The time scale applies to all records. The ordinate scale is the same in the upper ten curves (A to J) and is increased fourfold in the lower two curves (K and L). The number on each curve gives the depolarization in mV.

(continued from page 199)

By analogy with chemical kinetics, another way of expressing these transitions between closed and open gates is with the kinetic scheme:

$$1-n \underset{\beta_n}{\overset{\alpha_n}{\rightleftharpoons}} n$$

This scheme likewise illustrates that the forward rate constant applies to the closed gates, and the backward rate constant applies to the open gates (see Appendix 4).

26. **'In the resting state, defined by $V=0$, n has a resting value'** In a steady state, such as the resting potential, the net value of n does not change, such that $dn/dt=0$, and closed gates open and open gates close with equal frequency, such that $\alpha_n(1-n)=\beta_n n$. By solving for n (see Appendix 4), the proportion of gates open in the steady state can be expressed in terms of α_n and β_n. Although this equation, $n_0 = \alpha_{n0}/(\alpha_{n0}+\beta_{n0})$, is presented in the context of the resting potential, it holds true for any steady-state value of n.

27. **'If V is changed suddenly, α_n and β_n instantly take up values appropriate to the new voltage.'** Here H&H explicitly state a key inference, namely, that the rate constants α and β change instantaneously upon a change in voltage. Their reasoning is as follows. The conductance offset upon step repolarization to a new voltage is well fitted by an exponential decay. Because an exponential decay is defined by a constant τ and constant n_∞, the rates α and β must also be invariant while the voltage is held constant. Thus, because α and β do not change gradually during the voltage step, they must take on their new values immediately upon a change in voltage.

The proportion of gates open at rest is denoted by n_0; with a voltage step, the value of n relaxes to a new steady state, denoted by n_∞. The 'infinity' subscript refers to the new steady-state value, which can be calculated in the same way as n_0 (see note 26), giving **Equation 9**. The time course with which the proportion of open gates at V_0 approaches the value associated with the new voltage is set by the time constant, τ_n, of the exponential decay. Accordingly, **Equation 8** states that the number of open gates, n, at the new voltage changes over time, from n_0 to n_∞, following a single exponential, $1-exp(-t/\tau_n)$, scaled by the magnitude of change $(n_\infty-n_0)$. The expression for the time constant given in **Equation 10** is derived from the solution to the differential equation in **Equation 7** (see Appendix 4).

28. **'where $g_{K\infty}$ is the value which the conductance finally attains and g_{K0} is the conductance at $t=0$.'** The theory describes gating in terms of voltage-dependent rate constants, but the rate constants cannot be directly measured. The preceding equations, however, provide expressions for n_∞ and τ_n, which can be derived from the measurable parameter of conductance. Substituting **Equation 8** into **Equation 6** gives **Equation 11**, which expresses the conductance, g_K, as a function of $g_{K\infty}$, g_{K0}, and τ_n.

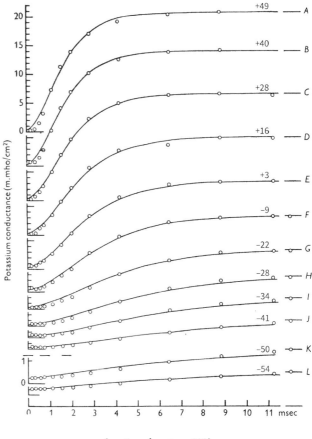

(continued on page 203)

eqn. (11) with a value of τ_n chosen to give the best fit. It will be seen that there is reasonable agreement between theoretical and experimental curves, except that the latter show more initial delay. Better agreement might have been obtained with a fifth or sixth power, but the improvement was not considered to be worth the additional complication.

The rate constants α_n *and* β_n. At large depolarizations $g_{K\infty}$ seems to approach an asymptote about 20–50% greater than the conductance at -100 mV.

TABLE 1. Analysis of curves in Fig. 3

Curve	V (mV) (1)	$g_{K\infty}$ (m.mho/cm²) (2)	n_∞ (3)	τ_n (msec) (4)	α_n (msec⁻¹) (5)	β_n (msec⁻¹) (6)
—	$(-\infty)$	$(24\cdot31)$	$(1\cdot000)$	—	—	—
A	-109	$20\cdot70$	$0\cdot961$	$1\cdot05$	$0\cdot915$	$0\cdot037$
B	-100	$20\cdot00$	$0\cdot953$	$1\cdot10$	$0\cdot866$	$0\cdot043$
C	-88	$18\cdot60$	$0\cdot935$	$1\cdot25$	$0\cdot748$	$0\cdot052$
D	-76	$17\cdot00$	$0\cdot915$	$1\cdot50$	$0\cdot610$	$0\cdot057$
E	-63	$15\cdot30$	$0\cdot891$	$1\cdot70$	$0\cdot524$	$0\cdot064$
F	-51	$13\cdot27$	$0\cdot859$	$2\cdot05$	$0\cdot419$	$0\cdot069$
G	-38	$10\cdot29$	$0\cdot806$	$2\cdot60$	$0\cdot310$	$0\cdot075$
H	-32	$8\cdot62$	$0\cdot772$	$3\cdot20$	$0\cdot241$	$0\cdot071$
I	-26	$6\cdot84$	$0\cdot728$	$3\cdot80$	$0\cdot192$	$0\cdot072$
J	-19	$5\cdot00$	$0\cdot674$	$4\cdot50$	$0\cdot150$	$0\cdot072$
K	-10	$1\cdot47$	$0\cdot496$	$5\cdot25$	$0\cdot095$	$0\cdot096$
L	-6	$0\cdot98$	$0\cdot448$	$5\cdot25$	$0\cdot085$	$0\cdot105$
—	(0)	$(0\cdot24)$	$(0\cdot315)$	—	—	—

Col. 1 shows depolarization in mV; col. 2, final potassium conductance; col. 3, $n_\infty = (g_{K\infty}/\bar{g}_K)^{\frac{1}{4}}$; col. 4, time constant used to compute curve; col. 5, $\alpha_n = n_\infty/\tau_n$; col. 6, $\beta_n = (1-n_\infty)/\tau_n$. The figure of 24·31 was chosen for $\bar{g}_K$ because it made the asymptotic value of n_∞ 5% greater than the value at -100 mV.

For the purpose of calculation we assume that $n = 1$ at the asymptote which is taken as about 20% greater than the value of $g_{K\infty}$ at $V = -100$ mV. These assumptions are somewhat arbitrary, but should introduce little error since we are not concerned with the behaviour of g_K at depolarizations greater than about 110 mV. In the experiment illustrated by Fig. 3, $g_{K\infty} = 20$ m.mho/cm² at $V = -100$ mV. $\bar{g}_K$ was therefore chosen to be near 24 m.mho/cm². This value was used to calculate n_∞ at various voltages by means of eqn. (6). α_n and β_n could then be obtained from the following relations which are derived from eqns. (9) and (10):

$$\alpha_n = n_\infty/\tau_n,$$
$$\beta_n = (1-n_\infty)/\tau_n.$$

The results of analysing the curves in Fig. 3 by this method are shown in Table 1.

An estimate of the resting values of α_n and β_n could be obtained from the decline in potassium conductance associated with repolarization. The procedure was essentially the same but the results were approximate because the

(continued from page 201)

This form is 'suitable for comparison with the experimental results,' shown in **Figure 3**. The conductances at different voltages (open symbols, derived from Paper 2) allow g_{K0} and $g_{K\infty}$ to be read off as the starting and ending points. From g_{K0} and $g_{K\infty}$ it is straightforward to calculate n_0 and n_∞ as $(g_{K0}/\bar{g}_K)^{1/4}$ and $(\bar{g}_{K\infty}/\bar{g}_K)^{1/4}$. The data can then be fitted by **Equation 11** (solid curves) after selecting τ_n by eye. These estimates of τ_n and n_∞ will later be used to calculate values of α_n and β_n appropriate to each voltage. With these values, H&H have a complete, quantitative description of the behavior of the potassium conductance under voltage-clamp, a description based entirely on the physical referent of four charged gating particles.

29. **'the improvement was not considered to be worth the additional complication.'** H&H pause in their theoretical development, as they did repeatedly in the experimental papers, to consider potential errors. Despite the good fit of the calculated line to the data points in **Figure 3**, H&H draw attention to the mismatch at the onset of the conductance increase. They note that the data might be better fitted by assuming more gating particles per channel and a correspondingly higher exponent for n, for example, by setting $g_K = \bar{g}_K n^6$, to introduce a further delay of the rise of the calculated curves. They return to the practical, however: since they must do all their calculations by hand, and since the equations are not derived from first principles, obtaining a more precise fit that is not grounded in a known physical reality is not worth the added computational cost.

In fact, the use of the fourth power, implying four gates, turned out to be consistent with the four voltage sensors on the tetrameric potassium channel. Later work indicated that the deviation that H&H note here—namely the delayed rise of potassium current relative to that predicted by the model—probably arises because each subunit can assume more than two states: after each voltage sensor moves from the resting to the activated position, an additional transition is still required for opening, delaying the onset of current (Bezanilla et al. 1994; Zagotta et al. 1994).

30. **'These assumptions are somewhat arbitrary, but should introduce little error . . . . α_n and β_n could then be obtained'** The maximal conductance $\bar{g}_K$ is needed to estimate n_∞. H&H point out that even with depolarizations of more than 100 mV from rest, the potassium conductance does not saturate, so g_K does not reach $\bar{g}_K$. They estimate that $\bar{g}_K$ is about 20% higher than the conductance that they measure at +40 mV *abs*. With this value of $\bar{g}_K$, H&H can calculate n_∞ at each voltage. With the τ_n values obtained from curve fitting the data in **Figure 3**, they can calculate the forward and backward rates at every potential (given in **Table 1**). Because these rates describe how fast the gates that prevent or permit conductance open and close at every voltage, the rates are as close to first principles as H&H can achieve.

resting value of the potassium conductance was not known with any accuracy when the membrane potential was high. Fig. 2 illustrates an experiment in which the membrane potential was restored to its resting value after a depolarization of 25 mV. It will be seen that both the rise and fall of the potassium conductance agree reasonably with theoretical curves calculated from eqn. (11) after an appropriate choice of parameters. The rate constants derived from these parameters were (in msec^{-1}): $\alpha_n = 0\cdot21$, $\beta_n = 0\cdot70$ when $V = 0$ and $\alpha_n = 0\cdot90$, $\beta_n = 0\cdot43$ when $V = -25$ mV.

In order to find functions connecting α_n and β_n with membrane potential we collected all our measurements and plotted them against V, as in Fig. 4. Differences in temperature were allowed for by adopting a temperature coefficient of 3 (Hodgkin *et al.* 1952) and scaling to 6° C. The effect of replacing sodium by choline on the resting potential was taken into account by displacing the origin for values in choline sea water by $+4$ mV. The continuous curves, which are clearly a good fit to the experimental data, were calculated from the following expressions:

$$\alpha_n = 0\cdot01 \, (V + 10) \bigg/ \left[\exp \frac{V + 10}{10} - 1 \right], \tag{12}$$

$$\beta_n = 0\cdot125 \, \exp \, (V/80), \tag{13}$$

where α_n and β_n are given in reciprocal msec and V is the displacement of the membrane potential from its resting value in mV.

These expressions should also give a satisfactory formula for the steady potassium conductance ($g_{K\infty}$) at any membrane potential (V), for this relation is implicit in the measurement of α_n and β_n. This is illustrated by Fig. 5, in which the abscissa is the membrane potential and the ordinate is $(g_{K\infty}/\bar{g}_K)^{\frac{1}{4}}$. The smooth curve was calculated from eqn. (9) with α_n and β_n substituted from eqns. (12) and (13).

Fig. 4 shows that β_n is small compared to α_n over most of the range; we therefore do not attach much weight to the curve relating β_n to V and have used the simplest expression which gave a reasonable fit. The function for α_n was chosen for two reasons. First, it is one of the simplest which fits the experimental results and, secondly, it bears a close resemblance to the equation derived by Goldman (1943) for the movements of a charged particle in a constant field. Our equations can therefore be given a qualitative physical basis if it is supposed that the variation of α and β with membrane potential arises from the effect of the electric field on the movement of a negatively charged particle which rests on the outside of the membrane when V is large and positive, and on the inside when it is large and negative. The analogy cannot be pressed since α and β are not symmetrical about $E = 0$, as they should be if Goldman's theory held in a simple form. Better agreement might

31. **'the membrane potential was restored to its resting value after a depolarization of 25 mV.'** The transition rates at the resting potential must be derived from steps *to* the resting potential, as in the repolarization of **Figure 2**. The decay phase, which follows a single exponential, reflects deactivation of the conductance, interpreted as the closing of any one of the four gates.

32. **'the simplest expression which gave a reasonable fit.'** H&H gather their calculations of α_n and β_n from all experiments and for all voltages, and plot them in **Figure 4**. They make use of their previous estimate of the Q_{10} of 3 to standardize the measurements to a (cold) value of 6°C by dividing the rate constants by $3^{(T-6)/10}$ for every temperature T. H&H also account for the hyperpolarization of the axon in choline seawater so that all voltage values are accurate. **Equations 12** and **13** provide a fit to the points over the voltage range of interest and permit interpolation of rate constants for any voltage value, including those between the experimental points. The expressions are similar in form to those of Eyring rate theory, which describe the rate constants of chemical reactions in which the reactants must overcome a high energy barrier to proceed from one state to another.

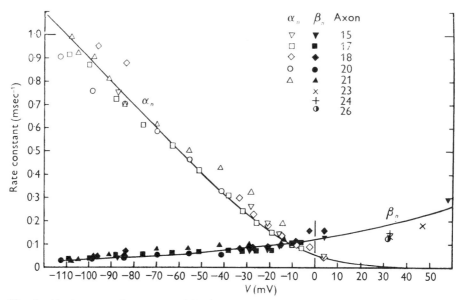

Fig. 4. Abscissa: membrane potential minus resting potential in sea water. Ordinate: rate constants determining rise (α_n) or fall (β_n) of potassium conductance at 6° C. The resting potential was assumed to be 4 mV higher in choline sea water than in ordinary sea water. Temperature differences were allowed for by assuming a Q_{10} of 3. All values for $V < 0$ were obtained by the method illustrated by Fig. 3 and Table 1; those for $V > 0$ were obtained from the decline of potassium conductance associated with an increase of membrane potential or from repolarization to the resting potential in choline sea water (e.g. Fig. 2). Axons 17–21 at 6–11° C, the remainder at about 20° C. The smooth curves were drawn from eqns. (12) and (13).

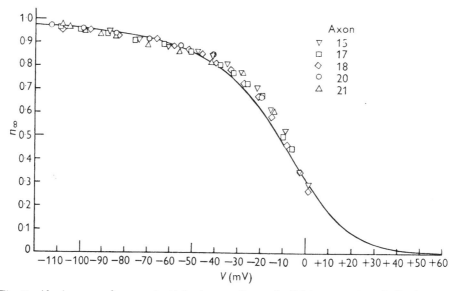

Fig. 5. Abscissa: membrane potential minus resting potential in sea water. Ordinate: experimental measurements of n_∞ calculated from the steady potassium conductance by the relation $n_\infty = \sqrt[4]{(g_{K\infty}/\bar{g}_K)}$, where $\bar{g}_K$ is the 'maximum' potassium conductance. The smooth curve is drawn according to eqn. (9).

33

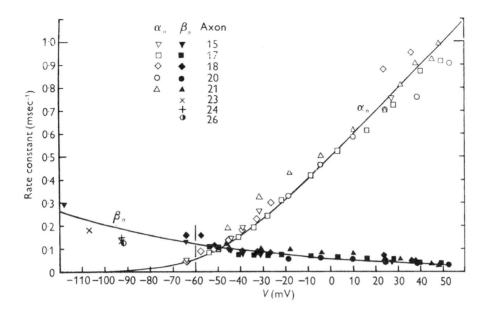

33. **Figure 5.** The n_∞ points were calculated from the fourth root of the measured $g_{K\infty}$ from Paper 2, Figure 10, normalized by the estimated $\bar{g}_K$ (see note 30 and Paper 2, note 64). The solid line is not a fit to the data but is the line resulting from **Equation 9**, in which n_∞ is given in terms of the rate constants.

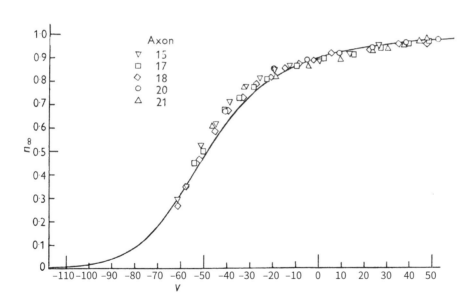

be obtained by postulating some asymmetry in the structure of the membrane, but this assumption was regarded as too speculative for profitable considera-tion.   34

The sodium conductance

There are at least two general methods of describing the transient changes in sodium conductance. First, we might assume that the sodium conductance 35 is determined by a variable which obeys a second-order differential equation. Secondly, we might suppose that it is determined by two variables, each of which obeys a first-order equation. These two alternatives correspond roughly to the two general types of mechanism mentioned in connexion with the nature of inactivation (pp. 502–503). The second alternative was chosen since it was simpler to apply to the experimental results.

The formal assumptions made are:

$$g_{Na} = m^3 h \bar{g}_{Na}, \tag{14}$$

$$\frac{dm}{dt} = \alpha_m (1-m) - \beta_m m, \tag{15}$$

$$\frac{dh}{dt} = \alpha_h (1-h) - \beta_h h, \tag{16}$$

where $\bar{g}_{Na}$ is a constant and the α's and β's are functions of V but not of t.

These equations may be given a physical basis if sodium conductance is assumed to be proportional to the number of sites on the inside of the membrane which are occupied simultaneously by three activating molecules but are not blocked by an inactivating molecule. m then represents the proportion of activating molecules on the inside and $1-m$ the proportion on the outside; h is the proportion of inactivating molecules on the outside and $1-h$ the proportion on the inside. α_m or β_h and β_m or α_h represent the transfer rate constants in the two directions.

Application of these equations will be discussed first in terms of the family of curves in Fig. 6. Here the circles are experimental estimates of the rise and fall of sodium conductance during a voltage clamp, while the smooth curves were calculated from eqns. (14)–(16).   36

The solutions of eqns. (15) and (16) which satisfy the boundary conditions $m = m_0$ and $h = h_0$ at $t = 0$ are

$$m = m_\infty - (m_\infty - m_0) \exp(-t/\tau_m), \tag{17}$$

$$h = h_\infty - (h_\infty - h_0) \exp(-t/\tau_h), \tag{18}$$

where $m_\infty = \alpha_m/(\alpha_m + \beta_m)$ and $\tau_m = 1/(\alpha_m + \beta_m)$,
             $h_\infty = \alpha_h/(\alpha_h + \beta_h)$ and $\tau_h = 1/(\alpha_h + \beta_h)$.

In the resting state the sodium conductance is very small compared with the value attained during a large depolarization. We therefore neglect m_0 if the

34. ***'this assumption was regarded as too speculative for profitable consideration.'*** This statement stands out for the transparency of its logic. In considering a physical basis for the expressions that describe the rate constants as a function of voltage, each plausible idea had shortcomings; only if additional unmeasurable conditions were fulfilled (e.g., membrane asymmetry) might these limitations be overcome. H&H present these ideas but acknowledge that the resulting hypotheses were not sufficiently substantiated to be worth further exploration. Nevertheless, the reader benefits from full disclosure of the experimentalists' thought process, should the option of experimental investigation arise in the future.

35. ***'two general methods of describing the transient changes in sodium conductance.'*** Since the sodium conductance first increases and then decreases, two processes must be taking place. H&H suggest that one variable (here, one gating particle) might do two things consecutively, or two variables (two gating particles) might each do one thing separately (see also note 12.) The latter, where the two processes are independent, is conceptually and computationally simpler. Hence H&H propose activation gates whose probability of being open is given by m (with $1 - m$ as the closed probability), which account for the rise of the sodium conductance, and inactivation gates whose open probability is given by h (with $1 - h$ as the closed probability), which account for the decay of the sodium conductance in the presence of continued depolarization. H&H apply the same form of equations to both the activation and inactivation gates of the sodium conductance as they used for the gates of the potassium conductance, though with different parameter names.

Because sodium channels open with a shorter lag than potassium channels, the rising phase of the conductance curves are fitted best by an exponential decay cubed, suggesting that three gates must open for the conductance to be active (rather than four in the case of the potassium conductance). The inactivation process can be well fitted by a single exponential decay, and it is therefore represented by a single particle. Further evidence for the existence of one inactivation gate comes from the measurements of the steady-state inactivation of the conductance, which was well described by the two-state model equation (see Paper 4, note 17). In a manner similar to the potassium n gates, the forward rate constants α_m and α_h act on the closed gates, and the backward rate constants β_m and β_h act on the open gates (see note 25).

Molecular studies later demonstrated that voltage-gated sodium channels have four domains, each with a voltage sensor. Three of these voltage sensors must move for the channel to open, while movement of the fourth domain, which occurs more slowly, is necessary for inactivation to proceed normally (Chanda & Bezanilla 2002). Thus, the m^3h formulation bears striking similarities to the structural attributes of sodium channels.

36. ***'circles are experimental estimates of the rise and fall of sodium conductance . . . while the smooth curves were calculated'*** The plots in ***Figure 6***, and the associated calculations, are analogous to those of ***Figure 3*** (see note 28). The points represent the conductances calculated from the voltage-clamped sodium currents from Paper 2. The solid lines are fits to ***Equations 14–16*** that provide estimates of m_0 and m_∞, h_0 and h_∞, and τ_m and τ_h.

depolarization is greater than 30 mV. Further, inactivation is very nearly complete if $V < -30$ mV so that h_∞ may also be neglected. The expression for the sodium conductance then becomes

$$g_{\mathrm{Na}} = g'_{\mathrm{Na}} \left[1 - \exp\left(-t/\tau_m\right)\right]^3 \exp\left(-t/\tau_h\right), \tag{19}$$

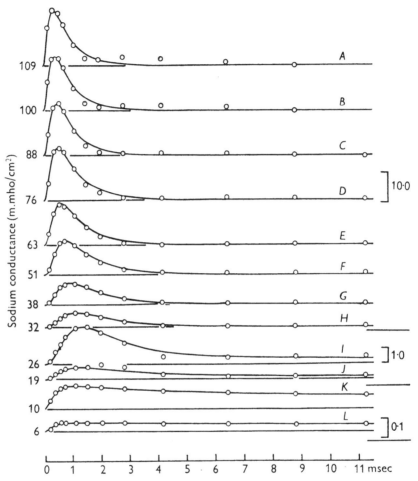

Fig. 6. Changes of sodium conductance associated with different depolarizations. The circles are experimental estimates of sodium conductance obtained on axon 17, temperature 6–7° C (cf. Fig. 3). The smooth curves are theoretical curves with parameters shown in Table 2; A to H drawn from eqn. 19, I to L from 14, 17, 18 with $\bar{g}_{\mathrm{Na}} = 70 \cdot 7$ m.mho/cm². The ordinate scales on the right are given in m.mho/cm². The numbers on the left show the depolarization in mV. The time scale applies to all curves.

where $g'_{\mathrm{Na}} = \bar{g}_{\mathrm{Na}} m_\infty^3 h_0$ and is the value which the sodium conductance would attain if h remained at its resting level (h_0). Eqn. (19) was fitted to an experimental curve by plotting the latter on double log paper and comparing it with a similar plot of a family of theoretical curves drawn with different ratios of τ_m to τ_h. Curves A to H in Fig. 6 were obtained by this method and gave the

PH. CXVII.

37. **'The expression for sodium conductance then becomes'** As they did for the potassium conductance, H&H recast **Equations 17** and **18** for m and h into an expression for the experimentally measured parameter, sodium conductance, to give **Equation 19**. This analysis is largely parallel to the analysis of the potassium conductance in **Equation 11**, but the sodium conductance is the product of an activation term that rises from 0 to 1 and an inactivation term that decays from 1 to 0 as the voltage ranges from the most hyperpolarized to the most depolarized potentials. This product is scaled by g'_{Na}, which is the maximal sodium conductance $\bar{g}_{Na}$ multiplied by m_∞^3 (the probability of finding all activation gates open at a particular potential), multiplied by h_0 (the probability that the inactivation gate was open at the *previous* voltage). Thus, g'_{Na} is the maximal sodium conductance that *could* occur with a voltage step, assuming that neither further inactivation nor recovery from inactivation takes place. The solid lines in **Figure 6** are the fits that arise from **Equation 19**, which permit estimation of τ_m and τ_h at each voltage. The estimated values are given in **Table 2**. As noted in the text, H&H use two different methods to fit the different curves, depending on which simplifying assumptions suit the behavior of m and h at each voltage.

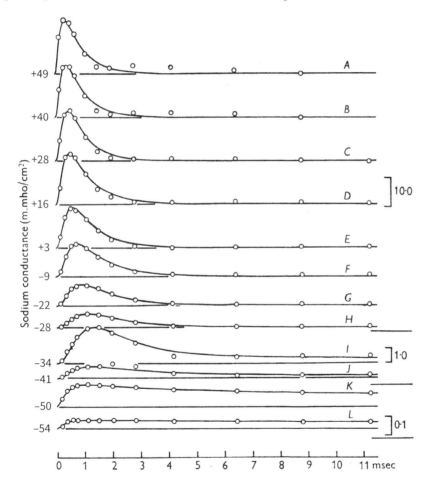

TABLE 2. Analysis of curves in Fig. 6

Curve	V (mV)	g'_{Na} (m.mho/cm^2)	m_∞	τ_m (msec)	α_m (msec^{-1})	β_m (msec^{-1})	τ_h (msec)	h_∞	α_h (msec^{-1})	β_h (msec^{-1})
—	($-\infty$)	(42·9)	(1·00)	—	—	—	—	—	—	—
A	−109	40·3	0·980	0·140	7·0	(0·14)	0·67	(0)	(0)	1·50
B	−100	42·6	0·997	0·160	6·2	(0·02)	0·67	(0)	(0)	1·50
C	−88	46·8	1·029	0·200	5·15	(−0·14)	0·67	(0)	(0)	1·50
D	−76	39·5	0·975	0·189	5·15	0·13	0·84	(0)	(0)	1·19
E	−63	38·2	0·963	0·252	3·82	0·15	0·84	(0)	(0)	1·19
F	−51	30·7	0·895	0·318	2·82	0·33	1·06	(0)	(0)	0·94
G	−38	20·0	0·778	0·382	2·03	0·58	1·27	(0)	(0)	0·79
H	−32	15·3	0·709	0·520	1·36	0·56	1·33	(0)	(0)	0·75
I	−26	7·90	0·569	0·600	0·95	0·72	(1·50)	(0·029)	(0·02)	(0·65)
J	−19	1·44	0·323	0·400	0·81	1·69	(2·30)	(0·069)	(0·03)	(0·40)
K	−10	0·13	0·145	0·220	0·66	3·9	(5·52)	(0·263)	(0·05)	(0·13)
L	6	0·046	0·103	0·200	0·51	4·5	(6·73)	(0·388)	(0·06)	(0·09)
—	(0)	(0·0033)	(0·042)	—	—	—	—	(0·608)	—	—

Values enclosed in brackets were not plotted in Figs. 7–10 either because they were too small to be reliable or because they were not independent measurements obtained in this experiment.

No annotations to this page.

values of g'_{Na}, τ_m and τ_h shown in Table 2. Curves I to L were obtained from eqns. (17) and (18) assuming that h_∞ and τ_h had values calculated from experiments described in a previous paper (Hodgkin & Huxley, 1952 c).

The rate constants α_m and β_m. Having fitted theoretical curves to the experimental points, α_m and β_m were found by a procedure similar to that used with α_n and β_n, i.e.

$$\alpha_m = m_\infty/\tau_m, \quad \beta_m = (1-m_\infty)/\tau_m,$$

the value of m_∞ being obtained from $\sqrt[3]{g'_{Na}}$ on the basis that m_∞ approaches unity at large depolarizations.

38

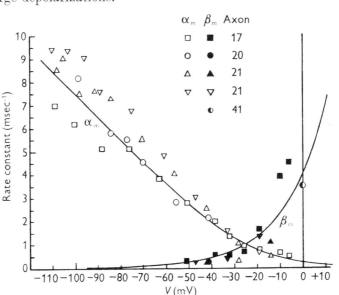

Fig. 7. Abscissa: membrane potential minus resting potential in sea water. Ordinate: rate 39 constants (α_m and β_m) determining initial changes in sodium conductance at 6° C. All values for $V < 0$ were obtained by the method illustrated by Fig. 6 and Table 2; the value at $V = 0$ was obtained from the decline in sodium conductance associated with repolarization to the resting potential. The temperature varied between 3 and 11° C and was allowed for by assuming a Q_{10} of 3. The smooth curves were drawn from eqns. (20) and (21).

Values of α_m and β_m were collected from different experiments, reduced to a temperature of 6° C by adopting a Q_{10} of 3 and plotted in the manner shown in Fig. 7. The point for $V = 0$ was obtained from what we regard as the most reliable estimate of the rate constant determining the decline of sodium conductance when the membrane is repolarized (Hodgkin & Huxley, 1952b, table 1, axon 41). The smooth curves in Fig. 7 were drawn according to the equations:

$$\alpha_m = 0\cdot1 \ (V + 25)\bigg/\left(\exp\frac{V+25}{10} - 1\right), \tag{20}$$

$$\beta_m = 4 \exp{(V/18)}, \tag{21}$$

where α_m and β_m are expressed in msec^{-1} and V is in mV.

38. *'the value of m_∞ being obtained from $^3\sqrt{g'_{Na}}$ on the basis that m_∞ approaches unity at large depolarizations.'* This approach is parallel to that used to find n_∞, but here the probability of finding the activation gate open in the steady state is the cube root of the peak sodium conductance. The proxy for maximum sodium conductance, g'_{Na}, is used as a normalization factor so that m_∞ saturates at 1.

39. *Figure 7.* As was done for the potassium conductance, the activation and deactivation rates α_m and β_m were calculated from the τ_m values obtained from curves fitted to the conductance data in *Figure 6*. Deactivation values at the resting potential ($V = 0$) were obtained in Paper 3 from exponential fits to the tail current elicited by repolarization to the resting potential. Once again, the lines allow the interpolation of rate constants at voltages for which measurements were not made. Temperature is again accounted for; Q_{10} is used to scale these rate constants to predict the values for 6°C. Note that the activation and deactivation rates for the m gating particles are considerably faster than those for the n gating particles in *Figure 4*. This difference, along with the additional delay of the potassium conductance introduced by raising n to the fourth power (compared to the third power for m), makes the sodium conductance activate more rapidly, as required for the generation of an action potential.

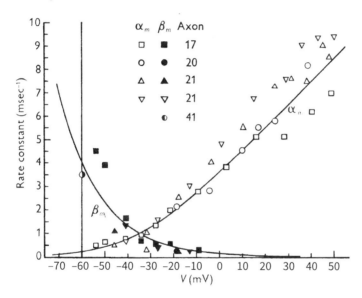

Fig. 8 illustrates the relation between m_∞ and V. The symbols are experimental estimates and the smooth curve was calculated from the equation   40

$$m_\infty = \alpha_m / (\alpha_m + \beta_m),\tag{22}$$

where α_m and β_m have the values given by eqns. (20) and (21).

The rate constants α_h and β_h. The rate constants for the inactivation process were calculated from the expressions   41

$$\alpha_h = h_\infty / \tau_h,$$
$$\beta_h = (1 - h_\infty) / \tau_h.$$

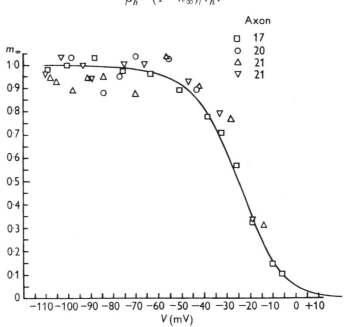

Fig. 8. Abscissa: membrane potential minus resting potential in sea water. Ordinate: m_∞ obtained by fitting curves to observed changes in sodium conductance at different depolarizations (e.g. Fig. 6 and Table 2). The smooth curve is drawn according to eqn. (22). The experimental points are proportional to the cube root of the sodium conductance which would have been obtained if there were no inactivation.

Values obtained by these equations are plotted against membrane potential in Fig. 9. The points for $V < -30$ mV were derived from the analysis described in this paper (e.g. Table 2), while those for $V > -30$ mV were obtained from the results given in a previous paper (Hodgkin & Huxley, 1952 c). A temperature coefficient of 3 was assumed and differences in resting potential were allowed for by taking the origin at a potential corresponding to $h_\infty = 0.6$.

The smooth curves in this figure were calculated from the expressions

$$\alpha_h = 0.07 \exp (V/20),\tag{23}$$

and

$$\beta_h = 1 \bigg/ \left(\exp \frac{V + 30}{10} + 1 \right).\tag{24}$$

40. *'The symbols are experimental estimates and the smooth curve was calculated'* Again parallel to the steady-state activation curve for potassium, in *Figure 8*, H&H plot m_∞ as the cube root of the measured amplitude of the peak sodium conductance at each voltage and superimpose a curve for m_∞ calculated from the rate constants.

41. *'The rate constants for the inactivation process were calculated'* H&H repeat the process for inactivation, with time constants for inactivation measured directly from depolarizations. The time constants, along with measurements of h_∞, permit the extraction of the rate constants, α_h and β_h, (*Equations 23* and *24*) which are fitted as before.

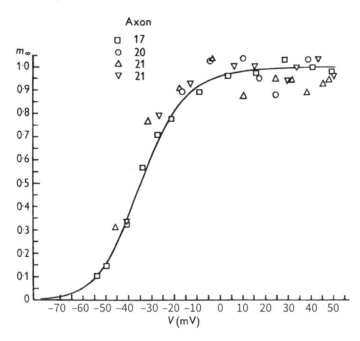

The steady state relation between h_∞ and V is shown in Fig. 10. The smooth curve is calculated from the relation

42

$$h_\infty = \alpha_h/(\alpha_h + \beta_h), \tag{25}$$

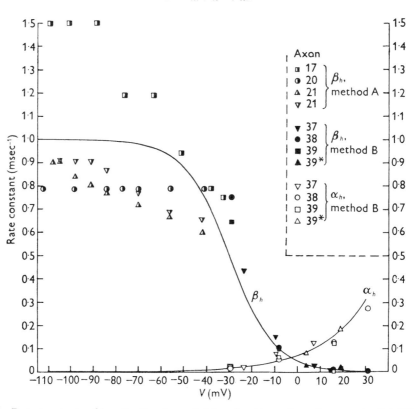

Fig. 9. Rate constants of inactivation (α_h and β_h) as functions of membrane potential (V). The smooth curves were calculated from eqns. (23) and (24). The experimental values of α_h and β_h were obtained from data such as those in Table 2 of this paper (method A) or from the values of τ_h and h_∞ given in Table 1 of Hodgkin & Huxley (1952c) (method B). Temperature differences were allowed for by scaling with a Q_{10} of 3. Axon 39 was at 19° C; all others at 3–9° C. The values for axons 37 and 39* were displaced by $-1\cdot5$ and -12 mV in order to give $h_\infty = 0\cdot6$ at $V = 0$.

with α_h and β_h given by eqns. (23) and (24). If $V > -30$ mV this expression approximates to the simple expression used in a previous paper (Hodgkin & Huxley, 1952 c), i.e.

$$h_\infty = 1 \Big/ \Big(1 + \exp \frac{V_h - V}{7}\Big),$$

where V_h is about -2 and is the potential at which $h_\infty = 0\cdot5$. This equation is 43 the same as that giving the effect of a potential difference on the proportion of negatively charged particles on the outside of a membrane to the total number of such particles on both sides of the membrane (see p. 503). It is therefore consistent with the suggestion that inactivation might be due to the

42. **'The smooth curve is calculated from the relation'** The data points in **Figure 9** represent the rate constants extracted from curve fitting as in **Figure 6** (method A) or the time constant of the onset of inactivation from Paper 4 (method B); the solid line is the fit to the points. Note that the inactivation rate appears to saturate at depolarized potentials. Although axon 17 shows an unusually rapid inactivation rate, the fitted line for β_h (the closing rate constant of the inactivation gate at depolarized potentials) provides a fit that adequately captures the mean response. The h_∞ curve is then calculated from the rate constants according to **Equation 25**, giving the solid line in **Figure 10**.

43. **'V_h is about −2 and is the potential where h_∞ is 0.5.'** The h_∞ curve can be described by the two-state equation derived from the Boltzmann principle (Paper 4, note 17). The parameter V_h is the voltage at which $h_\infty = \frac{1}{2}$, and is 2 mV more depolarized than rest. Thus, nearly half the sodium conductance is inactivated at the resting potential.

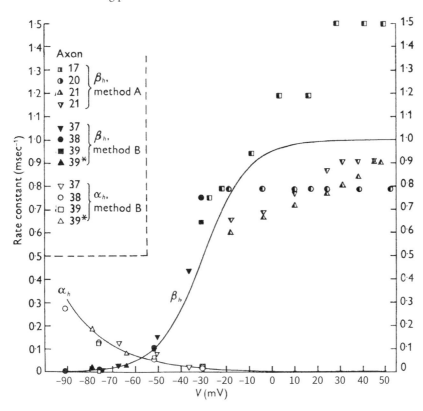

movement of a negatively charged particle which blocks the flow of sodium ions when it reaches the inside of the membrane. This is encouraging, but it must be mentioned that a physical theory of this kind does not lead to satisfactory functions for α_h and β_h without further *ad hoc* assumptions.   44

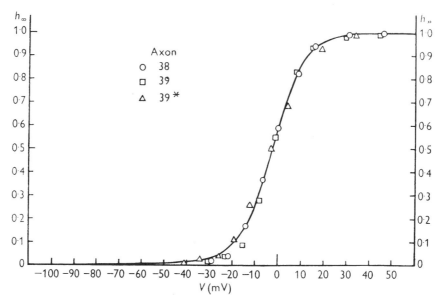

Fig. 10. Steady state relation between h and V. The smooth curve is drawn according to eqn. (25). The experimental points are those given in Table 1 of Hodgkin & Huxley (1952c). Axon 38 (5° C) as measured. Axon 39 (19° C) displaced -1.5 mV. Axon 39* (3° C, fibre in derelict state) displaced -12 mV. The curve gives the fraction of the sodium-carrying system which is readily available, as a function of membrane potential, in the steady state.

PART III. RECONSTRUCTION OF NERVE BEHAVIOUR   45

The remainder of this paper will be devoted to calculations of the electrical behaviour of a model nerve whose properties are defined by the equations which were fitted in Part II to the voltage clamp records described in the earlier papers of this series.

Summary of equations and parameters

We may first collect the equations which give the total membrane current I 46 as a function of time and voltage. These are:

$$I = C_M \frac{dV}{dt} + \bar{g}_K n^4 (V - V_K) + \bar{g}_{Na} m^3 h (V - V_{Na}) + \bar{g}_l (V - V_l), \tag{26}$$

where
$$dn/dt = \alpha_n(1-n) - \beta_n n, \tag{7}$$

$$dm/dt = \alpha_m(1-m) - \beta_m m, \tag{15}$$

$$dh/dt = \alpha_h(1-h) - \beta_h h, \tag{16}$$

44. *'This is encouraging, but it must be mentioned that a physical theory of this kind does not lead to satisfactory functions . . . without further ad hoc assumptions.'* H&H refer back to the fact that the model describes the probability of finding a particle in one of two states. Since this function fits the data extremely well, as shown in **Figure 10**, the physical manifestation of such a change of state might be that a charged particle moves to inactivate current flow. They comment, however, that even if such a molecular mechanism exists, it does not help much in determining rate constants.

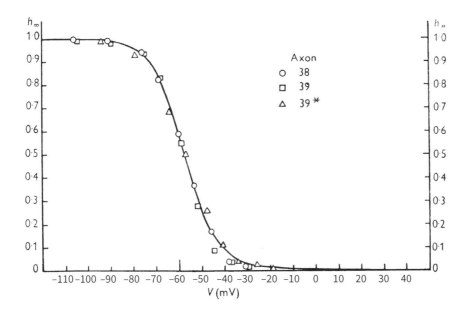

45. *'PART III. RECONSTRUCTION OF NERVE BEHAVIOR.'* This final section of H&H's opus is arguably the part for which they remain most widely known: the application of the Hodgkin-Huxley equations—the mathematical descriptions of sodium, potassium, and leak conductances recorded during voltage clamp—to predict, or, as they more exactly express it, to reconstruct, the action potential of the squid giant axon.

46. *'We may first collect the equations'* The sequence of equations reveals the underlying logic of this section. Reconstructing the changes in membrane voltage, including the action potential, requires knowing the membrane current that generates these changes. The total membrane current (**Equation 26**) consists of capacity current and ionic current; the latter can be broken down into individual sodium, potassium, and leak currents, each of which is the product of conductance and driving force. Each conductance is a proportion of the maximal conductance; H&H's evidence indicates that the proportion may be represented as the probability of finding a set of hypothetical gates open. The time course with which each gate opens and closes is determined by rate constants, which vary systematically with voltage. The conductances are therefore functions of time and voltage, which makes the currents also functions of time and voltage. Thus, by knowing the voltage and its recent history, it should be possible to calculate first how much of each current is flowing, and second how the currents will change with membrane potential and thereby generate voltage changes, including those recognizable as action potentials.

and

$$\alpha_n = 0\cdot01 \ (V+10)\bigg/\left(\exp\frac{V+10}{10}-1\right), \tag{12}$$

$$\beta_n = 0\cdot125 \ \exp\left(V/80\right), \tag{13}$$

$$\alpha_m = 0\cdot1 \ (V+25)\bigg/\left(\exp\frac{V+25}{10}-1\right), \tag{20}$$

$$\beta_m = 4 \ \exp\left(V/18\right), \tag{21}$$

$$\alpha_h = 0\cdot07 \ \exp\left(V/20\right), \tag{23}$$

$$\beta_h = 1\bigg/\left(\exp\frac{V+30}{10}+1\right). \tag{24}$$

Equation (26) is derived simply from eqns. (1)–(6) and (14) in Part II. The four terms on the right-hand side give respectively the capacity current, the current carried by K ions, the current carried by Na ions and the leak current, for 1 cm² of membrane. These four components are in parallel and add up to give the total current density through the membrane I. The conductances to K and Na are given by the constants $\bar{g}_K$ and $\bar{g}_{Na}$, together with the dimensionless quantities n, m and h, whose variation with time after a change of membrane potential is determined by the three subsidiary equations (7), (15) and (16). The α's and β's in these equations depend only on the instantaneous value of the membrane potential, and are given by the remaining six equations.

Potentials are given in mV, current density in μA/cm², conductances in m.mho/cm², capacity in μF/cm², and time in msec. The expressions for the α's and β's are appropriate to a temperature of $6\cdot3^\circ$ C; for other temperatures they must be scaled with a Q_{10} of 3.

The constants in eqn. (26) are taken as independent of temperature. The values chosen are given in Table 3, column 2, and may be compared with the experimental values in columns 3 and 4.

Membrane currents during a voltage clamp

Before applying eqn. (26) to the action potential it is well to check that it predicts correctly the total current during a voltage clamp. At constant voltage $dV/dt=0$ and the coefficients α and β are constant. The solution is then obtained directly in terms of the expressions already given for n, m and h (eqns. (8), (17) and (18)). The total ionic current was computed from these for a number of different voltages and is compared with a series of experimental curves in Fig. 11. The only important difference is that the theoretical current has too little delay at the sodium potential; this reflects the inability of our equations to account fully for the delay in the rise of g_K (p. 509). 47

'Membrane' and propagated action potentials

By a 'membrane' action potential is meant one in which the membrane potential is uniform, at each instant, over the whole of the length of fibre 48

47. *'Before applying eqn. (26) to the action potential it is well to check that it predicts correctly the total current during a voltage clamp.'* Recall that the conductance traces were calculated from experimentally measured voltage-clamped currents on the assumption of independent ionic fluxes in parallel with a near-perfect capacitor. These conductance traces were fitted to extract time constants (τ's), which were used to estimate rate constants based on the hypothesis that the proportion of open activation and inactivation gates (m, h, and n) changes with time and voltage. For convenience, it has become standard to refer to the gating particles as m-, h-, and n-gates, although H&H did not use this terminology (see note 14). The calculated rate constants were then fitted with functions of voltage so that values could be interpolated for all membrane potentials. Many well-justified assumptions were used in this process, most particularly the idea of physical gating particles based on mathematical fitting, but it remains possible that the accumulation of errors and approximations—or a still-concealed mistake in the basic model—might make the back-calculation of the original currents impossible or inaccurate. Here, H&H test whether the rate constants, extracted through these many steps of assumption and calculation, can be used in reverse to recreate the total voltage-clamped current. The data, shown in **Figure 11**, illustrate the success of the theory up to this point. The simulations do well at replicating the total membrane current, with the minor but noted flaw that the calculated potassium conductance rises more rapidly than the experimental data.

48. *'By a 'membrane' action potential is meant one in which the membrane potential is uniform'* The stationary membrane action potential only occurs under the experimental condition that HH&K developed for application of the voltage-clamp method. Recall that in the guard system that houses the squid axon, the long axial electrode wires maintain a significant stretch of the axon at a uniform potential. In Paper 1, Figures 8 and 9, HH&K verified that the axon could fire an action potential under these unusual conditions.

TABLE 3

Constant (1)	Value chosen (2)	Experimental values		Reference (5)
		Mean (3)	Range (4)	
C_M (μF/cm²)	1·0	0·91	0·8 to 1·5	Table 1, Hodgkin et al. (1952)
V_{Na} (mV)	−115	−109	−95 to −119	p. 455, Hodgkin & Huxley (1952a)
V_K (mV)	+12	+11	+9 to +14	Table 3, values for low temperature in sea water, Hodgkin & Huxley (1952b)
V_l (mV)	−10·613*	−11	−4 to −22	Table 5, Hodgkin & Huxley (1952b)
$\bar{g}_{Na}$ (m.mho/cm²)	120	{ 80, 160 }	{ 65 to 90, 120 to 260 }	Fully analysed results, Table 2† } Hodgkin & Huxley (1952a) Fresh fibres, p. 465†
$\bar{g}_K$ (m.mho/cm²)	36	34	26 to 49	p. 463, Hodgkin & Huxley (1952a)
$\bar{g}_l$ (m.mho/cm²)	0·3	0·26	0·13 to 0·50	Table 5, Hodgkin & Huxley (1952b)

* Exact value chosen to make the total ionic current zero at the resting potential ($V = 0$).

† The experimental values for $\bar{g}_{Na}$ were obtained by multiplying the peak sodium conductances by factors derived from the values chosen for α_m, β_m, α_h, and β_h.

49. ***Table 3*** gives the sources of values used in the model. At the resting potential, the ionic current must be zero or else the membrane potential would be changing, rather than resting. Thus, the leak conductance is set to make the ionic current zero at the resting potential.

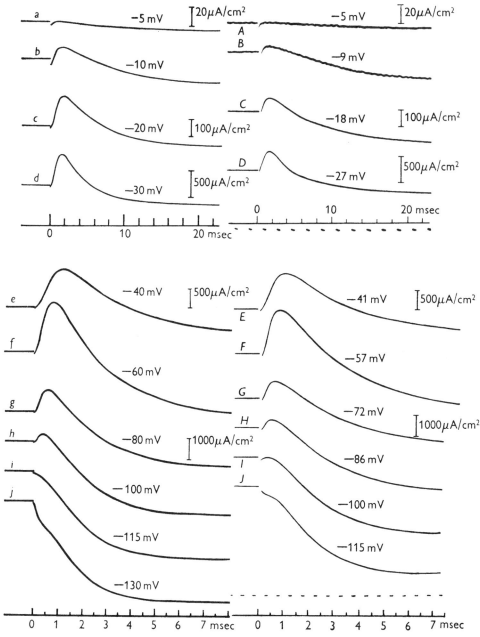

Fig. 11. Left-hand column: time course of membrane current during voltage clamp, calculated 50 for temperature of 4° C from eqn. (26) and subsidiaries and plotted on the same scale as the experimental curves in the right-hand column. Right-hand column: observed time course of membrane currents during voltage clamp. Axon 31 at 4° C; compensated feedback. The time scale changes between *d*, *D* and *e*, *E*. The current scale changes after *b*, *B*; *c*, *C*; *d*, *D* and *f*, *F*.

50. ***Figure 11*** shows the remarkable success of the equations at simulating currents (*left*) that match both the amplitude and time course of the total current measured under voltage clamp (*right*, see note 47).

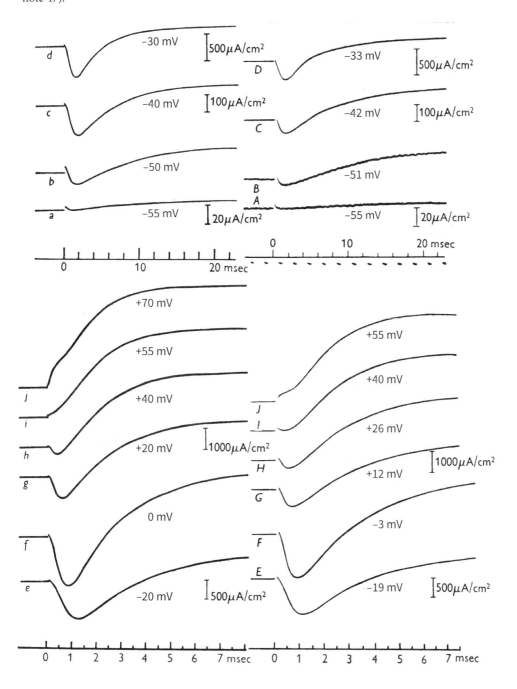

considered. There is no current along the axis cylinder and the net membrane current must therefore always be zero, except during the stimulus. If the stimulus is a short shock at $t=0$, the form of the action potential should be given by solving eqn. (26) with $I=0$ and the initial conditions that $V=V_0$ and m, n and h have their resting steady state values, when $t=0$.

The situation is more complicated in a propagated action potential. The fact that the local circuit currents have to be provided by the net membrane current leads to the well-known relation

$$i = \frac{1}{r_1 + r_2} \frac{\partial^2 V}{\partial x^2}, \tag{27}$$

where i is the membrane current per unit length, r_1 and r_2 are the external and internal resistances per unit length, and x is distance along the fibre. For an axon surrounded by a large volume of conducting fluid, r_1 is negligible compared with r_2. Hence

$$i = \frac{1}{r_2} \frac{\partial^2 V}{\partial x^2},$$

or

$$I = \frac{a}{2R_2} \frac{\partial^2 V}{\partial x^2}, \tag{28}$$

where I is the membrane current density, a is the radius of the fibre and R_2 is the specific resistance of the axoplasm. Inserting this relation in eqn. (26), we have

$$\frac{a}{2R_2} \frac{\partial^2 V}{\partial x^2} = C_M \frac{\partial V}{\partial t} + \bar{g}_K n^4 (V - V_K) + \bar{g}_{Na} m^3 h (V - V_{Na}) + \bar{g}_l (V - V_l), \tag{29}$$

the subsidiary equations being unchanged.

Equation (29) is a partial differential equation, and it is not practicable to solve it as it stands. During steady propagation, however, the curve of V against time at any one position is similar in shape to that of V against distance at any one time, and it follows that

$$\frac{\partial^2 V}{\partial x^2} = \frac{1}{\theta^2} \frac{\partial^2 V}{\partial t^2},$$

where θ is the velocity of conduction. Hence

$$\frac{a}{2R_2 \theta^2} \frac{\mathrm{d}^2 V}{\mathrm{d}t^2} = C_M \frac{\mathrm{d}V}{\mathrm{d}t} + \bar{g}_K n^4 (V - V_K) + \bar{g}_{Na} m^3 h (V - V_{Na}) + \bar{g}_l (V - V_l). \tag{30}$$

This is an ordinary differential equation and can be solved numerically, but the procedure is still complicated by the fact that θ is not known in advance. It is necessary to guess a value of θ, insert it in eqn. (30) and carry out the numerical solution starting from the resting state at the foot of the action potential. It is then found that V goes off towards either $+\infty$ or $-\infty$, according as the guessed θ was too small or too large. A new value of θ is

51. **'The situation is more complicated in a propagated action potential.'** In the 'membrane' or stationary action potential, the segment of the axon being measured is isopotential. Therefore, current and voltage vary only with time, and not distance, and are governed by the simple differential equation of **Equation 26**. In contrast, in the propagating action potential, current and voltage vary with both time and distance along the axon. The equations governing membrane voltage and current therefore become partial differential equations of these two variables.

52. **'the local circuit currents have to be provided by the net membrane current'** H&H had each independently provided evidence supporting the theory of local circuit currents (see Historical Background), Hodgkin in his early work on unmyelinated axons (Hodgkin 1937a, 1937b, 1939) and Huxley on myelinated axons (Huxley & Stämpfli 1949). The local circuit current theory stated that current flowed in loops: intracellularly down the axon cylinder, out across the membrane, longitudinally along a return path outside the axon, and into the membrane. By Ohm's law, current flows longitudinally wherever there is a longitudinal voltage difference.

H&H refer to the 'well-known' relationship between transmembrane current and the second spatial derivative of membrane voltage, which is derived from cable theory in Hodgkin and Rushton (1946). The relationship is illustrated in simplified form in *Figure A*. By Kirchoff's law (see Appendix 2.2), I_B, the current flowing across the membrane at point B, is equal to the difference between the currents I_{AB} and I_{BC} flowing along the axon. For example, if I_{AB} is smaller than I_{BC}, then I_B will be a net inward current of $I_{BC} - I_{AB}$. Otherwise, charge would accumulate at point B. By Ohm's law, I_{AB} and I_{BC} are proportional to the voltage differences between adjacent points on the membrane (equations below the schematic). I_B, therefore, becomes the *difference* between those two differences, which can be expressed as: $I_B = \Delta\Delta V_B / R$. In the limit, as points A, B, and C are brought progressively closer to one another, the difference of differences becomes the second derivative in **Equation 27**. Note that this analysis takes into account only the ionic (resistive) current; capacitive current has been omitted. As a result, the equations shown in *Figure A* hold only for the case when the membrane potential is not changing in time. The relationship between transmembrane current and the second spatial derivative of voltage still holds in the dynamic condition of propagation, but the equations become more complex.

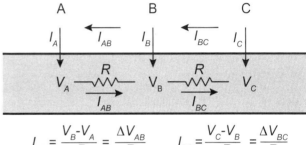

$$I_{AB} = \frac{V_B - V_A}{R} = \frac{\Delta V_{AB}}{R} \qquad I_{BC} = \frac{V_C - V_B}{R} = \frac{\Delta V_{BC}}{R}$$

$$I_B = I_{BC} - I_{AB} = \frac{\Delta V_{BC} - \Delta V_{AB}}{R} = \frac{\Delta\Delta V_B}{R}$$

Figure A

In H&H's formulation, R and I in the difference equations of *Figure A* become internal resistance per unit length, r_2, and membrane current per unit length, i. The larger the cross-sectional area of the axon, the larger the area available for longitudinal current flow. As a result, the longitudinal resistance of the axon varies inversely with the square of the axon diameter. To make the equations generalizable, the membrane current per unit length, i, is converted to current per unit surface area of the axon, I (current density), by dividing it by the perimeter of the axon, $2\pi a$, where a is the axonal radius. Similarly, r_2 can be converted to specific resistance R_2 (units of $\Omega \cdot$ cm) by multiplying by the cross-sectional area of the axon, πa^2. In the expression for I given in **Equation 28**, substituting $1/r_2$ with $\pi a^2/R_2$ and then dividing by $2\pi a$ simplifies to $a/2R_2$ as the coefficient on the voltage term. (As noted, r_1 is neglected since it is so small relative to r_2.) By substituting **Equation 28** into **Equation 26**, which applied only to a stationary action potential, H&H arrive at **Equation 29**, which generalizes to the propagating action potential.

53. **'During steady propagation, however, the curve of V against time at any position is similar in shape to that of V against distance at any one time'** The differential equation that describes the stationary action potential is complicated to solve by hand; adding the dimension of length makes **Equation 29** into a partial differential equation, for which attaining a numerical solution is far

(continued on page 231)

then chosen and the procedure repeated, and so on. The correct value brings V back to zero (the resting condition) when the action potential is over.

The solutions which go towards $\pm \infty$ correspond to action potentials travelling slower than normal under a travelling anode or faster than normal under a travelling cathode. We suspect that a system which tends to $-\infty$ for all values of θ after an initial negative displacement of V is one which is incapable of propagating an action potential.

NUMERICAL METHODS

Membrane action potentials

Integration procedure. The equations to be solved are the four simultaneous first-order equations (26), (7), (15), and (16) (p. 518). After slight rearrangement (which will be omitted in this description) these were integrated by the method of Hartree (1932–3). Denoting the beginning and end of a step by t_0 and t_1 ($= t_0 + \delta t$) the procedure for each step was as follows: 54

(1) Estimate V_1 from V_0 and its backward differences.

(2) Estimate n_1 from n_0 and its backward differences.

(3) Calculate $(dn/dt)_1$ from eqn. 7 using the estimated n_1 and the values of α_n and β_n appropriate to the estimated V_1.

(4) Calculate n_1 from the equation

$$n_1 - n_0 = \frac{\delta t}{2} \left\{ \left(\frac{dn}{dt} \right)_0 + \left(\frac{dn}{dt} \right)_1 - \frac{1}{12} \left[\Delta^2 \left(\frac{dn}{dt} \right)_0 + \Delta^2 \left(\frac{dn}{dt} \right)_1 \right] \right\};$$

$\Delta^2 (dn/dt)$ is the second difference of dn/dt; its value at t_1 has to be estimated.

(5) If this value of n_1 differs from that estimated in (2), repeat (3) and (4) using the new n_1. If necessary, repeat again until successive values of n_1 are the same.

(6) Find m_1 and h_1 by procedures analogous to steps (2)–(5).

(7) Calculate $\bar{g}_K n_1^4$ and $\bar{g}_{Na} m_1^3 h_1$.

(8) Calculate $(dV/dt)_1$ from eqn. 26 using the values found in (7) and the originally estimated V_1.

(9) Calculate a corrected V_1 by procedures analogous to steps (4) and (5). This result never differed enough from the original estimated value to necessitate repeating the whole procedure from step (3) onwards.

The step value had to be very small initially (since there are no differences at $t = 0$) and it also had to be changed repeatedly during a run, because the differences became unmanageable if it was too large. It varied between about 0·01 msec at the beginning of a run or 0·02 msec during the rising phase of the action potential, and 1 msec during the small oscillations which follow the spike.   55

Accuracy. The last digit retained in V corresponded to microvolts. Sufficient digits were kept in the other variables for the resulting errors in the change of V at each step to be only occasionally as large as $1\,\mu V$. It is difficult to estimate the degree to which the errors at successive steps accumulate, but we are confident that the overall errors are not large enough to be detected in the illustrations of this paper.

Temperature differences. In calculating the action potential it was convenient to use tables giving the α's and β's at intervals of 1 mV. The tabulated values were appropriate to a fibre at 6·3° C. To obtain the action potential at some other temperature T' °C the direct method would be to multiply all α's and β's by a factor $\phi = 3^{(T'-6\cdot3)/10}$, this being correct for a Q_{10} of 3. Inspection of eqn. 26 shows that the same result is achieved by calculating the action potential at 6·3° C with a membrane capacity of $\phi C_M\,\mu F/cm^2$, the unit of time being $1/\phi$ msec. This method was adopted since it saved recalculating the tables.

(continued from page 229)

more difficult. H&H, however, note a feature of the system that greatly simplifies the solution to **Equation 29**: A propagating action potential moves along the axon at a constant rate, such that the same voltage waveform will emerge at successive sites on the axon, but delayed in time by the distance divided by the speed of propagation, θ. Therefore, H&H can measure the action potential at one site on the axon and use the knowledge that the waveform of the action potential plotted against *time* has the identical shape to the waveform plotted against *distance*.

To arrive at the expression $\partial^2 V/\partial x^2 = (1/\theta^2)(\partial^2 V/\partial t^2)$, note that the first derivative of voltage with respect to distance ($\partial V/\partial x$) is, by the chain rule for differentiation, equal to $\partial V/\partial t \times \partial t/\partial x$, where $\partial t/\partial x$ is the reciprocal of conduction velocity, $1/\theta$. Therefore, $\partial V/\partial x$ is equal to $(1/\theta)(\partial V/\partial t)$. The second derivative of voltage with respect to distance, $\partial^2 V/\partial x^2$, then becomes $\partial(\partial V/\partial t)/\partial x \times \partial t/\partial x$, or $(1/\theta^2)(\partial^2 V/\partial t^2)$. This expression can be substituted into **Equation 29** to give **Equation 30**. Although **Equation 30** is far easier to solve, θ is unknown. H&H therefore must iteratively guess at the propagation velocity during their numerical integration of the equations (see note 58). The equations for the propagating action potential were solved numerically by computer more than a decade later (Cooley & Dodge 1966), which facilitated a more detailed exploration of the predictions of the equations under a wide range of conditions.

54. **'The procedure for each step was as follows'** At this point, H&H have set up the equations that must be solved to calculate the total membrane current flowing at any time and at any voltage (see note 46), and they have obtained estimates of all the constant terms necessary to provide numerical values of current. Recall that when the voltage is changed slightly, all the rate constants that govern n, m, and h will change instantaneously; the gates will begin to move with specific rates toward a new steady-state (that they may not have time to reach before the voltage changes further); the sodium and potassium conductance will change; current will flow; charge will be deposited on the membrane capacitance; the voltage will change; and the process will begin again. The challenge for H&H in reconstructing the action potential is to introduce an initial voltage change from rest, and to use the rate constants to calculate dn/dt, dm/dt, dh/dt, and the steady-state n, m, and h values; from there to find the conductance; to calculate the ionic current (equal and opposite to the capacity current); and finally to determine the voltage change, δV, that occurs in a tiny period of time, δt. This new voltage can then be plotted. With the new voltage value, the rate constants change again and the process is repeated. These calculations are described more fully in Appendix 5.

55. **'The step value . . . varied between about 0.01 msec at the beginning of a run or 0.02 msec during the rising phase of the action potential, and 1 msec during the small oscillations which follow the spike.'** If a time step is long enough for the voltage to change greatly, the resulting changes in rate constants associated with intermediate voltages will not be accounted for adequately, and inaccuracies in the calculated membrane potential trajectory will occur and become compounded with each new step. Therefore, small time steps—on the order of 10–30 µs—must be used when the voltage is changing rapidly, but larger time steps may be used when the voltage is changing slowly (see Appendix 5). Each value was calculated (by Huxley) on a mechanical calculator that required entering numbers and turning a crank many times for each addition, subtraction, multiplication, or division (see Appendix 5). For this reason, H&H sought any simplification that would increase computational efficiency without introducing significant errors.

Propagated action potential

Equations. The main equation for a propagated action potential is eqn. (30). Introducing   56
a quantity $K = 2R_2\theta^2 C_M/a$, this becomes

$$\frac{\mathrm{d}^2 V}{\mathrm{d}t^2} = K\left\{\frac{\mathrm{d}V}{\mathrm{d}t} + \frac{1}{C_M}\left[\bar{g}_K n^4(V - V_K) + \bar{g}_{Na} m^3 h\,(V - V_{Na}) + \bar{g}_l(V - V_l)\right]\right\}. \tag{31}$$

The subsidiary equations (7), (15) and (16), and the α's and β's, are the same as for the membrane equation.

Integration procedure. Steps (1)–(7) were the same as for the membrane action potential. After that the procedure was as follows:

(8) Estimate $(\mathrm{d}V/\mathrm{d}t)_1$ from $(\mathrm{d}V/\mathrm{d}t)_0$ and its backward differences.

(9) Calculate $(\mathrm{d}^2 V/\mathrm{d}t^2)_1$ from eqn. (31), using the values found in (7) and the estimated values of V_1 and $(\mathrm{d}V/\mathrm{d}t)_1$.

(10) Calculate a corrected $(\mathrm{d}V/\mathrm{d}t)_1$ by procedures analogous to steps (4) and (5).

(11) Calculate a corrected V_1 by a procedure analogous to step (4), using the corrected $(\mathrm{d}V/\mathrm{d}t)_1$.

(12) If necessary, repeat (9)–(11) using the new V_1 and $(\mathrm{d}V/\mathrm{d}t)_1$, until successive values of V_1 agree.

Starting conditions. In practice it is necessary to start with V deviating from zero by a finite amount ($0\cdot1$ mV was used). The first few values of V, and hence the differences, were obtained as follows. Neglecting the changes in g_K and g_{Na}, eqn. (31) is   57

$$\frac{\mathrm{d}^2 V}{\mathrm{d}t^2} = K\left\{\frac{\mathrm{d}V}{\mathrm{d}t} + \frac{g_0}{C_M}\,V\right\},$$

where g_0 is the resting conductance of the membrane. The solution of this equation is $V = V_0 e^{\mu t}$, where μ is a solution of

$$\mu^2 - K\mu - Kg_0/C_M = 0. \tag{32}$$

When K has been chosen, μ can thus be found and hence V_1, V_2, etc. ($V_0 e^{\mu t_1}$, $V_0 e^{\mu t_2}$, etc.).

After several runs had been calculated, so that K was known within fairly narrow limits, time was saved by starting new runs not from near $V = 0$ but from a set of values interpolated between corresponding points on a run which had gone towards $+\infty$ and another which had gone towards $-\infty$.

Choice of K. The value of K chosen for the first run makes no difference to the final result, but the nearer it is to the correct value the fewer runs will need to be evaluated. The starting value was found by inserting in eqn. (32) a value of μ found by measuring the foot of an observed action potential.

Calculation of falling phase. The procedure outlined above is satisfactory for the rising phase and peak of the action potential but becomes excessively tedious in the falling phase and the   58 oscillations which follow the spike. A different method, which for other reasons is not applicable in the earlier phases, was therefore employed. The solution was continued as a membrane action potential, and the value of $\mathrm{d}^2 V/\mathrm{d}t^2$ calculated at each step from the differences of $\mathrm{d}V/\mathrm{d}t$. From these it was possible to derive an estimate of the values (denoted by z) that $\mathrm{d}^2 V/\mathrm{d}t^2$ would have taken in a propagated action potential. The membrane solution was then re-calculated using the following equation instead of eqn. (31):

$$\frac{\mathrm{d}V}{\mathrm{d}t} = -\frac{1}{C_M}\{\bar{g}_K n^4(V - V_K) + \bar{g}_{Na} m^3 h\,(V - V_{Na}) + \bar{g}_l\,(V - V_l)\} + \frac{z}{K}. \tag{33}$$

This was repeated until the z's assumed for a particular run agreed with the $\mathrm{d}^2 V/\mathrm{d}t^2$'s derived from the same run. When this is the case, eqn. (33) is identical with eqn. (31), the main equation for the propagated action potential.

56. *'The main equation for a propagated action potential is eqn. (30).'* After an appropriate value for conduction velocity, θ, is found, it is rolled into the constant K, along with specific internal resistance (R_2), capacitance (C), and axon diameter (a) in **Equation 31**. A calculation comparable to the one for the stationary action potential can then be undertaken.

57. *'The first few values of V, and hence the differences, were obtained as follows.'* In the propagated action potential, the initial depolarization at any point on the axon is driven by a continuous current flowing intracellularly from upstream regions of the axon; the amount of this current depends on both the magnitude and the propagation velocity of the action potential. H&H mimic the initial depolarization with a small voltage change of 0.1 mV. Since little voltage-gated conductance is activated with a deviation of 0.1 mV from rest, they can ignore sodium and potassium conductances. Doing so simplifies **Equation 31** so that it only contains the resting conductance, g_0 (g_{leak} plus the values of g_{Na} and g_K at the resting potential), which makes the second order differential equation simpler to solve. This simplification is important because they must estimate the constant K, which depends on conduction velocity, θ, and which therefore determines the amount of current that will drive subsequent depolarization. If the initial guess of K is incorrect, successive voltage values (calculated in their steps 8–12) will tend toward the meaningless extremes of $+\infty$ or $-\infty$. Only an accurate estimate of K, arrived at by successive approximation, will generate the correct action potential waveform. From K, H&H can solve for θ.

58. *'becomes excessively tedious in the falling phase'* Because the repolarization phase of the action potential is less rapid than the rise, the voltage changes very little when successive time steps are tiny. Therefore, H&H used a method of successive approximation: they estimated where the next voltage value would be based on the calculated *membrane* action potential waveform and then used that value to estimate an additional term, z. The term z/K was then added to the (simpler, first order) differential equation that defines the membrane action potential; the propagated action potential equation reduces to this value if z is correct. The values were compared with the second derivative values that had been calculated at previous voltages to verify the estimate of z. The rest of the repolarization phase could then be computed simply from the first-order differential equation.

RESULTS

RESULTS

Membrane action potentials

Form of action potential at 6° C. Three calculated membrane action potentials, with different strengths of stimulus, are shown in the upper part of Fig. 12.   59 Only one, in which the initial displacement of membrane potential was 15 mV, is complete; in the other two the calculation was not carried beyond the middle of the falling phase because of the labour involved and because the solution   60

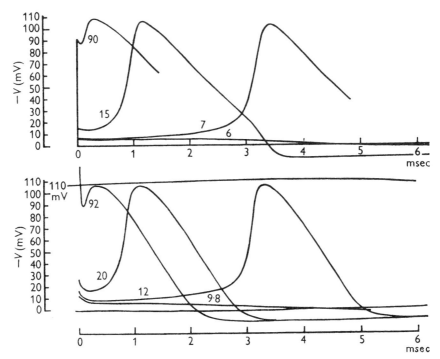

Fig. 12. Upper family: solutions of eqn. (26) for initial depolarizations of 90, 15, 7 and 6 mV (calculated for 6° C). Lower family: tracings of membrane action potentials recorded at 6° C from axon 17. The numbers attached to the curves give the shock strength in mμcoulomb/cm². The vertical and horizontal scales are the same in both families (apart from the slight curvature indicated by the 110 mV calibration line). In this and all subsequent figures depolarizations (or negative displacements of V) are plotted upwards.

had become almost identical with the 15 mV action potential, apart from the displacement in time. One solution for a stimulus just below threshold is also shown.

The lower half of Fig. 12 shows a corresponding series of experimental membrane action potentials. It will be seen that the general agreement is good, as regards amplitude, form and time-scale. The calculated action potentials do, however, differ from the experimental in the following respects: 61 (1) The drop during the first 0·1 msec is smaller. (2) The peaks are sharper.

59. **'Three calculated membrane action potentials, with different strengths of stimulus, are shown'** At last! The voltage changes are calculated from the initial voltage change, which is calculated from the charge applied as $V = Q/C$, and the new rate constants associated with that new voltage. The apparently continuous lines in the top panel of **Figure 12** stem from the calculation of each voltage change over the time steps of 10–20 µs, through the procedures described above, and the waveforms are strikingly similar to the experimentally measured action potential recorded with comparable stimuli, in the bottom panel. Thus, the voltage-dependent rate constants (kinetics) and voltage-dependent steady-state conductances (amplitudes) extracted from voltage-clamp recordings indeed provide sufficient information to recreate the action potential. The word *sufficient* is used here in its scientific sense: it is not necessary to hypothesize additional variables to account for the phenomenon (but see H&H's analysis below, note 61). Note, in this and subsequent plots of action potentials, the value 0 on the *y*-axis is the resting potential (approximately –60 mV *abs*), and depolarizations are upwards.

60. **'because of the labour involved'** Again, since each point had to be literally cranked out by hand on the mechanical calculator, defining the short-latency and long-latency action potential beyond the first half of the repolarization, which looked just like that of the middle-latency action potential, was unlikely to yield enough additional information to be worth the effort.

61. **'The calculated action potentials do, however, differ from the experimental'** It is interesting to note the extent to which H&H analyze the mismatches, down to the relatively slower onset of depolarization (the 'drop') in the first 100 µs. On the one hand, these are simple acknowledgments of the imperfections of the reconstructed action potentials. On the other hand, the later explanation that the error can be accounted for by specific, quantifiable deviations between the simulated and recorded currents lends support to the fundamental hypothesis that the current magnitude and kinetics control the shape of the action potential waveform (see notes 97 and 98).

(3) There is a small hump in the lower part of the falling phase. (4) The ending of the falling phase is too sharp. The extent to which these differences are the result of known shortcomings in our formulation will be discussed on pp. 542–3.

The positive phase of the calculated action potential has approximately the correct form and duration, as may be seen from Fig. 13 in which a pair of curves are plotted on a slower time scale.   62

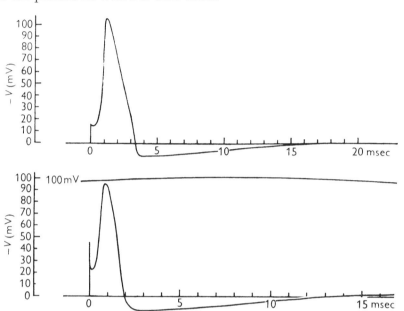

Fig. 13. Upper curve: solution of eqn. (26) for initial depolarization of 15 mV, calculated for 6° C. Lower curve: tracing of membrane action potential recorded at 9·1° C (axon 14). The vertical scales are the same in both curves (apart from curvature in the lower record). The horizontal scales differ by a factor appropriate to the temperature difference.

Certain measurements of these and other calculated action potentials are collected in Table 4.

Form of action potential at 18·5° C. Fig. 14 shows a comparison between a calculated membrane action potential at 18·5° C and an experimental one at 20·5° C. The same differences can be seen as at the low temperature, but,   63 except for the initial drop, they are less marked. In both the calculated and the experimental case, the rise of temperature has greatly reduced the duration of the spike, the difference being more marked in the falling than in the rising phase (Table 4), as was shown in propagated action potentials by Hodgkin & Katz (1949).

The durations of both falling phase and positive phase are reduced at the higher temperature by factors which are not far short of that (3·84) by which the rate constants of the permeability changes are raised ($Q_{10} = 3\cdot0$). This is the justification for the differences in time scale between the upper and lower parts in Figs. 13 and 14.

62. **'a pair of curves are plotted on a slower time scale.'** The curves in *Figure 13* are shown on two different time bases. The simulated (upper) action potential was computed with all rate constants set to their values at 6°C. The experimental (lower) action potential was recorded at a warmer temperature. The waveforms of the action potentials are remarkably similar. At warmer temperatures, the action potential becomes briefer and has a lower amplitude, as a result of the temperature-sensitivity of the rate constants.

63. **'a comparison between a calculated membrane action potential at 18.5°C and an experimental one at 20.5°C.'** For the simulation in *Figure 14*, the rate constants were scaled with a Q_{10} of 3, by multiplying by the factor $3^{(T_2 - T_1)/10}$ to obtain the values appropriate for action potential waveforms closer to room temperature. The match is again very good, with the exceptions noted. To compensate for the faster kinetics generated by the two-degree higher temperature in the experimental record, H&H again compress the time base of the computed record.

TABLE 4.　Characteristics of calculated action potentials

Type of action potential	Temperature (°C)	Stimulus	Spike height (mV)	Amplitude of positive phase (mV)	Peak conductance (m.mho/cm²)	Duration of rising phase, 20 mV to peak (msec)	Duration of falling phase, peak to $V=0$ (msec)	Duration of positive phase (msec)	Interval from peak of potential to peak of conductance (msec)	Max. rate of rise (V/sec)
Propagated	18·5	—	90·5	9·7	32·6	0·252	0·67	5·20	−0·016	431
Membrane	18·5	15 mV depolarization	96·8	10·5	30·7	0·275	0·61	5·09	+0·012	564
Membrane	6·3	100 mV depolarization	108·8	—	45·5	—	—	—	+0·16	—
Membrane	6·3	90 mV depolarization	108·5	—	44·8	—	—	—	+0·15	—
Membrane	6·3	15 mV depolarization	105·4	11·2	37·0	0·59	2·21	14·15	+0·15	311
Membrane	6·3	7 mV depolarization	102·1	—	33·4	0·62	—	—	+0·16	277
Membrane	6·3	Anode break	112·1	11·2	53·4	0·50	2·54	14·4	+0·14	414

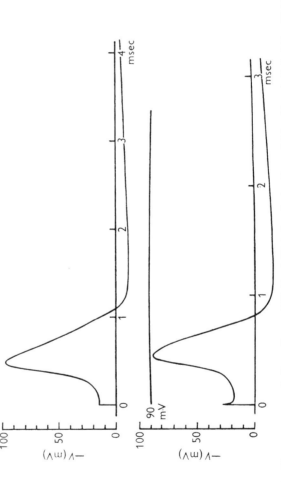

Fig. 14.　Upper curve: solution of eqn. (26) for initial depolarization of 15 mV, calculated for 18·5° C. Lower curve: tracing of membrane action potential recorded at 20·5° C (axon 11). Vertical scales are similar. Horizontal scales differ by a factor appropriate to the temperature difference.

64. *Table 4*. Key parameters of the calculated action potentials are given. Points with no values are the cases where the computations were stopped because they were too laborious.

Propagated action potential

Form of propagated action potential. Fig. 15 compares the calculated propagated action potential, at 18·5° C, with experimental records on both fast and slow time bases. As in the case of the membrane action potential, the only   65 differences are in certain details of the form of the spike.

Velocity of conduction. The value of the constant K that was found to be needed in the equation for the propagated action potential (eqn. 31) was 10·47 msec⁻¹. This constant, which depends only on properties of the membrane,

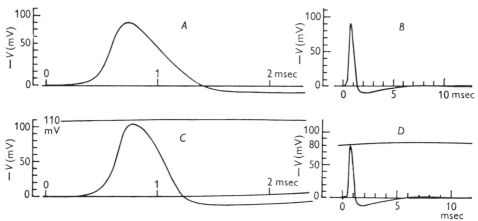

Fig. 15. *A*, solution of eqn. (31) calculated for K of 10·47 msec⁻¹ and temperature of 18·5° C. *B*, same solution plotted on slower time scale. *C*, tracing of propagated action potential on same vertical and horizontal scales as *A*. Temperature 18·5° C. *D*, tracing of propagated action potential from another axon on approximately the same vertical and horizontal scales as *B*. Temperature 19·2° C. This axon had been used for several hours; its spike was initially 100 mV.

determines the conduction velocity in conjunction with the constants of the   66 nerve fibre considered as a cable. The relation is given by the definition of K (p. 524), from which

$$\theta = \sqrt{(Ka/2R_2C_M)}, \tag{34}$$

where θ = conduction velocity, a = radius of axis cylinder, R_2 = specific resistance of axoplasm, and C_M = capacity per unit area of membrane.

The propagated action potential was calculated for the temperature at which the record C of Fig. 15 was obtained, and with the value of C_M ($1\cdot0\,\mu\text{F/cm}^2$) that was measured on the fibre from which that record was made. Since θ, a and R_2 were also measured on that fibre, a direct comparison between calculated and observed velocities is possible. The values of a and R_2 were $238\,\mu$ and $35\cdot4\ \Omega$. cm respectively. Hence the calculated conduction velocity is

$$(10470 \times 0\cdot0238/2 \times 35\cdot4 \times 10^{-6})^{\frac{1}{2}}\ \text{cm/sec} = 18\cdot8\ \text{m/sec}.$$

The velocity found experimentally in this fibre was $21\cdot2$ m/sec.

65. *'Fig. 15 compares the calculated propagated action potential, at 18.5°C, with experimental records on both fast and slow time bases.'* In many ways, **Figure 15** is the culmination of what H&H (and indeed the rest of the field) had been trying to accomplish since the 1930s: to understand the physical basis of the propagating action potential, at relatively physiological temperature, well enough to reconstruct it from a quantitative knowledge of the underlying variables. They have succeeded extraordinarily well. H&H explicitly note that the experimental action potential in *D*, recorded at room temperature, was obtained only at the end of several hours of experiments after its amplitude had become reduced. Recall that the experimental action potentials had been recorded in Plymouth in 1948 and the analysis was done in Cambridge starting in 1950. Therefore, the only action potentials that H&H would have had available to compare with the calculated records were those previously photographed from their oscilloscope screen under comparable conditions.

66. *'This constant, which depends only on properties of the membrane, determines the conduction velocity'* For the propagating action potential, the total membrane current density, *I*, was expressed as the second derivative of voltage with respect to time, scaled by a constant term, *K*, composed of the conduction velocity, θ, as well as other axon properties (see notes 52 and 56). The calculations that led to the reconstruction of the propagating action potential required iterative estimation of a realistic value for *K*, which turned out to be 10.47/msec. From this value and the other axonal parameters of capacitance, diameter, and axial resistance, the value of θ can be calculated and compared to experimental measurements of the rate of propagation of an action potential (conduction velocity). These values, of 18.8 m/sec and 21.2 m/sec, turn out to agree well, again providing an additional check on the validity not only of the rate constant values that H&H have extracted but also of the conceptualization of the basis for the action potential that they have put forth.

As an aside, since the duration of the depolarizing phase of the propagated action potential is about 0.5 msec, and the propagation speed is about 20 m/sec, the physical length of the action potential—the distance between points on the axon that are beginning and completing depolarization at any given time—can be calculated to be about 10 mm.

Impedance changes

Time course of conductance change. Cole & Curtis (1939) showed that the impedance of the membrane fell during a spike, and that the fall was due to a great increase in the conductance which is in parallel with the membrane capacity. An effect of this kind is to be expected on our formulation, since the entry of Na^+ which causes the rising phase, and the loss of K^+ which causes the falling phase, are consequent on increases in the conductance of the membrane to currents carried by these ions. These component conductances are evaluated during the calculation, and the total conductance is obtained by adding them and the constant 'leak conductance', $\bar{g}_l$.

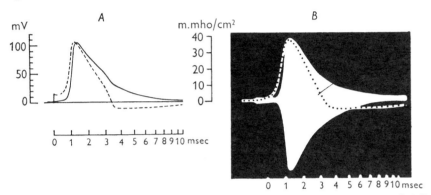

Fig. 16. *A*, solution of eqn. (26) for initial depolarization of 15 mV at a temperature of 6° C. The broken curve shows the membrane action potential in mV; the continuous curve shows the total membrane conductance ($g_{Na} + g_K + \bar{g}_l$) as a function of time. *B*, records of propagated action potential (dotted curve) and conductance change reproduced from Cole & Curtis (1939). The time scales are the same in *A* and *B*.

Fig. 16 *A* shows the membrane potential and conductance in a calculated membrane action potential. For comparison, Fig. 16 *B* shows superposed records of potential and impedance bridge output (proportional to conductance change), taken from Cole & Curtis's paper. The time scale is the same in *B* as in *A*, and the curves have been drawn with the same peak height. It will be seen that the main features of Cole & Curtis's record are reproduced in the calculated curve. Thus (1) the main rise in conductance begins later than the rise of potential; (2) the conductance does not fall to its resting value until late in the positive phase; and (3) the peak of the conductance change occurs at nearly the same time as the peak of potential. The exact time relation between the peaks depends on the conditions, as can be seen from Table 4.

We chose a membrane action potential for the comparison in Fig. 16 because the spike duration shows that the experimental records were obtained at about 6° C, and our propagated action potential was calculated for 18·5° C. The conductance during the latter is plotted together with the potential in Fig. 17. The same features are seen as in the membrane action potential, the delay

PH. CXVII.

67. **'Cole & Curtis (1939) showed that the impedance of the membrane fell during a spike . . . due to a great increase in the conductance which is in parallel with the membrane capacity.'** The well-known experiment by Cole and Curtis (1939) is reproduced in *Figure 16B*. They measured axonal impedance (resistance and capacitance) during the course of an action potential (see Historical Background) with a Wheatstone bridge (see Appendix 3.4). The circuit involves two known constant resistances, one known adjustable resistance, and one unknown resistance. A voltage is applied, and the variable resistor is adjusted until the output voltage is zero. This procedure is known as 'balancing the bridge.' The unknown resistance can then be can be calculated from the values of the known resistances.

Cole and Curtis used the squid axon as the unknown impedance and applied a high-frequency oscillating voltage, which allowed them to measure both resistance and capacitance. In the experiment shown, the output voltage starts at zero with the bridge balanced, at which point the resting impedance of the axon can be measured. When an action potential is elicited, the bridge goes out of balance, and the voltage signal grows in amplitude as a result of the changing membrane impedance. During this period, the individual cycles of the high-frequency oscillatory signal merge together into the white region seen in the figure. Cole and Curtis measured the amplitude and phase of the oscillations and found that the membrane resistance, but not the membrane capacitance, changed appreciably during the action potential. The envelope of the positive phases of oscillation is roughly proportional to the change in conductance (inverse of resistance) during the action potential. The action potential itself is superimposed on the conductance signal.

68. **'the main features of Cole & Curtis's record are reproduced in the calculated curve.'** In calculating the voltage changes that give rise to the membrane (stationary) action potential, H&H necessarily obtained measurements of the sodium, potassium, and leak conductances at every point in the waveform. They summed all those conductances (regardless of the direction in which current is flowing) to obtain the solid line in *Figure 16A*; the action potential voltage trace is superimposed as the dashed line. H&H draw attention to the similarity between the calculated total conductance change and the values measured by Cole and Curtis, both in their general time course and their relationship to the associated voltage waveform, providing another independent check on the validity of H&H's model. Note that the rate constants in the simulation had values appropriate to 6°C, because H&H were able to infer from Cole and Curtis's action potential duration that their experiments must have been done near that temperature.

between the rise of potential and the rise of conductance being even more marked.

Absolute value of peak conductance. The agreement between the height of the conductance peak in Fig. 16 *A* and the half-amplitude of the bridge output in Fig. 16 *B* is due simply to the choice of scale. Nevertheless, our calculated action potentials agree well with Cole & Curtis's results in this respect. [69] These authors found that the average membrane resistance at the peak of the impedance change was $25\ \Omega.\text{cm}^2$, corresponding to a conductance of $40\ \text{m.mho/cm}^2$. The peak conductances in our calculated action potentials ranged from 31 to 53 m.mho/cm² according to the conditions, as shown in Table 4.

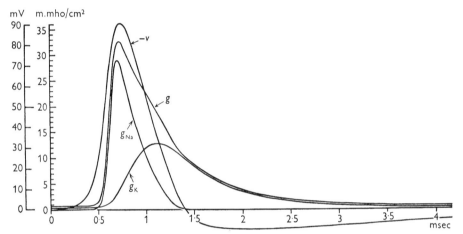

Fig. 17. Numerical solution of eqn. (31) showing components of membrane conductance (g) during propagated action potential ($-V$). Details of the analysis are as in Fig. 15.

Components of conductance change. The manner in which the conductances to Na⁺ and K⁺ contribute to the change in total conductance is shown in Fig. 17 for the calculated propagated action potential. The rapid rise is due [70] almost entirely to sodium conductance, but after the peak the potassium conductance takes a progressively larger share until, by the beginning of the positive phase, the sodium conductance has become negligible. The tail of raised conductance that falls away gradually during the positive phase is due solely to potassium conductance, the small constant leak conductance being of course present throughout.

Ionic movements

Time course of ionic currents. The time course of the components of membrane current carried by sodium and potassium ions during the calculated propagated spike is shown in Fig. 18 *C*. The total ionic current contains also [71] a small contribution from 'leak current' which is not plotted separately.

Two courses are open to current which is carried into the axis cylinder by ions crossing the membrane: it may leave the axis cylinder again by altering

69. *'our calculated action potentials agree well with Cole & Curtis's results in this respect.'* Cole and Curtis were studying propagating action potentials in their 1939 study (see Historical Background), so it is of interest to compare the magnitudes of H&H's calculated individual ionic conductances with Cole and Curtis's experimentally measured total conductance during the propagated action potential. The peaks of the summed conductances calculated by H&H (shown in ***Table 4***) indeed bracket the mean value measured in the earlier study.

70. *'The manner in which the conductances to Na⁺ and K⁺ contribute to the change in total conductance is shown in Fig. 17 for the calculated propagated action potential.* Note again that this is the most physiological condition that H&H simulate. In this iconic figure, the action potential is labeled as $-V$ (owing to their conventions), the total conductance is given by line g, and the individual ionic conductances are labeled as g_{Na} and g_K. The sodium conductance is transient, with a time course not unlike that of the conductance evoked by a step depolarization. The potassium conductance during an action potential is quite different from that activated by a step: it falls rapidly after reaching its maximal value, and its peak is considerably smaller than the peak sodium conductance (see Paper 1, note 56). These differences from the conductance activated by step depolarizations result from the brevity of the depolarization, which is a consequence of the potassium conductance itself. As H&H point out, the leak conductance, which is not plotted, would simply be a horizontal line (just above the x-axis), since its amplitude is constant, regardless of voltage.

71. *'components of membrane current carried by sodium and potassium ions during the calculated propagated spike'* After successfully describing the voltage change during the propagated action potential, and then showing the conductances that gave rise to that propagated action potential, H&H illustrate the currents during that action potential (***Figure 18***). By definition, these currents cannot be measured in voltage clamp, since, during propagation, the voltage is varying across space as well as changing with time.

Note that the sodium and potassium flux overlap considerably: the sodium current is maximal during the rising phase but lasts well into the falling phase, while the potassium current peaks at the start of the falling phase and its long duration is largely responsible for the afterhyperpolarization of the membrane voltage. Thus, the two currents act against one another during the repolarization phase. Such overlap does not occur in all neurons (Carter & Bean 2009). The energetic inefficiency of the action potential in the squid giant axon may be an acceptable cost, given the role of the axon in the escape reflex.

the charge on the membrane capacity, or it may turn either way along the axis cylinder making a net contribution, I, to the local circuit current. The magnitudes of these two terms during steady propagation are $-C_M\, \mathrm{d}V/\mathrm{d}t$ and $(C_M/K)\, \mathrm{d}^2V/\mathrm{d}t^2$ respectively, and the manner in which the ionic current is divided between them at the different stages of the spike is shown in Fig. 18 B. It will be seen that the ionic current is very small until the potential is well beyond the threshold level, which is shown by Fig. 12 A to be about 6 mV.

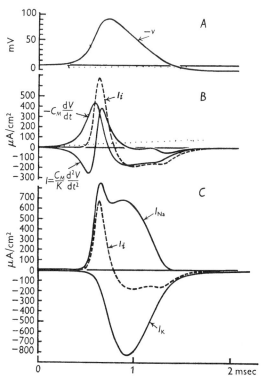

Fig. 18. Numerical solution of eqn. (31) showing components of membrane current during propagated action potential. A, membrane potential ($-V$). B, ionic current (I_i), capacity current $\left(-C_M\dfrac{\mathrm{d}V}{\mathrm{d}t}\right)$ and total membrane current $\left(I=\dfrac{C_M}{K}\dfrac{\mathrm{d}^2V}{\mathrm{d}t^2}\right)$. C, ionic current (I_i), sodium current (I_{Na}) and potassium current (I_K). The time scale applies to all the curves. Details of the analysis are as in Fig. 15.

During this period the current for charging the membrane capacity comes almost entirely from local circuits. The fact that the ionic current does not become appreciable as soon as the threshold depolarization is passed is due partly to the smallness of the currents reached in any circumstances near the threshold, and partly to the delay with which sodium conductance rises when the potential is lowered.

Total movements of ions. The total entry of sodium and loss of potassium can be obtained by integrating the corresponding ionic currents over the whole

72. **'Two courses are open to current which is carried into the axis cylinder by crossing the membrane: it may leave the axis cylinder again . . . , or it may turn either way along the axis cylinder'**
In the membrane action potential, all charges that flow into any point along the axon settle on the membrane and discharge the capacitance at that point, thereby changing the local membrane voltage (see Paper 1, notes 13, 44, and 60). For the propagating action potential, the incoming charges can either discharge the local capacitance, generating a capacity current ('leave the axis cylinder'), or they can spread longitudinally down the axon as axial current and depolarize the membrane at a distance, participating in the loops of current flow that form local circuits. The current flowing down the axon is given in **Equation 30** as $(a/2R_2\theta^2)(\mathrm{d}^2V/\mathrm{d}t^2)$. Since the constant K was subsequently defined as $2R_2\theta^2C_M/a$, this expression reduces by substitution to $(C_M/K)(\mathrm{d}^2V/\mathrm{d}t^2)$.

73. **'the manner in which the ionic current is divided between them at the different stages of the spike is shown in Fig. 18B.'** The calculated propagating action potential is shown again in **Figure 18A**, with **Figure 18C** illustrating the individual sodium and potassium currents, as well as their sum I_i. The middle panel splits I_i into its component parts, the axial current, I, and the capacity current $-C_M\mathrm{d}V/\mathrm{d}t$. H&H point out that the inward ionic current (plotted upward) becomes appreciable only well after the onset of the rising phase of the action potential. Note that the capacity current and the ionic current differ for the propagating action potential, although they would be equal and opposite for a stationary action potential. In the propagating case, the initial local depolarization, even beyond threshold, comes from axial current flowing from upstream, depolarized portions of the axon. The downward deflections for I correspond to current that would be locally depolarizing while the upward deflections would be locally hyperpolarizing.

impulse. This has been done for the four complete action potentials that we calculated, and the results are given in Table 5. It will be seen that the results at 18·5° C are in good agreement with the values found experimentally by Keynes (1951) and Keynes & Lewis (1951), which were obtained at comparable temperatures.   74

Ionic fluxes. The flux in either direction of an ion can be obtained from the net current and the equilibrium potential for that ion, if the independence principle (Hodgkin & Huxley, 1952*a*) is assumed to hold. Thus the outward   75 flux of sodium ions is $I_{Na}/(\exp (V - V_{Na}) F/RT - 1)$, and the inward flux of potassium ions is $-I_K/(\exp (V_K - V) F/RT - 1)$. These two quantities were evaluated at each step of the calculated action potentials, and integrated over the whole impulse. The integrated flux in the opposite direction is given in each case by adding the total net movement. The results are given in Table 5, where they can be compared with the results obtained with radioactive tracers by Keynes (1951) on *Sepia* axons. It will be seen that our theory predicts too little exchange of Na and too much exchange of K during an impulse. This discrepancy will be discussed later.

Refractory period

Time course of inactivation and delayed rectification. According to our theory, there are two changes resulting from the depolarization during a spike which make the membrane unable to respond to another stimulus until a certain time has elapsed. These are 'inactivation', which reduces the level to which

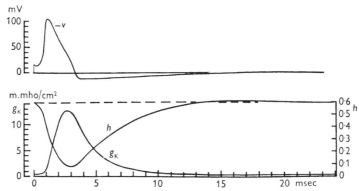

Fig. 19. Numerical solution of eqn. (26) for initial depolarization of 15 mV and temperature of 6° C. Upper curve: membrane potential, as in Fig. 13. Lower curves show time course of g_K and h during action potential and refractory period.

the sodium conductance can be raised by a depolarization, and the delayed rise in potassium conductance, which tends to hold the membrane potential   76 near to the equilibrium value for potassium ions. These two effects are shown in Fig. 19 for the calculated membrane action potential at 6° C. Both curves reach their normal levels again near the end of the positive phase, and finally

74. **'the results at 18.5°C are in good agreement with the values found experimentally'** Keynes (1951) and Keynes & Lewis (1951) had used radioactive sodium and potassium to measure the quantity of each ion that flowed into and out of the axons of *Sepia* (cuttlefish) during an action potential. H&H integrate the total calculated current density (coulombs/sec/cm^2) over time (sec) to get total charge transfer density (coulombs/cm^2), which can be translated into μμmole/cm^2 (picomole/cm^2) by dividing by Faraday's constant (96,485 coulombs/mole). These net ionic movements are given in **Table 5**. The match between the predictions of the model and the measurements made with a completely different method provides further support for the validity of H&H's approach.

75. **'The flux in either direction of an ion can be obtained . . . if the independence principle . . . is assumed to hold.'** The estimates of charge transfer above are from the *net* current, but ions can move both into and out of the cell through the active conductance mechanism (in modern terms, through open channels). Far from the equilibrium potential, most ionic flux will be in one direction, but near equilibrium potential, substantial flux will happen in both directions, even though the net current is close to zero. H&H refer back to the independence principle that they derived and validated previously (see Paper 2, note 75) and to the equation that later became known as the GHK current equation (see Paper 2, notes 52 and 80). They use the equations to calculate the total movement of each ion in each direction (one-way fluxes) and report that their theory underestimates sodium movement and overestimates potassium movement reported in previous work by Keynes (given in **Table 5**).

76. **'two changes . . . make the membrane unable to respond to another stimulus. . . . These are "inactivation" . . . and the delayed rise in potassium conductance'** Up to this point, H&H have shown that their equations can reproduce the primary electrical behavior of the axon, the action potential, as well as the impedance changes and charge transfer that occur during the action potential. They now consider several additional behaviors of the axon, which had been observed experimentally but were not yet understood, and they test whether these responses can also be explained by the theory. These include the absolute and relative refractory period, threshold instability, anode break excitation, accommodation, and subthreshold oscillations.

As the heading of this section indicates, inactivation and the delayed potassium current both contribute to the refractory period, defined as the period after an action potential during which a second action potential either requires a stronger stimulus or cannot be elicited at all. **Figure 19** illustrates not only the membrane action potential and potassium conductance that have been shown before, but also the value of *h*, the proportion of inactivation gates that are open (scale on the right *y*-axis). During the action potential, *h* falls from 0.6 to less than 0.1, and takes about 12 msec to return to its resting value (see Paper 4, note 24). At the same time, the potassium conductance, which is elevated during the action potential, takes about 10 msec to return to its resting value, producing the afterhyperpolarization, which H&H refer to as 'the positive phase.' Therefore, for several milliseconds after the spike, any depolarization that might normally bring the axon to threshold must work against a reduction in the available sodium conductance and an elevated potassium conductance.

TABLE 5. Ionic movements during an impulse. All values are expressed in $\mu\mu$mole/cm² and represent the excess over the corresponding movement in the resting state. In the theoretical cases the integration is taken as far as the 3rd intersection with the base line after the spike; it is begun in case (1) when $V = 0\cdot1$ mV; (2) and (3) at the stimulus; (4) when $V = 0$ before the spike. Experimental data from Keynes (1951) for row 6 and from Keynes & Lewis (1951) for rows 5 and 7.

Type of action potential	Temp. (°C)	Stimulus (mV)	Sodium			Potassium		
			Influx	Outflux	Net entry	Influx	Outflux	Net loss
Theoretical (*Loligo*):								
1 Propagated	18·5	—	5·42	1·09	4·33	1·72	5·98	4·26
2 Membrane	18·5	15	5·01	1·02	3·99	1·71	5·78	4·07
3 Membrane	6·3	15	19·30	4·84	14·46	6·17	20·49	14·32
4 Membrane	6·3	Anode break	26·61	9·45	17·16	6·64	23·41	16·77
Experimental:								
5 Propagated (*Loligo*)	22	—	—	—	3·5	—	—	3·0
6 Propagated (*Sepia*)	14	—	10·3	6·6	3·7	0·39	4·7	4·3
7 Propagated (*Sepia*)	22	—	—	—	3·8	—	—	3·6

77. ***Table 5.*** As H&H state, the calculated net sodium and potassium flux is in good agreement with that measured with radioactive tracers (compare calculated and experimental data for the propagated action potential in *Loligo*, for 'net entry' and 'net loss.') The discrepancy with studies of influx and efflux by Keynes on *Sepia* axons is discussed at the very end of the paper (see note 98).

settle down after a heavily damped oscillation of small amplitude which is not seen in the figure.

Responses to stimuli during positive phase. We calculated the responses of the membrane when it was suddenly depolarized by 90 mV at various times during the positive phase of the membrane action potential at 6° C. These are shown by the upper curves in Fig. 20. After the earliest stimulus the

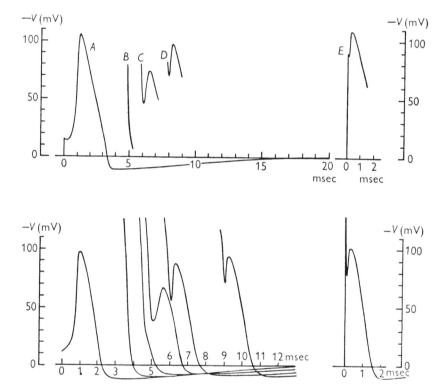

Fig. 20. Theoretical basis of refractory period. Upper curves: numerical solutions of eqn. (26) for temperature of 6° C. Curve *A* gives the response to 15 mμcoulomb/cm² applied instantaneously at $t=0$. Curve *E* gives the response to 90 mμcoulomb/cm² again applied in the resting state. Curves *B* to *D* show effect of applying 90 mμcoulomb/cm² at various times after curve *A*. Lower curves: a similar experiment with an actual nerve, temperature 9° C. The voltage scales are the same throughout. The time scales differ by a factor appropriate to the temperature difference.

membrane potential falls again with hardly a sign of activity, and the membrane can be said to be in the 'absolute refractory period'. The later [78] stimuli produce action potentials of increasing amplitude, but still smaller than the control; these are in the 'relative refractory period'. Corresponding [79] experimental curves are shown in the lower part of Fig. 20. The agreement is good, as regards both the duration of the absolute refractory period and the changes in shape of the spike as recovery progresses.

78. **'After the earliest stimulus . . . the membrane can be said to be in the "absolute refractory period."'** For a brief time after an action potential, no depolarizing stimulus, no matter how large, can trigger another action potential; the membrane potential simply falls toward rest when the stimulus is withdrawn. This period is the absolute refractory period. The upper curves in **Figure 20** illustrate that the model replicates the refractory periods shown in the lower, experimental curves, as a consequence of the kinetics of inactivation of the sodium conductance and deactivation of the potassium conductance. As indicated in **Figure 19**, during the absolute refractory period (curve *B*) the sodium conductance is more than 75% inactivated, and potassium conductance is still elevated. As a result, evoked sodium current can never exceed the potassium current. The net ionic current therefore can never become inward, and the conditions required to initiate action potential cannot be fulfilled.

79. **'The later stimuli produce action potentials of increasing amplitude, but still smaller than the control; these are in the "relative refractory period."'** For a while after the absolute refractory period, action potentials can be evoked, but only by stimuli that are larger than normally needed. This period, during which threshold is elevated, is the relative refractory period. The later, small action potentials (curves *C* and *D*) illustrate that the amplitude of the spike depends on the availability of the sodium conductance relative to the potassium conductance. They also show that the phrase 'all-or-none' as applied to action potentials only refers to the *occurrence* of the action potential: the membrane either does (all) or does not (none) reach threshold and generate a spike. The absolute *magnitude* of the resulting action potential, however, can vary.

Excitation

Our calculations of excitation processes were all made for the case where the membrane potential is uniform over the whole area considered, and not for the case of local stimulation of a whole nerve. There were two reasons for this: first, that such data from the squid giant fibre as we had for comparison were obtained by uniform stimulation of the membrane with the long electrode; and, secondly, that calculations for the whole nerve case would have been extremely laborious since the main equation is then a partial differential equation.

Threshold. The curves in Figs. 12 and 21 show that the theoretical 'membrane' has a definite threshold when stimulated by a sudden displacement of membrane potential. Since the initial fall after the stimulus is much less marked in these than in the experimental curves, it is relevant to compare the lowest point reached in a just threshold curve, rather than the magnitude of the original displacement. In the calculated series this is about 6 mV and in the experimental about 8 mV. This agreement is satisfactory, especially as the 80 value for the calculated series must depend critically on such things as the leak conductance, whose value was not very well determined experimentally.

The agreement might have been somewhat less good if the comparison had been made at a higher temperature. The calculated value would have been much the same, but the experimental value in the series at 23° C shown in Fig. 8 of Hodgkin *et al.* (1952) is about 15 mV. However, this fibre had been stored for 5 hr before use and was therefore not in exactly the same state as those on which our measurements were based.

Subthreshold responses. When the displacement of membrane potential was less than the threshold for setting up a spike, characteristic subthreshold responses were seen. One such response is shown in Fig. 12, while several 81 are plotted on a larger scale in Fig. 21*B*. Fig. 21*A* shows for comparison the corresponding calculated responses of our model. The only appreciable differences, in the size of the initial fall and in the threshold level, have been mentioned already in other connexions.

During the positive phase which follows each calculated subthreshold response, the potassium conductance is raised and there is a higher degree of 'inactivation' than in the resting state. The threshold must therefore be raised in the same way as it is during the relative refractory period following a spike. This agrees with the experimental findings of Pumphrey, Schmitt & Young (1940).

Anode break excitation. Our axons with the long electrode in place often gave anode break responses at the end of a period during which current was made to flow inward through the membrane. The corresponding response of 82 out theoretical model was calculated for the case in which a current sufficient

80. *'the theoretical "membrane" has a definite threshold. . . . In the calculated series this is about 6 mV and in the experimental about 8 mV.'* The next topics addressed are threshold and subthreshold responses. Hodgkin had investigated the elusive subthreshold responses since his early days working on nerve in the mid 1930s (see Historical Background). In the model, as in the experiments, threshold—the voltage beyond which an action potential is elicited and beneath which the membrane returns to rest—is 'definite' for a fixed set of initial conditions. Examples of depolarizations to near-threshold voltages at high resolution are shown in ***Figure 21***, and show the 'satisfactory' match between experimental and calculated records. Threshold is estimated as −54 mV *abs.* by the model and −52 mV *abs.* from the data.

81. ***'When the displacement of membrane potential was less than the threshold for setting up a spike, characteristic subthreshold responses were seen.'*** In ***Figure 21***, subthreshold responses are not only clearly demonstrated, but fully explained as the result of voltage-gated sodium currents that activate, but not sufficiently to overcome the leak and potassium currents that eventually bring the membrane potential back to rest. H&H note that such subthreshold responses will be more likely whenever the balance between depolarizing sodium and repolarizing potassium currents is tipped toward the latter, as in the relative refractory period (see note 79). Under these conditions, threshold will be raised.

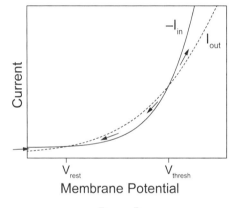

Figure B

Figure B schematizes the relationship between the magnitudes of inward and outward current as a function of voltage in the context of small, rapid displacements from rest. Inward current is mostly voltage-gated sodium current plus a small inward leak at potentials more depolarized than the leak reversal potential (−50 mv *abs.*); outward current is mostly voltage-gated potassium current plus a small leak at potentials more hyperpolarized than −50 mV *abs*. The curves in the figure are derived from H&H's equations for n_∞ and m_∞ as functions of voltage, and from $\bar{g}_K$, $\bar{g}_{Na}$, and g_L (***Table 3***). h is set to its resting value of 0.6, and maintains this value at all potentials, since it responds more slowly to sudden depolarization than do n and m.

The inward (solid line) and outward (dotted line) currents oppose each other, and each current becomes larger with depolarization, but with different voltage-dependences. Their relative magnitudes give rise to two crossing points—the resting potential (V_{rest}) and threshold (V_{thresh}). At these two voltages, therefore, the membrane is at equilibrium because the inward and outward currents sum to zero, but the two equilibria differ in character. At rest, any small displacement leads to net current flow in a direction that shifts the membrane potential back *toward* its resting value: hyperpolarization makes net inward current flow, and depolarization makes net outward current flow (arrows). The resting potential is therefore a stable equilibrium. In contrast, with any displacement from threshold, the net current flow moves the membrane potential *away* from threshold: toward rest for hyperpolarization, toward the peak of the action potential for depolarization. Hence threshold is an unstable equilibrium. The key point is that the *discontinuous* 'all-or-none' behavior at threshold arises from the crossing points between the two *continuous* current-voltage relationships governed by basic biophysical principles.

82. *'axons . . . often gave anode break responses at the end of a period during which current was made to flow inward through the membrane.'* The term 'anode break' comes from early work in which anodal current was applied extracellularly, hyperpolarizing the membrane. The offset ('break') of this current often led to an action potential (Hodgkin 1951). In H&H's experiments, hyperpolarization through the internal electrode, which evokes passive inward current, has the same effect. Anode break responses are also referred to as rebound excitation or disinhibitory action potentials.

to bring the membrane potential to 30 mV above the resting potential was suddenly stopped after passing for a time long compared with all the time-constants of the membrane. To do this, eqn. (26) was solved with $I = 0$ and the initial conditions that $V = +30$ mV, and m, n and h have their steady state values for $V = +30$ mV, when $t = 0$. The calculation was made for a temperature 83

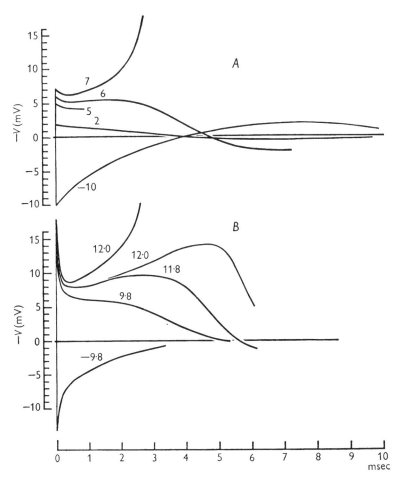

Fig. 21. A, numerical solutions of eqn. (26) for 6° C. The numbers attached to the curves give the initial depolarization in mV (also the quantity of charge applied in mμcoulomb/cm²). B, response of nerve membrane at 6° C to short shocks; the numbers show the charge applied in mμcoulomb/cm². The curves have been replotted from records taken at low amplification and a relatively high time-base speed.

of 6·3° C. A spike resulted, and the time course of membrane potential is plotted in Fig. 22A. A tracing of an experimental anode break response is shown in Fig. 22B; the temperature is 18·5° C, no record near 6° being available. It will be seen that there is good general agreement. (The oscillations after the positive phase in Fig. 22B are exceptionally large; the response of

83. *'eqn. (26) was solved with I = 0 and the initial conditions that V = +30 mV, and m, n and h have their steady state values for V = +30 mV, when t = 0.'* For the calculation, no current was applied ($I = 0$), and all the gates were set at open probabilities n_∞, m_∞, and h_∞ for 30 mV more hyperpolarized than rest (−90 mV *abs.*). This condition would set the sodium and potassium conductances near zero, but would allow nearly complete recovery from inactivation of the sodium conductance, as h would increase from about 0.6 at rest to near 1 (see ***Figure 10***).

this axon to a small constant current was also unusually oscillatory as shown in Fig. 23.)

The basis of the anode break excitation is that anodal polarization decreases the potassium conductance and removes inactivation. These effects persist for an appreciable time so that the membrane potential reaches its resting value with a reduced outward potassium current and an increased inward sodium current. The total ionic current is therefore inward at $V = 0$ and the membrane undergoes a depolarization which rapidly becomes regenerative.

84

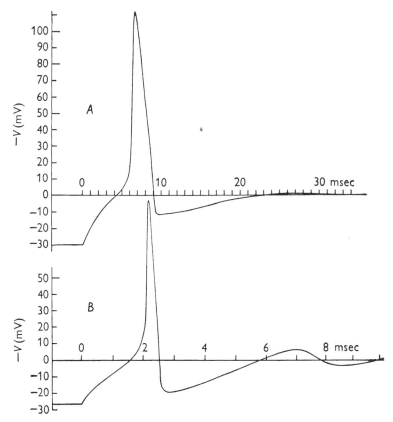

Fig. 22. Theoretical basis of anode break excitation. A, numerical solution of eqn. (26) for boundary condition $-V = -30$ mV for $t < 0$; temperature 6° C. B, anode break excitation following sudden cessation of external current which had raised the membrane potential by 26·5 mV; giant axon with long electrode at 18·5° C. Time scales differ by a factor appropriate to the temperature difference.

Accommodation. No measurements of accommodation were made nor did we make any corresponding calculations for our model. It is clear, however, that the model will show 'accommodation' in appropriate cases. This may be shown in two ways. First, during the passage of a constant cathodal current through the membrane, the potassium conductance and the degree of inactivation will rise, both factors raising the threshold. Secondly, the steady state

85

84. *'The basis of the anode break excitation is that anodal polarization decreases the potassium conductance and removes inactivation.'* As a result of the large increase in the availability of the sodium conductance from holding near −90 mV *abs.*, when the hyperpolarizing step is turned off and the membrane potential returns toward rest, the depolarization activates the sodium conductance. Normally, the amount of sodium current at rest does not outweigh the resting potassium conductance, but in this case, with *h* close to 1, it has nearly doubled. Threshold is therefore reduced to a value below the usual resting potential and an action potential is generated. The comparison between the calculated and experimental responses is shown in **Figure 22**.

The basis of anode break, as well as the refractory periods, is schematized in *Figure C*. As in *Figure B*, the curves approximate the magnitudes of summed inward current (solid line) and summed outward current (dotted line) that would be elicited by steps to any voltage, given a fixed degree of inactivation of the sodium conductance. With the resting availability of sodium current, the resting potential and threshold are represented by the points at which curve 1 crosses the outward current (see note 81). The three other solid curves approximate the shifts in the inward current that result from changes in baseline sodium current inactivation. With prolonged hyperpolarization, sodium current recovers from inactivation, increasing the total inward current flowing upon 'anode break' or relief of hyperpolarization (curve 2). As noted by H&H, hyperpolarization also deactivates potassium current and outward current is therefore reduced (not shown). As a result, inward current exceeds outward current at any potential. A stable equilibrium point no longer exists, and an action potential occurs. In contrast, during the absolute refractory period immediately after an action potential, inactivation is increased, and inward current is reduced such that it does not exceed outward current even with applied depolarization (curve 3). Later, during the relative refractory period, when sodium current has partially recovered from inactivation, a crossing point is restored and an action potential can be produced (curve 4), but threshold is elevated relative to curve 1.

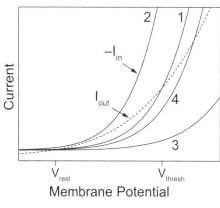

Figure C

85. *'the model will show "accommodation" in appropriate cases.'* Accommodation is the gradual decrease in action potential firing frequency during a steady depolarizing current injection. H&H anticipate that the ionic currents active under these conditions would likely resemble those measured during the relative refractory period. In both cases, a depolarization toward threshold occurs before full recovery from inactivation of the sodium conductance or complete deactivation of the potassium conductance. As a consequence, threshold is raised, and the action potential, if it occurs, will be of smaller magnitude. Successive action potentials will have progressively higher thresholds and longer interspike intervals, which together form the signature of accommodation. Subsequent work in different cell types has demonstrated that, in addition to the rapid phase of inactivation of sodium current described by H&H, slower phases of inactivation as well as gradual activation of voltage- and calcium-dependent potassium currents also contribute to accommodation.

ionic current at all strengths of depolarization is outward (Fig. 11), so that an applied cathodal current which rises sufficiently slowly will never evoke a regenerative response from the membrane, and excitation will not occur.   86

Oscillations

In all the calculated action potentials and subthreshold responses the membrane potential finally returns to its resting value with a heavily damped oscillation. This is well seen after subthreshold stimuli in Figs. 21 A and 24, but the action potentials are not plotted on a slow enough time base or with a large enough vertical scale to show the oscillations which follow the positive phase.

The corresponding oscillatory behaviour of the real nerve could be seen after   87 a spike or a subthreshold short shock, but was best studied by passing a small constant current through the membrane and recording the changes of membrane potential that resulted. The current was supplied by the long internal electrode so that the whole area of membrane was subjected to a uniform current density. It was found that when the current was very weak the potential changes resulting from inward current (anodal) were almost exactly similar to those resulting from an equal outward current, but with opposite sign. This is shown in Fig. 23 B and C, where the potential changes are about ± 1 mV. This symmetry with weak currents is to be expected from our equations, since they can be reduced to a linear form when the displacements of all the variables from their resting values are small. Thus, neglecting products, squares and higher powers of δV, δm, δn and δh, the deviations of V, m, n and h from their resting values (0, m_0, n_0 and h_0 respectively), eqn. (26)   88 (p. 518) becomes

$$\delta I = C_M \frac{\mathrm{d}\delta V}{\mathrm{d}t} + \bar{g}_K n_0^4 \delta V - 4\bar{g}_K n_0^3 V_K \delta n + \bar{g}_{Na} m_0^3 h_0 \delta V$$
$$- 3\bar{g}_{Na} m_0^2 h_0 V_{Na} \delta m - \bar{g}_{Na} m_0^3 V_{Na} \delta h + \bar{g}_l \delta V. \qquad (35)$$

Similarly, eqn. (7) (p. 518) becomes

$$\frac{\mathrm{d}\delta n}{\mathrm{d}t} = \frac{\partial \alpha_n}{\partial V} \delta V - (\alpha_n + \beta_n)\delta n - n_0 \frac{\partial (\alpha_n + \beta_n)}{\partial V} \delta V,$$

or

$$(p + \alpha_n + \beta_n)\delta n = \left\{ \frac{\partial \alpha_n}{\partial V} - n_0 \frac{\partial (\alpha_n + \beta_n)}{\partial V} \right\} \delta V, \qquad (36)$$

where p represents $\mathrm{d}/\mathrm{d}t$, the operation of differentiating with respect to time.

The quantity δn can be eliminated between eqns. (35) and (36). This process is repeated for δm and δh, yielding a fourth-order linear differential equation with constant coefficients for δV. This can be solved by standard methods for any particular time course of the applied current density δI.

86. *'an applied cathodal current which rises sufficiently slowly will never evoke a regenerative response from the membrane, and excitation will not occur.'* With a slow enough depolarization ('applied cathodal current'), the activation of the potassium conductance, even though it is relatively slow, will take place as the sodium conductance inactivates. Under these conditions, the inward current never exceeds the outward current, and the axon will never generate an action potential. Slow ramp-like depolarizations can therefore have a negative effect on excitability, emphasizing the point that the rate, as well as the magnitude, of depolarization dictates the probability of action potential generation.

87. *'In all the calculated action potentials . . . the membrane potential finally returns to its resting value with a heavily damped oscillation. . . . The corresponding oscillatory behavior of the real nerve could be seen'* After the action potential, oscillations of the membrane potential are present in both the calculated and experimental records. During this period, h is high enough that another action potential will not be produced. The afterhyperpolarization allows the sodium conductance to recover slightly, however, so that the return toward the resting potential activates some sodium current (by increasing m) which makes the voltage overshoot rest. This small, slow depolarization activates the potassium conductance (by increasing n), which repolarizes the membrane below rest, but to a lesser extent than the initial afterhyperpolarization. The cycle continues, but with a lower amplitude during each successive cycle.

88. *'the deviations of V, m, n and h from their resting values'* The goal here is to estimate the small change in membrane current resulting from very small changes in voltage around rest, which in turn drives opposing (restorative) changes in voltage, leading to oscillations. Because the changes in membrane potential and conductance are small, H&H can approximate the full differential equation, **Equation 26**, with the difference equation, **Equation 35**. This approximation, obtained by linearization around an operating point, estimates discrete small changes, δ, which represent the changes in V, m, n, and h that result from small perturbations in the membrane potential. With the expressions in **Equation 36**, δm, δn, δh can be made to drop out of the approximation. After omitting negligibly small terms, what remains is a simplified, fourth-order linear equation, which can be solved explicitly to give the change in voltage with time. The oscillations that follow an action potential can then be replicated both experimentally and computationally with small applied depolarizations or hyperpolarizations of the membrane, as shown in **Figure 23**.

Fig. 23 *A* shows the response of the membrane to a constant current pulse calculated in this way. The constants in the equations are chosen to be appropriate to a temperature of 18·5° C so as to make the result comparable with the tracings of experimental records shown in *B* and *C*. It will be seen that the calculated curve agrees well with the records in *B*, while those in *C*, obtained from another axon, are much less heavily damped and show a higher

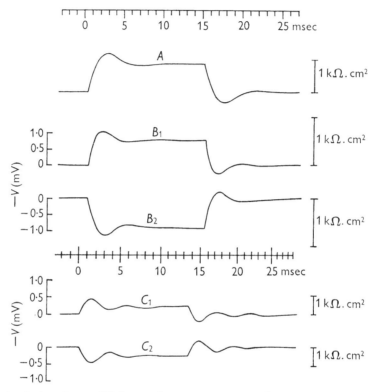

Fig. 23. *A*, solution of eqn. (35) for small constant current pulse; temperature 18·5° C; linear approximation. The curve shows $\delta V / \delta I$ (plotted upwards) as a function of time. *B*, changes in membrane potential associated with application of weak constant currents of duration 15 msec and strength $\pm 1\cdot49\,\mu\text{A/cm}^2$. B_1, cathodic current; B_2, anodic current. Depolarization is shown upward. Temperature 19° C. *C*, similar records from another fibre enlarged to have same time scale. Current strengths are $\pm 0\cdot55\,\mu\text{A/cm}^2$. Temperature 18° C. The response is unusually oscillatory.

frequency of oscillation. A fair degree of variability is to be expected in these respects since both frequency and damping depend on the values of the components of the resting conductance. Of these, g_{Na} and g_{K} depend critically on the resting potential, while $\bar{g}_l$ is very variable from one fibre to another.

Both theory and experiment indicate a greater degree of oscillatory behaviour than is usually seen in a cephalopod nerve in a medium of normal ionic com- 89 position. We believe that this is largely a direct result of using the long internal

89. *'Both theory and experiment indicated a greater degree of oscillatory behavior than is usually seen'* H&H note that the long internal electrode kept the stretch of axonal membrane all at one voltage (isopotential), which exaggerated the oscillations. Under normal conditions, the voltage across the length of membrane would vary, so the current could flow along the axon rather than changing the local voltage, which would dampen the oscillations. Hence, the observed oscillations can be accounted for quantitatively by the biophysical properties of the membrane and its conductances, but here they are magnified by the space clamp of the experimental preparation and may not be so large under physiological conditions (see note 91).

electrode. If current is applied to a whole nerve through a point electrode, neighbouring points on the membrane will have different membrane potentials and the resulting currents in the axis cylinder will increase the damping.

The linear solution for the behaviour of the theoretical membrane at small displacements provided a convenient check on our step-by-step numerical procedure. The response of the membrane at 6·3° C to a small short shock was 90 calculated by this means and compared with the step-by-step solution for an initial depolarization of the membrane by 2 mV. The results are plotted in Fig. 24. The agreement is very close, the step-by-step solution deviating in the direction that would be expected to result from its finite amplitude (cf. Fig. 21).

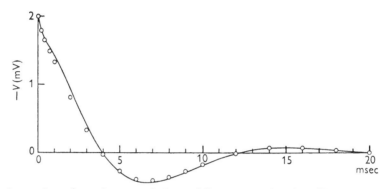

Fig. 24. Comparison of step-by-step solution and linear approximation. Eqn. (26), temperature 6° C; initial displacement of $-V = 2$ mV. Continuous line: step-by-step solution. Circles: linear approximation with same initial displacement.

As pointed out by Cole (1941), the process underlying oscillations in membrane potential must be closely connected with the inductive reactance observed with alternating currents. In our theoretical model the inductance is due partly to the inactivation process and partly to the change in potassium conductance, the latter being somewhat more important. For small displace- 91 ments of the resting potential the variations in potassium current in 1 cm² of membrane are identical with those in a circuit containing a resistance of 820 Ω in series with an inductance which is shunted by a resistance of 1900 Ω. The value of the inductance is 0·39 H at 25° C, which is of the same order as the 0·2 H found by Cole & Baker (1941). The calculated inductance increases 3-fold for a 10° C fall in temperature and decreases rapidly as the membrane potential is increased; it disappears at the potassium potential and is replaced by a capacity for $E > E_{\mathrm{K}}$.

DISCUSSION

The results presented here show that the equations derived in Part II of this paper predict with fair accuracy many of the electrical properties of the squid giant axon: the form, duration and amplitude of spike, both 'membrane' 92

90. *'The linear solution for the behaviour of the theoretical membrane at small displacements pro-*
vided a convenient check on our step-by-step numerical procedure.' H&H had two methods of
solving **Equation 26**: (1) the linear approximation of **Equations 35** and **36**, which works well for
small displacements of membrane voltage, and (2) the complete system of equations that they used
for the step-by-step calculation of the action potentials and other voltage responses throughout the
paper. In **Figure 24**, they illustrate that both methods give nearly identical results, lending further
credibility to their numerical approach.

91. *'the inductance is due partly to the inactivation process and partly to the change in potassium*
conductance' An inductor is a circuit element containing a coil that generates a voltage when current flowing through it is changed. An inductor and a capacitor in parallel form an LC circuit, which
acts as a resonator and therefore can give rise to oscillations (see Appendix 2.6). When Cole and
colleagues observed oscillations, they hypothesized that some aspect of the membrane acted as an
inductor (Curtis & Cole 1940; Cole & Baker 1941; Cole & Curtis 1942). H&H indicate that their
model—which includes only a capacitor (the membrane), voltage-dependent resistors (the ionic
conductances), and batteries (the driving forces)—generates oscillations. The interplay between the
onset/recovery from inactivation of the sodium conductance and the activation/deactivation of the
potassium conductance *mimics* an inductance. Rather than having the fixed *phase* lag between voltage and current characteristic of an inductor, however, the conductances produce a *time* lag owing
to the rate constants associated with gate opening and closure. Thus, the oscillations arise via an
entirely different mechanism than in a true inductor.

92. *'the equations derived in Part II of this paper predict with fair accuracy many of the electrical*
properties of the squid giant axon' This may be the greatest understatement of the entire series of
five papers, especially given the subsequent list of properties that are predicted.

and propagated; the conduction velocity; the impedance changes during the spike; the refractory period; ionic exchanges; subthreshold responses; and oscillations. In addition, they account at least qualitatively for many of the phenomena of excitation, including anode break excitation and accommodation. This is a satisfactory degree of agreement, since the equations and constants were derived entirely from 'voltage clamp' records, without any adjustment to make them fit the phenomena to which they were subsequently applied. Indeed any such adjustment would be extremely difficult, because in most cases it is impossible to tell in advance what effect a given change in one of the equations will have on the final solution.

The agreement must not be taken as evidence that our equations are anything more than an empirical description of the time-course of the changes in permeability to sodium and potassium. An equally satisfactory description of the voltage clamp data could no doubt have been achieved with equations of very different form, which would probably have been equally successful in 93 predicting the electrical behaviour of the membrane. It was pointed out in Part II of this paper that certain features of our equations were capable of a physical interpretation, but the success of the equations is no evidence in favour of the mechanism of permeability change that we tentatively had in mind when formulating them.

The point that we do consider to be established is that fairly simple permeability changes in response to alterations in membrane potential, of the kind deduced from the voltage clamp results, are a sufficient explanation of the 94 wide range of phenomena that have been fitted by solutions of the equations.

Range of applicability of the equations

The range of phenomena to which our equations are relevant is limited in two respects: in the first place, they cover only the short-term responses of the 95 membrane, and in the second, they apply in their present form only to the isolated squid giant axon.

Slow changes. A nerve fibre whose membrane was described by our equations would run down gradually, since even in the resting state potassium leaves and sodium enters the axis cylinder, and both processes are accelerated by activity. This is no defect in describing the isolated squid giant axon, which does in fact run down in this way, but some additional process must take place in a nerve in the living animal to maintain the ionic gradients which are the immediate source of the energy used in impulse conduction.

After-potentials. Our equations give no account of after-potentials, apart from the positive phase and subsequent oscillations.

Conditions of isolated giant axon. There are many reasons for supposing that the resting potential of the squid giant axon is considerably lower after isolation than when it is intact in the living animal. Further evidence for this view

93. *'The agreement must not be taken as evidence that our equations are anything more than an empirical description. . . . An equally satisfactory description . . . could no doubt have been achieved with equations of a very different form'* H&H again emphasize that the equations are empirical, and hence to some extent arbitrary, because the variables lack referents for which molecular evidence is available. In later memoirs (C&D p. 303; Huxley 2000), both Hodgkin and Huxley stated that they fully expected the equations to be superseded relatively rapidly with more biologically grounded mathematics.

94. *'The point that we do consider to be established is that fairly simple permeability changes in response to alterations in membrane potential . . . are a sufficient explanation'* The one conclusion that H&H emphasize is that permeability, or conductance, is a function of membrane potential; thus, *voltage* is the independent variable. Hence, the terms 'voltage-gated conductances' and, later, 'voltage-gated channels' enter the literature and the collective neurophysiological consciousness.

95. *'limited in two respects'* After including discussions of error throughout the papers, H&H end with a discussion of limitations. First, they draw attention to the relatively short time scales on which all the measurements were made and raise the possibility that additional variables might emerge with longer measurements; indeed, longer depolarizations were later found to produce longer-lasting 'slow inactivation' of sodium and potassium conductances. Second, they do not assume that their discoveries would apply to nerve cells and neurons beyond the squid giant axon, much less universally. They also acknowledge that the physical state of the preparation might have skewed some values.

is provided by the observation (Hodgkin & Huxley, 1952c) that the maximum inward current that the membrane can pass on depolarization is increased by previously raising the resting potential by 10–20 mV by means of anodally directed current. Our equations could easily be modified to increase the resting potential (e.g. by reducing the leak conductance and adding a small outward current representing metabolic extrusion of sodium ions). We have not made any calculations for such a case, but certain qualitative results are evident from inspection of other solutions. If, for instance, the resting potential were raised (by 12 mV) to the potassium potential, the positive phase and subsequent oscillations after the spike would disappear, the rate of rise of the spike would be increased, the exchange of internal and external sodium in a spike would be increased, the membrane would not be oscillatory unless depolarized, and accommodation and the tendency to give anode break responses would be greatly reduced. Several of these phenomena have been observed when the resting potential of frog nerve is raised (Lorente de Nó, 1947), but no corresponding information exists about the squid giant axon.

Applicability to other tissues. The similarity of the effects of changing the concentrations of sodium and potassium on the resting and action potentials of many excitable tissues (Hodgkin, 1951) suggests that the basic mechanism of conduction may be the same as implied by our equations, but the great differences in the shape of action potentials show that even if equations of the same form as ours are applicable in other cases, some at least of the parameters must have very different values.   96

Differences between calculated and observed behaviour

In the Results section, a number of points were noted on which the calculated behaviour of our model did not agree with the experimental results. We shall now discuss the extent to which these discrepancies can be attributed to known shortcomings in our equations. Two such shortcomings were pointed out in Part II of this paper, and were accepted for the sake of keeping the equations simple. One was that the membrane capacity was assumed to behave as a 'perfect' condenser (phase angle 90°; p. 505), and the other was that the equations governing the potassium conductance do not give as much delay in the conductance rise on depolarization (e.g. to the sodium potential) as was observed in voltage clamps (p. 509).   97

The assumption of a perfect capacity probably accounts for the fact that the initial fall in potential after application of a short shock is much less marked in the calculated than in the experimental curves (Figs. 12 and 21). Some of the initial drop in the experimental curves may also be due to end-effects, the guard system being designed for the voltage clamp procedure but not for stimulation by short shocks.

96. *'the great differences in the shape of action potentials show that even if equations of the same form as ours are applicable . . . some at least of the parameters must have very different values.'* H&H anticipate that even if different neurons indeed have voltage-gated sodium and potassium conductances, the rates that govern those conductances must differ across cells, generating sodium and potassium conductances with distinct voltage-dependence and kinetics, which would yield different action potential waveforms. In this way, they anticipate diversity of ionic conductances, within and across species, which would become a focus of study in subsequent decades.

97. *'membrane capacity was assumed to behave as a perfect condenser . . . equations . . . do not give as much delay in the conductance rise on depolarization . . . as was observed.'* H&H return to the two known points in which their model deviated from the experimental measurements, but which they tolerated for computational ease. Regarding the question of capacitance, they state that the apparatus might have exaggerated the initial depolarization, so the error might have been at the level of the measurement rather than the assumption about the membrane.

The inadequacy of the delay in the rise of potassium conductance has several effects. In the first place the falling phase of the spike develops too early, 98 reducing the spike amplitude slightly and making the peak too pointed in shape (p. 525). In the membrane action potentials these effects become more marked the smaller the stimulus, since the potassium conductance begins to rise during the latent period. This causes the spike amplitude to decrease more in the calculated than in the experimental curves (Fig. 12).

The low calculated value for the exchange of internal and external sodium ions is probably due to this cause. Most of the sodium exchange occurs near the peak of the spike, when the potential is close to the sodium potential. The early rise of potassium conductance prevents the potential from getting as close to the sodium potential, and from staying there for as long a time, as it should.

A check on these points is provided by the 'anode break' action potential. 99 Until the break of the applied current, the quantity n has the steady state value appropriate to $V = +30\ \mathrm{mV}$, i.e. it is much smaller than in the usual resting condition. This greatly increases the delay in the rise of potassium conductance when the membrane is depolarized. It was found that the spike height was greater (Table 4), the peak was more rounded, and the exchange of internal and external sodium was greater (Table 5), than in an action potential which followed a cathodal short shock.

The other important respect in which the model results disagreed with the experimental was that the calculated exchange of internal and external potassium ions per impulse was too large. This exchange took place largely during the positive phase, when the potential is close to the potassium potential and the potassium conductance is still fairly high. We have no satisfactory explanation for this discrepancy, but it is probably connected with the fact that the value of the potassium potential was less strongly affected by changes in external potassium concentration than is required by the Nernst equation. 100

SUMMARY   101

1. The voltage clamp data obtained previously are used to find equations which describe the changes in sodium and potassium conductance associated with an alteration of membrane potential. The parameters in these equations were determined by fitting solutions to the experimental curves relating sodium or potassium conductance to time at various membrane potentials.

2. The equations, given on pp. 518–19, were used to predict the quantitative behaviour of a model nerve under a variety of conditions which corresponded to those in actual experiments. Good agreement was obtained in the following cases:

(*a*) The form, amplitude and threshold of an action potential under zero membrane current at two temperatures.

(*b*) The form, amplitude and velocity of a propagated action potential.

98. *'The inadequacy of the delay in the rise of potassium conductance has several effects.'* The errors that H&H tally all indicate that the calculation generates slightly too much potassium current at short latencies. The discrepancies between the experimental and the calculated action potentials can largely be attributed to this excessive activation of potassium current, which also accounts for the overestimation of potassium efflux relative to the radioactive tracer measurements published by Keynes the previous year. The curtailing of the action potential might also account for the smaller inward sodium flux. Since the net ionic movements were in closest agreement for the squid axon, however, it is also possible that species differences might partly explain the discrepancies; indeed, the intracellularly recorded *Sepia* action potential reproduced in Hodgkin's 1951 review article (recorded by Weidmann 1951) appears to have about a 1-msec halfwidth at ~15°C. The longer action potential relative to the squid raises the possibility that the potassium flux might indeed be somewhat lower in *Sepia* (see note 77).

99. *'A check on these points is provided by the "anode break" action potential.'* Hyperpolarizing the membrane, which brings the value of n much closer to zero than its resting value, delayed the activation of potassium conductance upon depolarization, and also broadened, rounded, and magnified the action potential as H&H had anticipated. Regarding why $n^4 \bar{g}_K$ was not quite successful in mimicking the slow onset of the potassium conductance, many possibilities exist. H&H consider the possibility that n^5 or n^6 might have worked better, but given present knowledge that a potassium channel has four voltage sensors, such a modification might be empirically handy but not biologically accurate. Among the most likely reasons for the discrepancy is that each potassium channel subunit can assume more than two states (see note 29).

100. *'the value of the potassium potential was less strongly affected by changes in external potassium concentration than is required by the Nernst equation.'* H&H reported that changing extracellular potassium led to a smaller shift in E_K than predicted by the Nernst equation (see Paper 3, note 50). They propose that, during those experiments, potassium might have accumulated at the membrane, shifting E_K in the depolarized direction, decreasing driving force, and reducing total potassium flux. This effect, if it occurred, might be most pronounced during the afterhyperpolarization. During the calculated action potential, however, E_K was held constant. Consequently, the calculated potassium current might have a larger driving force than the experimental current, which would account for why the calculated flux exceeded the experimental flux. In later work, Frankenhaeuser and Hodgkin (1956) repeatedly stimulated action potentials and quantified the changes in afterhyperpolarization, which they attributed to the accumulation of extracellular potassium in the restricted space between the axonal membrane and the Schwann cells. This region came to be known as the Frankenhaeuser-Hodgkin space.

101. *'Summary'* At the outset of Paper 5, H&H stated that in the absence of molecular knowledge of the conductance mechanism, any quantitative description might be of limited value. Nevertheless, their extensive experiments and theoretical framework placed constraints on the model and allowed them to develop equations that successfully replicated a wide range of the electrical behaviors of the membrane. Several aspects of the model were molecularly prescient, such as the anticipation of gating current, the relation of the n^4 term to the four voltage sensing domains of the potassium channel tetramer, and the relation of the $m^3 h$ term to the three voltage-sensing domains that gate sodium channel opening, with movement of the fourth domain being permissive for inactivation. Even the errors that they identified presaged later discoveries, such as of the restricted diffusion of extracellular ions in the Schwann cell clefts. Many exceptions and extension have indeed been found to what came to be known as 'Hodgkin-Huxley kinetics,' such as the coupling of activation and inactivation, and the existence of intermediate closed and open states. Despite these revisions, the H&H model remains widely used and still serves as an instructive approximation of the activity of excitable cells for many applications in physiology.

(*c*) The form and amplitude of the impedance changes associated with an action potential.

(*d*) The total inward movement of sodium ions and the total outward movement of potassium ions associated with an impulse.

(*e*) The threshold and response during the refractory period.

(*f*) The existence and form of subthreshold responses.

(*g*) The existence and form of an anode break response.

(*h*) The properties of the subthreshold oscillations seen in cephalopod axons.

3. The theory also predicts that a direct current will not excite if it rises sufficiently slowly.

4. Of the minor defects the only one for which there is no fairly simple explanation is that the calculated exchange of potassium ions is higher than that found in *Sepia* axons.

5. It is concluded that the responses of an isolated giant axon of *Loligo* to 102 electrical stimuli are due to reversible alterations in sodium and potassium permeability arising from changes in membrane potential.

REFERENCES

COLE, K. S. (1941). Rectification and inductance in the squid giant axon. *J. gen. Physiol.* **25**, 29–51.

COLE, K. S. & BAKER, R. F. (1941). Longitudinal impedance of the squid giant axon. *J. gen. Physiol.* **24**, 771–788.

COLE, K. S. & CURTIS, H. J. (1939). Electric impedance of the squid giant axon during activity. *J. gen. Physiol.* **22**, 649–670.

GOLDMAN, D. E. (1943). Potential, impedance, and rectification in membranes. *J. gen. Physiol.* **27**, 37–60.

HARTREE, D. R. (1932–3). A practical method for the numerical solution of differential equations. *Mem. Manchr lit. phil. Soc.* **77**, 91–107.

HODGKIN, A. L. (1951). The ionic basis of electrical activity in nerve and muscle. *Biol. Rev.* **26**, 339–409.

HODGKIN, A. L. & HUXLEY, A. F. (1952*a*). Currents carried by sodium and potassium ions through the membrane of the giant axon of *Loligo*. *J. Physiol.* **116**, 449–472.

HODGKIN, A. L. & HUXLEY, A. F. (1952*b*). The components of membrane conductance in the giant axon of *Loligo*. *J. Physiol.* **116**, 473–496.

HODGKIN, A. L. & HUXLEY, A. F. (1952*c*). The dual effect of membrane potential on sodium conductance in the giant axon of *Loligo*. *J. Physiol.* **116**, 497–506.

HODGKIN, A. L., HUXLEY, A. F. & KATZ, B. (1949). Ionic currents underlying activity in the giant axon of the squid. *Arch. Sci. physiol.* **3**, 129–150.

HODGKIN, A. L., HUXLEY, A. F. & KATZ, B. (1952). Measurement of current-voltage relations in the membrane of the giant axon of *Loligo*. *J. Physiol.* **116**, 424–448.

HODGKIN, A. L. & KATZ, B. (1949). The effect of temperature on the electrical activity of the giant axon of the squid. *J. Physiol.* **109**, 240–249.

KEYNES, R. D. (1951). The ionic movements during nervous activity. *J. Physiol.* **114**, 119–150.

KEYNES, R. D. & LEWIS, P. R. (1951). The sodium and potassium content of cephalopod nerve fibres. *J. Physiol.* **114**, 151–182.

LORENTE DE NÓ, R. (1947). A study of nerve physiology. *Stud. Rockefeller Inst. med. Res.* **131**, **132**.

PUMPHREY, R. J., SCHMITT, O. H. & YOUNG, J. Z. (1940). Correlation of local excitability with local physiological response in the giant axon of the squid (*Loligo*). *J. Physiol.* **98**, 47–72.

102. **'It is concluded'** H&H end their *tour de force* with a mild but precise statement that electrical excitability in the squid giant axon results from voltage-gated conductances. The discoveries that underlie this simple conclusion, however, completely transformed the understanding of cellular excitability in particular and bioelectricity in general. As John W. Moore—a postdoc with Kenneth Cole in 1952—once quipped, it took the rest of the field about a decade to catch up.

Epilogue

But the field did catch up, and the following decades saw tremendous advances in physiology that built directly on the discoveries of H&H. Ultimately, what began as a basic scientific inquiry in a fragile invertebrate with a fortuitously oversized axon would provide the basis for the development of a vast array of biomedical research fields. The studies that the H&H papers made possible would not only yield immeasurable insights into brain and muscle function but also identify, explain, and alleviate medical conditions as diverse as epilepsies, ataxias, myotonias, arrhythmias, and pain.

In the late 1950s, calcium was identified as another ion that entered excitable cells via voltage-gated conductances, setting the stage in the next decade for the recognition of calcium ions as the ultimate transducers of electricity into physiological action, including muscle contraction and vesicular fusion.

The 1960s saw the growth of biophysical pharmacology, as specific conductances were found to be blocked by natural toxins and synthetic compounds. Sodium current was abolished by tetrodotoxin; potassium current was blocked and modulated by tetraethylammonium ions and other quaternary ammonium ions; and voltage-gated calcium current was reduced by divalent cations. Natural toxins and synthetic blockers helped reveal the complex links of voltage-gating to biochemical processes and also offered insights into the gating mechanisms of conductances and the permeation properties of pores that earned the name of 'ion channels.'

The 1970s yielded the first direct measurements of gating currents, confirming H&H's deduction that voltage-dependent conductances became activated as a consequence of charged particles moving in response to the transmembrane voltage. The decade also held the discovery that ion channels were proteins that could be isolated, purified, and analyzed. Currents through single ion channels were recorded for the first time, and the unitary openings and closures of ion channels permitted insights into ion translocation rates as well as protein conformational changes on the time scale of microseconds.

The 1980s brought the molecular revolution to electrophysiology as ion channel proteins became identified by their genes, starting with the Shaker potassium channel. A wealth of structure-function studies in expression systems vastly increased knowledge of how specific properties of ion channel proteins resulted from particular amino acid sequences. The refinement of the patch-clamp technique allowed high-quality electrophysiological recordings of voltage- and ligand-gated channels in small neurons and at tiny synapses of vertebrate species. The diversity of ion channels, as well as their susceptibility to modulation by kinases and phosphatases, began to be appreciated.

The 1990s continued the explosion of discoveries in synaptic physiology and neural plasticity, and initiated the detailed exploration of ion channels in dendrites and axons of mammalian neurons. The awareness grew that ion channel proteins did not work in isolation but formed complexes with auxiliary subunits and other proteins that linked them to other cell biological processes. Imaging methods, such as Förster (fluorescence) resonance energy transfer, allowed ion channel gating to be observed directly, and calcium imaging turned the action potential itself into a visual signal. The decade culminated with the crystallization of the first potassium channel protein, initiating an enormous cascade of structural studies visualizing the atomic structure of ion channels, which continue to the present day.

In the twenty-first century, the reductionist discoveries of biophysical electrophysiology served to make systems-level questions far more tractable. The reach and rigor of integrative physiology expanded as optogenetics harnessed a knowledge of ion channel gating to alter the activity of excitable cells at the will of the investigator. Thus, a half-century after Hodgkin and Huxley's new technologies, elegant experimentation, painstaking analyses, and brilliant insights had revealed the electrical secrets of ionic conductances, the conductances themselves were transformed into novel tools, which are now helping to unlock the remaining mysteries of physiology.[2]

---

[2] Further reading: *Hille, B. (2001) Ion channels of Excitable Membranes. Sinauer. Ashcroft, F. M. (1999) Ion Channels and Disease: Channelopathies. Academic Press.*

Appendices

The 1952 Hodgkin-Huxley papers draw on a wide range of interdisciplinary scientific knowledge. Most people who are motivated to read the original articles have had at least some training in calculus, physics, chemistry, and biology, but not everyone has maintained a ready facility with all these subjects after their initial period of study. Additionally, not everyone has been formally exposed to the relevant intersections between these disciplines or the specialties within them, such as biophysics, physical chemistry, and electronics, each with the associated mathematics.

The following five appendices have been written to provide a refresher and/or formalization of ideas relevant to the H&H papers for those who may need them, starting with the fundamentals and continuing directly toward specific electrophysiological concepts. The first three provide a review and explanation of general principles of mathematics (Appendix 1), electricity and simple circuits (Appendix 2), and electronic circuits pertinent to electrophysiological applications (Appendix 3). The final two present information specifically relevant to the Hodgkin-Huxley equations, covering the derivation of rate constants (Appendix 4) and numerical methods (Appendix 5). Although the material in the last two appendices is included in the papers and annotations, it is synthesized here into a more systematic and cohesive presentation.

Appendix 1

MATHEMATICAL PRINCIPLES

1.1 Independent and dependent variables. A complicating aspect of studying the action potential is the mutual dependence of the underlying variables, voltage and current: when current flows across the membrane, the transmembrane voltage changes, which changes the current flow, which changes the voltage, etc. Although this interrelated nature of current and voltage is precisely what generates the action potential, it is problematic from the experimentalist's perspective: most experimentation requires an independent variable that can be controlled, so that the investigator can measure the resulting changes of dependent variables. The extra challenge in studies of the action potential was that initially it was not even clear what the independent variable actually was—did current control voltage or did voltage control current? Or did some currents control other currents? Any of these possibilities appeared reasonable at the outset. One of the key achievements of the voltage-clamp technique was to control the membrane potential, making voltage into the independent variable in the experiment. H&H could then measure the consequent transmembrane ionic flux. The key word in the previous sentence is that flux indeed turned out to be 'consequent'—current flowed as a result of the voltage step: it *depended* directly on voltage, and voltage was ultimately identified as the biophysical independent variable as well as the experimental one.

The relationship between independent (controlled) variables and dependent (measured) variables can be illustrated graphically, with the independent variable plotted on the x-axis and the dependent variable plotted on the y-axis. It is customary to refer to the dependent variable first; that is, to speak of 'y versus x.' For this reason, plots of current versus voltage are called *I-V* curves. The name indicates that the first variable, current (I), is expressed as a function of the second variable, voltage (V).

1.2 Linearity. The simplest function is the linear equation $y = x$, which describes a straight line that passes through the origin. One can 'read' this function by translating it into the statement that y takes on whatever value x takes. A straight line can be modified on the graph in two ways, and still remain a straight line. The first way is that its steepness or 'slope' can be altered. If every y value has double the value of every x, the slope of the line is 2, and the equation is $y = 2x$, whereas if every y has half the value of every x, the slope is ½, and the equation is $y = x/2$. Slope can be expressed as the constant factor a, so the general equation becomes $y = ax$, which defines a line of slope a that passes through the origin (*Figure A*).

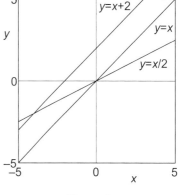

Figure A

This most basic of algebraic relations is of extraordinary importance for electrophysiology, owing to Ohm's law, $V = IR$. As stated above, in the context of the voltage clamp, I is the dependent variable, so it is more relevant to express this equation as $I = V/R$ or, since conductance is the inverse of resistance, as $I = gV$. On a graph, I is the y variable, V is the x variable and g is the slope. Thus, Ohm's law is represented by a straight line going through the origin, and the value of the slope provides a measure of the conductance (or the inverse of the resistance). Or stating the converse, *if* a plot of measured current against applied voltage defines a straight line, *then* the system under investigation obeys Ohm's law. This is precisely the reasoning applied by H&H in the approach to Paper 3: the linearity of the current-voltage relationship for tail current demonstrates that ionic currents obey Ohm's law.

The second way in which a straight line can vary on the graph is that it may be shifted with respect to the origin. One can define this shift by adding a constant value, b, to every y value, so that the equation for the line becomes the familiar $y = ax + b$, where b is the y-intercept. An equivalent way to define the shift is to subtract (or add) a constant value x_0 (usually pronounced 'x-nought') from every x value, so that the equation for the line becomes $y = a(x - x_0)$. In this formulation, x_0 is equal to $-b/a$. Thus, the equation $y = x + 2$ describes a straight line with slope of 1 and an x-intercept of -2 (*Figure A*).

This latter approach is most convenient in the context of plotting voltage-clamped currents. The simple equation written above, $I = gV$, states that current is zero where the voltage is zero. Both experiment and theory, however, indicate that this is not necessarily the case. Instead, current through an ion-selective channel goes to zero at the equilibrium potential for the permeant ion(s), which is often a nonzero value—thus, the straight line defined by Ohm's law requires a horizontal shift along the x-axis. Rewriting in terms of the relevant quantities, the equation becomes $I = g(V - V_0)$, where V_0 is the reversal potential for the permeant ion.

In addition to equations being described as linear, physical systems can also be described as linear based on the relationship of the dependent y variable (output) to the independent x variable (input). For linear systems with multiple inputs, the output is proportional to a weighted sum of the inputs: $y = a_1 x_1 + a_2 x_2 + b$. The inputs do not interact to augment or diminish each other's effect on the output; instead they are *independent*.

The straightforward nature of the mathematics—adding and subtracting—greatly simplifies experimental analysis of any system that turns out to be linear. In the case of excitable cells, despite the many nonlinearities of the conductances—such as their dependence on voltage, time, and space—the linear summation of *currents* at any given voltage makes their manipulation and quantification more tractable. For instance, voltage-clamped currents can be isolated pharmacologically: First, stimulus-evoked current can be measured before and after the application of a blocking compound. Next, since the output (the total current) is the sum of the inputs (the blocker-sensitive and the blocker-insensitive current), the blocker-sensitive current can be revealed by subtracting the current in the blocker from the total current. In a non-voltage-clamped membrane, however, sodium and potassium current flow are *not* independent. The flow of sodium current changes the membrane potential, which changes the amount of potassium current that flows, which also changes the membrane potential, etc. As a consequence, the voltage changes induced by sodium flux *cannot* be summed with the voltage changes produced by potassium flux to obtain the voltage change produced by flux of both ions. Nor can action potential waveforms recorded with and without a blocking compound be subtracted to quantify the contribution of the blocker-sensitive current on the action potential.

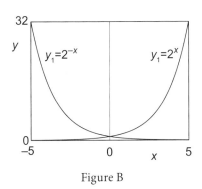

Figure B

1.3 Exponential functions. An exponential increase results when a quantity is repeatedly multiplied by a constant factor, where that factor is greater than 1. For instance, if a value y_0 that is initially equal to the amount A is doubled once, the value y_1 will be equal to $A \times 2^1$, and can be plotted for $x = 1$. If that value is doubled a second time, y_2 becomes $A \times 2^2$, for $x = 2$. A third doubling gives y_3 of $A \times 2^3$. This progressively increasing value of y can be expressed as the function $y = A \times 2^x$. Such a function, which has a x value as the exponent on a constant term greater than 1, defines an **exponential growth** function. Note that although the *proportionate* increase on every step is identical, the *absolute* difference between successive values gets larger and larger as x increases; as a consequence, exponential growth functions rapidly attain enormous values (*Figure B*).

Following a similar principle, an exponential decrease is produced when a quantity is repeatedly multiplied by a constant factor of less than 1. For an example parallel to the one given above, if the value y_0 with the amount A is halved once, y_1 will be $A \times (\frac{1}{2})^1$, which can also be written as $A \times 2^{-1}$. If the value is halved a second time, y_2 becomes $A \times (\frac{1}{2})^2$, or $A \times 2^{-2}$. A third halving gives y_3 of $A \times (\frac{1}{2})^3$, or $A \times 2^{-3}$. This progressively decreasing value of y can be expressed as the function $y = A \times (\frac{1}{2})^x$, or $A \times 2^{-x}$ (*Figure B*). Such a function, which has a negative x value as the exponent on a constant term greater than 1, defines an **exponential decay**. Although the *proportionate* decrease on every step is identical, the *absolute* difference between successive values gets smaller and smaller as x increases; consequently, the value of the function gets closer and closer to zero (approaching zero asymptotically), although formally, it never actually reaches zero.

Interesting mathematical outcomes arise when the constant that is raised to the power of x is Euler's constant, e, which has an approximate value of 2.71828. The exponential growth function $\mathbf{y=e^x}$, also written $y=\exp(x)$, is the only function y whose derivative dy/dx is equal to itself: $dy/dx=e^x$. The corresponding exponential decay function, $\mathbf{y=e^{-x}}$, also written $y=\exp(-x)$, is the only function whose derivative is the *negative* of itself: $dy/dx=-e^{-x}$. This latter point turns out to be especially important, because the equation $y+dy/dx=0$ is the simplest possible **first-order linear differential equation** first-order because its terms include a function and its first derivative—and it has the exponential decay function e^{-x} as a solution.

Many natural processes, especially those that evolve over time, are first-order processes that can be described by some form of an exponential decay with e as its base value. How this outcome arises and some of its implications are discussed in Appendix 2.3, but here, the focus is on what the exponential decay looks like and what its parameters represent.

The shape of the function $y=e^{-x}$ can be defined by evaluating it at a few key points. At $x=-\infty$, the exponent is an infinite positive value and y is likewise infinite. At $x=+\infty$, the exponent is an infinite negative value and y reaches 0. Two additional points help define the specific curve $y=e^{-x}$ in a way that is meaningful for thinking about scientific phenomena. At $x=0$, e is raised to a power of 0 and y is therefore 1. At $x=1$, the function has the value e^{-1} or $1/e$, which is $1/2.71828$ or about 0.37. In other words, at $x=1$, the function has decayed to 37% of the value it had at $x=0$. In many biophysical processes, a value decays over time (where the x value is time, t) from a fixed amplitude with certain kinetics. The initial amplitude, of course, does not necessarily equal 1, but it can be always described as 100% of its starting value A. The equation can therefore be scaled to start from a value of A at time zero, making the equation $y=Ae^{-t}$. Because the time course of different processes can vary, it is convenient to state how much time must pass before the quantity decays to $1/e$ or 37% of its initial value (equivalently, before it decays *by* 63%); in other words, how much time must pass before the exponent has a value of 1. This time can be found by expressing the exponent as t divided by a constant factor, called the **time constant**, τ (tau), which has units of seconds. The equation becomes

$$y = Ae^{-t/\tau},$$

or equivalently,

$$y = A\exp\left(\frac{-t}{\tau}\right).$$

When $t=\tau$ sec, the exponent is 1, and y is 37% of the initial value A (*Figure C*). For phenomena in which the x value is not time, the constant can be generically referred to as the **decay constant**.

An exponential decay is therefore fully defined by its initial amplitude at an x value of zero and its decay constant, just as a straight line can be defined by its slope and intercept. For temporal phenomena, a short or brief time constant (low numerical value) is indicative of a process with a rapid rate (high value). A long time constant (high value) is indicative of a process with a slow rate (low value). The **rate constant**, k, is thus the inverse of the time constant ($k=1/\tau$), and when it is convenient to talk in terms of rate constants, the exponential decay function can be rewritten as

$$y = Ae^{-kt},$$

or equivalently,

$$y = A\exp(-kt).$$

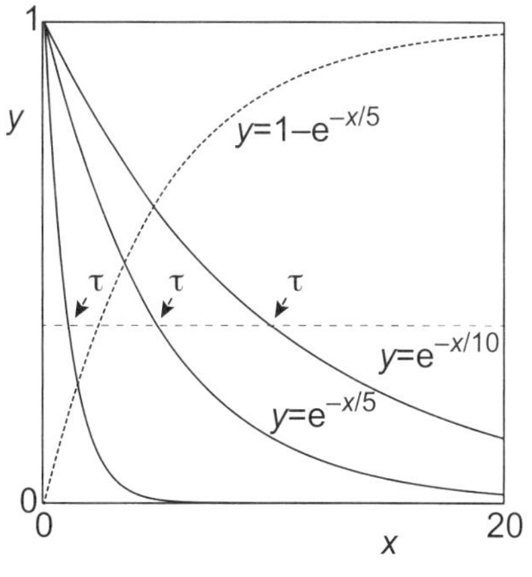

Figure C

Not all processes decay asymptotically to zero; sometimes a steady-state or equilibrium value remains. This value can be called y_{ss} (for steady state) or equivalently, y_{∞} (indicating that the value is reached at long times, where $t=\infty$). The equation can then be expressed as

$$y = Ae^{-t/\tau} + y_{ss}.$$

Note that at time zero this function has a value of $A+y_{ss}$.

An important conceptual point is that an exponential 'decay' defines the *shape* of the function—the absolute change in the y value is smaller at each fixed interval—but it does *not* necessarily mean that the value of the function is dropping. Quantities can 'decay' exponentially to yield *more* of the quantity represented on the y-axis. To avoid confusion, such a process can be equivalently referred to as an 'exponential relaxation.' An example from H&H is the recovery from inactivation by sodium currents. After a depolarizing stimulus that completely inactivates the sodium conductance, the current gradually recovers and approaches 100% of its initial value. Thus, instead of falling from 1 to 0, the function describing recovery from inactivation progresses from 0 to 1. Despite its increasing value, it is still formally a 'decay,' because the absolute increment at any time interval is smaller than in the preceding, equivalent time interval. This function is expressed by subtracting the exponential decay $e^{-t/\tau}$ from 1 and then scaling by the total amplitude:

$$y = A(1 - e^{-t/\tau}).$$

This function starts at 0 and grows with time to a value asymptotically approaching A (*Figure C*). At a time equal to the time constant τ, the value of y is $(1 - 0.37)$ of A, or 63% of A. Another common instance in electrophysiology of a value that can 'decay up' (relax) according to this equation is the passive charging of an RC circuit (like the membrane) to generate a voltage (see Appendix 2.3).

1.4 Saturating functions. A saturating function is one in which the y value initially grows as the x value increases; with increasing x values, however, the rate of growth decreases, and the y value asymptotically approaches a maximum. The equation referred to in the annotations as the two-state equation based on the Boltzmann principle, which is of particular relevance to biophysical processes, and it is a special case of a saturating function. In addition to saturating at high x values, this equation asymptotically approaches a minimum y value for low values of x; it saturates at both extremes.

The equation for such a saturating function is

$$y = \frac{1}{1 + e^{-x}}.$$

This equation can be understood by breaking it down into its components, beginning with the exponential term in the denominator (*Figure D*):

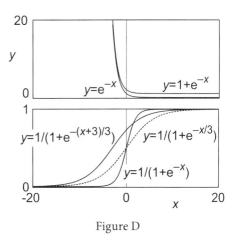

Figure D

1. As described in section 1.3, over the range $x = -\infty$ to ∞, the exponential decay $y = e^{-x}$ goes from ∞ to 0, with a y value of 1 at $x = 0$.

2. Adding 1 to the function gives $y = (1 + e^{-x})$, and shifts the decay up by +1 across the full range. The function now goes from ∞ to 1, with a y value of 2 at $x = 0$.

3. Inverting this function gives $y = 1/(1 + e^{-x})$, which goes from 0 to 1, with a y-value of ½ at $x = 0$.

The curve can vary in three parameters while maintaining its basic form (*Figure D*):

1. The curve can be made shallower or steeper by dividing the x value by a constant term, k, called the slope factor, so that $y = 1/(1 + e^{-x/k})$. The larger the k value, the shallower the slope. At the foot of the curve, before the y value saturates, an e-fold increase in the y value is obtained whenever the x value increases by k. In the dotted curve of *Figure D*, $k = 3$.

2. The curve can be shifted left or right on the x-axis, expressed by subtracting a constant $x_{1/2}$ from the x value, so that $y = 1/(1 + e^{-(x-x_{1/2})/k})$. When the x value is equal to $x_{1/2}$, the y value will be ½. The $x_{1/2}$ constant is called the 'value at half-maximum,' often shortened to the 'half-max value.' In the leftmost curve of *Figure D*, $x_{1/2} = -3$.

3. The maximum value of 1 can be scaled by multiplying by the maximal value of y that is attained, y_{max}.

This final equation becomes

$$y = \frac{y_{max}}{1 + e^{-(x-x_{1/2})/k}},$$

equivalently written

$$y = \frac{y_{max}}{1 + \exp\left(-(x - x_{1/2})/k\right)}.$$

Like a straight line that can be uniquely defined by its slope and intercept, or an exponential decay that can be uniquely defined by its amplitude and time constant, this equation can be uniquely defined by its half-max value, its slope factor, and its maximal value, which is often normalized to 1 (as in H&H's h_∞ curve).

This equation is often referred to as a sigmoid or 'S-shaped' function with the reference being made to the top part of the 'S.' An ambiguity of this terminology is that other functions are also described as S-shaped (with reference to the concave-up followed by concave-down shape of the curve at the initial rise), such as the product of exponentials that describes the rising phase of potassium conductance. Note, however, that the function described here is symmetrical (identical if flipped vertically and horizontally) whereas the product of exponentials is not.

This equation can be referred to as the 'two-state equation,' as it describes the equilibrium probability of finding a physical particle in one of two distinct states, when the transition between the two states requires the application (or removal) of a fixed amount of energy (Boltzmann's principle). In H&H, a single particle directly controls the measured conductance in the context of inactivation, in which the gate (the 'particle') can be in *either* an open *or* a closed configuration, making the conductance either available or inactivated (see Paper 4, Figure 5). In this case, the x variable is membrane potential, and the half-max value becomes $V_{1/2}$, the potential at which the conductance is half-inactivated. The slope factor k indicates that an e-fold change in availability occurs at the foot of the curve with every k millivolts of depolarization.

It is important to note that H&H do not use this equation to fit the *activation* of conductance versus voltage curves (see Paper 2, Figures 9 and 10) because they deduce that multiple particles must move to activate the conductance—three for the sodium conductance and four for the potassium conductance. The plots of m_∞ and n_∞, which represent the steady-state 'activated' or 'open' probability of a *single* gating particle (for the sodium and potassium conductance, respectively), would indeed be expected to follow the two-state equation. Nevertheless, because intermediate closed states (e.g., with only one open gate or one activated voltage-sensing domain) are often short-lived and therefore negligible, experimentally recorded activation data can be mimicked to a first approximation by a two-state equation. It has therefore become standard practice to fit this equation to conductance-voltage curves, as

$$g = \frac{g_{max}}{1 + e^{-(V-V_{1/2})/k}},$$

where g is the conductance of a specific ion at any voltage, and g_{max} is the maximal conductance to that ion. Normalizing this curve gives

$$\frac{g}{g_{max}} = \frac{1}{1 + e^{-(V-V_{1/2})/k}}.$$

The availability (inactivation) curve, which saturates at negative voltages rather than positive voltages, and is the same curve flipped on its vertical axis, can be obtained by inverting the sign on the exponent to give

$$\frac{g}{g_{max}} = \frac{1}{1 + e^{(V-V_{1/2})/k}}.$$

Even when the fitted curves deviate slightly from the data, the parameters obtained from such fits are convenient for comparison of current properties, for example, the half-max voltages of activation and/or slope factors of different potassium channels. Depending on the application, therefore, the approximation of a fit of the two-state equation may or may not be adequate.

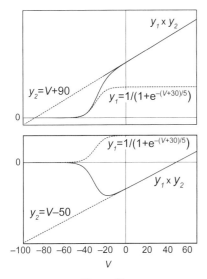

Figure E

1.5 The current-voltage relation. The two-state equation for conductance also dictates the shape of the current-voltage relation for an ion channel. One of the key points that emerges throughout the H&H papers is that current is conductance multiplied by driving force, for example, $I_{Na} = g_{Na}(V - V_{Na})$. Inspection of this equation shows that it is the product of a straight line describing driving force, $I = g_{max}(V - V_{Na})$, and the normalized conductance-voltage curve, which can be well-approximated by the two-state equation for g_{Na}/g_{max} given above (see section 1.4). At the most hyperpolarized potentials, the *I-V* curve for a voltage-gated channel activated by depolarization has a value of 0 because of the zero value of the two-state equation; at the intermediate potentials, it curves as a consequence of the increasing conductance; and at the most depolarized potentials, where the two-state equation saturates to a value of 1, it linearizes as the straight line is multiplied by 1. In this way, the various forms of the *I-V* curve for a voltage-gated channel, and indeed any ion channel, can be accounted for as the product of a straight line and a two-state equation. Two examples are shown in *Figure E* for hypothetical voltage-gated currents that are both half-activated at −30 mV, but reverse at either −90 mV (top) or +50 mV (bottom).

Appendix 2

ELECTRICAL PRINCIPLES AND CIRCUIT ELEMENTS

This and the following appendix provide a brief review of fundamental electrical and electronic principles as they pertain to electrophysiology, targeted to readers without extensive background in physics. Instead of leading with equations to explain phenomena, each section leads with ideas likely to be accessible to those with some exposure to laboratory experiments and/or the scientific literature involving electrophysiology. Some sections place the theory in the context of experimental protocols or conventions that are common in published research. For purposes of simplicity and relevance, some of the definitions and ideas presented below do not fully encompass the broader fields of nonbiological electricity and electronics.

2.1 Quantities, units, and physical relationships. A quantity is a physical quality, property, or phenomenon that can assume a numerical value—that is, it can be quantified.

The quantity **charge**, Q, is measured in units of **coulombs**, **C**. Charge can be positive or negative. Charges of opposite polarity attract one another; charges of the same polarity repel. In an aqueous solution, like the environment of a cell, many chemical elements exist in ionized form, such that they bear a net positive or negative charge. For instance, when dissolved in water, sodium exists as the monovalent cation Na^+, potassium exists as the monovalent cation K^+, and chlorine exists as the monovalent anion, Cl^- (chloride).

The quantity **voltage**, V (or E), is measured in units of **volts**, **V**. Simply stated, a voltage results when positive and negative charges are separated. The term **potential difference**, or **potential**, is also used to refer to voltage, with a definition that can be thought of as intuitive: when opposite charges are separated, they have the *potential* to come back together again, and will do so, should the opportunity arise. The more charges that are separated, the greater this potential, such that voltage is proportional to the amount of charge that is separated. More formally stated, separating opposite charges requires energy to overcome the attractive force between them. That energy is stored in the electric field (the force per unit charge) between the charges, and it becomes potential energy that can be recovered when the charges recombine. Voltage, or potential, therefore refers to the energy required to move charges against the attractive force, or the energy recovered when charges move with the attractive force; the more charges that are involved, the more energy will be required (or recovered). The unit, volt, is therefore joules/coulomb (energy/charge).

A separation of charges can be maintained by a nonconductive barrier, which in the context of cells is often a membrane. In electrical terms, this membrane acts as a capacitor, a device that can store charge. The quantity **capacitance**, C, quantifies this ability to store charge (i.e., to accumulate positive and negative charges on opposite sides of the barrier) and is measured in units of **farads**, **F**.

A capacitor in an electrical circuit is composed of two conductive 'plates' separated by an insulating substance, called a dielectric material, that prevents charges from passing through it. In depictions of circuits, a capacitor is illustrated as a pair of parallel lines of equal length representing the plates of the capacitor (*Figure A*). In cells, the membrane separating the conductive intracellular and extracellular fluids is composed of a lipid bilayer, in which the hydrophobic tails form the dielectric substance that impedes the passage of ions. The membrane thus acts as a capacitor. The greater the surface area of the cell, the more space exists for the storage of ions, and the greater the capacitance. Thus, capacitance is proportional to the area of the plates—in this case, the area of the membrane, A. Written mathematically, $C \propto A$.

Figure A

Capacitance is also dependent on the composition of the dielectric material ($C \propto \varepsilon$) and inversely proportional to the distance between the plates ($C \propto 1/d$); however, for lipid bilayers, the material and distance (the membrane thickness) tend to be constant. To a first approximation, therefore, the specific capacitance, C_M, or capacitance per unit area of membrane, can be treated as a biological constant, $1\,\mu F/cm^2$ (or $0.01\,pF/\mu m^2$), which H&H measured during their experiments.

The capacitor has the *ability* (capacity) to store charge, thereby keeping positive and negative charges separate, but no voltage actually exists across it unless charges accumulate or 'are deposited' on the capacitor plates. Mathematically, charge, voltage, and capacitance are related by the equation $\boldsymbol{Q = CV}$. In a physiological context, therefore, when ions (charges) are separated across the cell membrane (the capacitor), a voltage (or **membrane potential**, V_m) is established. The larger the capacitor (the larger the cell), the more charge must be deposited to generate a given voltage. For this reason, a large region of membrane, such as a neuronal soma, is often referred to as a 'capacitative load.'

In the absence of action potential firing, the quantity of ions (charge) separated across the cell membrane is sufficient to make a membrane potential (voltage) of about -60 mV in neurons or -90 mV in skeletal muscle fibers. The stable value is called the **resting membrane potential** or simply **resting potential** (V_{rest}). The amount of charge that must be separated to generate such a voltage can be calculated from $Q = CV$. In a spherical cell about the size of a mammalian neuronal soma ($\sim 20\,\mu m$ somatic diameter), the surface area is approximately $1200\,\mu m^2$. Given the specific membrane capacitance of $1\,\mu F/cm^2$, this area corresponds to a capacitance of ~ 12 pF. Therefore, the charge required to generate a voltage of about 100 mV is about 1.2×10^{-12} coulombs. Dividing by Faraday's constant ($\sim 10^5$ coulombs/mole) and multiplying by Avogadro's number (6.02×10^{23} ions/mole) indicates that about 10^7 ions must be separated across the membrane (about 8000 per square micron). An ion such as potassium has an intracellular concentration of about 150 mM in a mammal (about three times lower than in a saltwater organism like squid), which corresponds to approximately 10^{12} potassium ions in that same ~ 20-μm spherical cell body. Therefore, the *proportion* of ions that must cross the membrane to change the voltage by ~ 100 mV is extremely low, only about 1 in 100,000. When ions cross the membrane to generate a single action potential, the changes in intracellular and extracellular sodium, potassium, and chloride concentrations tend to be negligible, although exceptions exist when volumes are small, or concentrations are low.

The quantity **current**, I, is defined as the movement of charge. Current is measured in units of **amperes (amps)**, $\boldsymbol{A}$, which is charge per unit time (coulombs/sec). The flow of current depends on the characteristics of the substance(s) through which charges must pass. Generally, substances are described in terms of the extent to which they *impede* current flow, giving rise to the quantity **impedance**. Impedance can take multiple physical forms.

As described above, the dielectric material between the plates of a capacitor does not permit charge to pass through it; hence, capacitance is one form of impedance. Although the dielectric material prevents the direct flow of charges from one plate to another, a capacitor can pass current when the voltage is changing. Instead of flowing *through* the dielectric, charges accumulate on one plate of the capacitor, generating an electric field that repels like charges from the other plate and leaves opposite charges behind (*Figure B*). Thus, current flows, although the charges that are deposited onto one plate are physically distinct from those that leave the other. H&H call this current the **capacity current**, I_c (see also section 2.5); the term more commonly in use today is *capacitive current*. Since current is the flow of charge, dQ/dt, the expression for capacity current, I_c, can be arrived at by differentiating the relation $Q = CV$ to obtain $I_c = C(dV/dt)$.

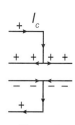

Figure B

In contrast, other substances allow individual charges to pass directly through them. The quantity **resistance**, R, measured in units of **ohms** (Ω) represents the difficulty with which current flows through a substance; hence resistance is another form of impedance. Current flows through metals very easily (low resistance), through saline solutions and protein pores less easily (moderate resistance), and through lipid bilayers hardly at all (high resistance). These substances all function as resistors in electronic circuits. In circuit diagrams, a resistor is drawn as a zigzag line, representing the impediment to current flow (*Figure C*).

The amount of current that flows through a resistor therefore depends on the magnitude of the resistance: $I \propto R$. Additionally, the amount of current that flows de-

Figure C

pends on the magnitude of the voltage applied across that resistance: $I \propto V$. These dependences are expressed as **Ohm's law**, commonly written as

$$I = \frac{V}{R}$$

or

$$V = IR$$

The reciprocal of resistance, **1/R**, gives the quantity **conductance**, **g** or **G**, which therefore is a measure of the ease with which a substance permits charge to permeate it. Conductance was originally measured in units of **mhos**—ohms spelled backward, used in the H&H papers—and was sometimes given the upside-down symbol ℧, but these units have been superseded by units of **siemens**, **S**. Ohm's law can be equivalently written by substituting a conductance term for the inverse of resistance: $I = gV$.

In electrophysiology, it can be convenient to use the different formulations of Ohm's law at different times depending on the quantities of interest. The versions with I as the dependent variable are useful for discussing measurements of current (as in voltage clamp), whereas $V = IR$ is preferred when voltage is reported (as in natural cellular activity). The equation is generally written in terms of g when fluxes through specific ion channels are of interest, and in terms of R when properties of the entire cell are considered without distinguishing channel types.

In H&H's terms, the *membrane* is said to conduct current, although, as described above, the lipid bilayer is itself a poor conductor because of its hydrophobic inner leaflets; instead, it is the proteinaceous ion channel pores embedded in the lipid that conduct current when open and resist current flow when closed. Ion channels therefore act as resistors in an equivalent circuit of the membrane. For ion channels that are made to open or close, or are 'gated', by specific stimuli such as voltage, the resistance changes depending on the stimulus. Those ion channels are therefore more accurately depicted as **variable resistors**, shown as a zigzag line with an arrow (resembling a needle on a circular gauge) through it (*Figure D*). The current flowing through the resistance of the membrane is called ionic current, I_i.

Thus, in the context of an excitable cell, current either flows as capacity current, in which ions are deposited on one side of the membrane and repel like charges on the opposite side of the membrane, or as ionic current, in which ions pass through the membrane via ion channel proteins embedded in the lipid bilayer.

A third component of impedance is **inductance**, discussed in section 2.6.

Figure D

2.2 Circuit elements. This section describes simple circuit elements as a basis for understanding amplifiers as they pertain to electrophysiology. In electronics, devices are constructed of **voltage sources** (such as batteries, power supplies, coils, or membranes) and **impedances** (resistors, capacitors, inductors, etc., including ion channel proteins), usually in service of converting electricity into another useful form (e.g., to do mechanical or chemical work or to generate light or heat). The voltage sources and impedances are connected into **circuits** through which **current carriers** (such as electrons or ions) flow.

A voltage source, such as a battery, uses energy to supply current through the resistance (impedance) of a circuit. Ideally, a voltage source can maintain a fixed voltage between its two terminals regardless of what other elements are in the circuit. In accordance with Ohm's law, $V = IR$, if the resistance of the circuit is high, little current will flow; if the resistance is low, a large current will flow. By extension, if the two terminals of an ideal voltage source were connected directly together—or 'shorted' to give a resistance of 0 Ω—the current would be infinite. In reality, voltage sources are not ideal (see Appendix 3.1), but it is useful to consider an ideal voltage source to introduce some concepts relevant to electronic circuits.

Most circuits contain more than one form of resistance; current flows through these resistances when powered by a voltage source. In circuit diagrams, a voltage source is usually depicted as four parallel lines of alternating short and long length, representing the cells of a battery (*Figure E*).

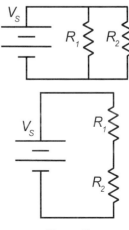

Figure E

Current in a circuit can flow via whatever route is available. If current can flow through *either* element A *or* element B, the elements can be said to be in *parallel* (*Figure E, top*). This situation contrasts with one in which current must flow through element A *and* element B, in which case the elements are in *series* (*Figure E, bottom*). When resistors are connected in series, the total resistance is the sum of all the resistances:

$$R_{total} = R_1 + R_2 + \cdots + R_n,$$

that is, *resistors in series sum linearly*. When resistors are connected in parallel, the total resistance is the inverse of the sum of the inverses of all the resistances:

$$R_{total} = \frac{1}{\frac{1}{R_1} + \frac{1}{R_2} + \cdots + \frac{1}{R_n}}.$$

A simple way of thinking about the latter point is to convert the resistances into conductances:

$$g_{total} = g_1 + g_2 + \cdots + g_n,$$

that is, *conductances in parallel sum linearly*. Regardless of how many branch points a circuit may have, the current flowing into any point in a circuit is always the same as the current flowing out of that point; this idea is known as **Kirchoff's law**.

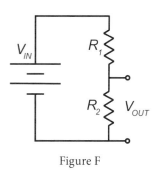

Figure F

Voltage dividers. With two resistors, R_1 and R_2, connected in series to a battery (*Figure F*), the total voltage drop, termed the input voltage, V_{IN}, occurs across both resistors. It is possible, however, to measure the voltage difference not only across the whole circuit (which is equal to the voltage of the source that powers it), but also across any resistance within the circuit. In fact, with any pair of resistors, it is often useful to describe the voltage drop across the second resistance, R_2. Here, this voltage is termed the output voltage, V_{OUT}, as it can be used as a voltage source for another part of the circuit. If the two resistors in series have equal value, half the voltage drop occurs across the first resistor, and half the drop is over the second, so $V_{OUT} = \frac{1}{2}V_{IN}$. With resistors of unequal value, the voltage drop will be greater over the resistance of higher value. To quantify, recall that the current flow, I, through each resistor is necessarily the same. According to Ohm's law, $I = V_{IN}/(R_1 + R_2)$. Because the output voltage, V_{OUT}, across the second resistor, R_2, must be IR_2, it can be expressed as

$$V_{OUT} = V_{IN} \times \frac{R_2}{R_1 + R_2}.$$

This circuit element, of two resistors in series, which split the total voltage drop into an input and output voltage, is called a **voltage divider**.

2.3 RC circuits. H&H deduced that the membrane contains both capacitative and resistive elements, which they could illustrate as a circuit diagram (see Paper 5, Figure 1). Such a circuit produces distinctive voltage responses. In a circuit consisting of a resistor alone, the application of a current through the resistor yields an instantaneous voltage of a magnitude determined by Ohm's law: $V = IR$. In a circuit such as the membrane, however, the time course of the voltage change induced by the current source, I, will be slowed (*Figure G*). Because the resistance and capacitance are in parallel, the total applied current, I_{total}, is the sum of the capacity current, I_c, and the resistive current, I_R. If a current step is applied at time $t = 0$, *no* current initially flows through the resistor, because the voltage starts at 0, and $I_R = V/R$. As a result, at $t = 0$, *all* the current flows as capacity current (I_c). As charge accumulates on the plates, however, the capacitor builds up a voltage, and resistive current begins to flow. Because $V = Q/C$ and $I_c = dQ/dt$, the rate of the voltage change, dV/dt, is

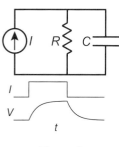

Figure G

I_c/C. As the voltage builds, the current through the resistor, I_R, increases, while the current through the capacitor, I_c, falls by a corresponding amount, so that I_{total} remains constant. Consequently, dV/dt, which depends on I_c, also falls, and the voltage asymptotically approaches its final value of $V = I_{total}R$ (Figure G).

If the applied current is terminated, the capacitor discharges. Under these conditions, the total applied current is zero. Therefore, $I_c + I_R = 0$. Because charge is capacitance times voltage, $Q = CV$, and current is the time derivative of charge, $I_c = dQ/dt$, it follows that $I_c = C(dV/dt)$. Recall also that $I_R = V/R$. Therefore, $C(dV/dt) + V/R = 0$, or

$$\frac{dV}{dt} + \frac{V}{RC} = 0.$$

This is a simple first-order linear differential equation (see Appendix 1.3). Solving the equation requires finding a function, $V(t)$, that can be divided by RC to give a value that is equal and opposite to its own derivative, dV/dt, so that the sum will be zero. Recall that the derivative of e^{kx} is ke^{kx}, so the derivative of e^{-x} is $-e^{-x}$. Hence, e^{-x} is a function for which the sum of the function and its derivative is zero. $V(t)$ must therefore be of a form related to e^{-t}. Taking this idea a step further, the derivative of $e^{-t/RC}$ is $(-1/RC) \times e^{-t/RC}$. The solution to the equation therefore becomes

$$V(t) = V_0 \exp\left(\frac{-t}{RC}\right),$$

where RC is the time constant τ in seconds.

Generalizing, in response to any square step current, I, the voltage across an RC circuit will change with an *exponentially decaying* time course, from its initial value V_0 at $t = 0$ to a final value V_∞, so that the total change in voltage is ultimately $(V_\infty - V_0)$, which is IR. The voltage as a function of time can therefore be expressed as

$$V(t) = V_0 + (V_\infty - V_0)\left[1 - \exp\left(\frac{-t}{\tau}\right)\right].$$

Experimentally, it is often convenient to probe the properties of the RC circuit of the membrane by the injection of square pulses of current. In this context, I_r is synonymous with the ionic current I_i, since ions carry the current through the resistor-like ion channels of the membrane. By measuring the magnitude and kinetics of the voltage change for a current pulse of a particular amplitude and making use of the relations $(V_\infty - V_0) = IR$ and $\tau = RC$, the membrane resistance and capacitance can be calculated. A higher capacitance will produce a slower voltage change; a higher resistance will produce a slower and larger voltage change.

2.4 DC and AC signals. A steady voltage applied across a resistor generates a steady current, called a DC signal (also abbreviated d.c. in H&H). Note that although the abbreviation 'DC' stands for 'direct current,' it is commonly applied to either a voltage or a current, with the emphasis that the signal is unvarying. In contrast, a sinusoidally varying voltage or current, such as the 50- or 60-Hz voltage that comes from standard electrical power outlets, is called an AC signal, where 'AC' is the abbreviation for 'alternating current.' Note that the sign on the current (and the voltage) flips as the current changes direction.

Resistors respond to an alternating voltage with an alternating current that is *in phase* with the voltage. For example, for a sinusoidal voltage with an amplitude of 1 V (2-V peak-to-peak magnitude), the current through a 1000 Ω resistor is also a sinusoid, with an amplitude of 1 mA (*Figure H*). Since current is proportional to voltage, as defined by Ohm's law, $I = V/R$, it peaks simultaneously with the voltage, such that the two sinusoids are in phase with one another. The amplitude of the current is the same at 1 Hz or 1000 Hz. For a resistor, therefore, the peak current resulting from an AC signal is independent of the frequency of the sinusoid.

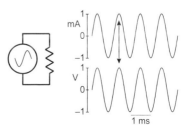

Figure H

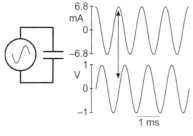

Figure I

Capacitors respond differently to a sinusoidal voltage. As noted above, $Q = CV$. Therefore, since $I = dQ/dt$, the capacity current can be given as $I_c = C(dV/dt)$ (i.e., the current is proportional to the *derivative* of the voltage). The derivative of a sinusoid, $\sin(x)$, is another sinusoid that is shifted 90 degrees in phase, $\cos(x)$ (*Figure I*). The current is therefore maximal when the *rate of change* of voltage is maximal, which occurs when the voltage crosses zero (arrows).

For a sinusoidal voltage waveform of frequency f and amplitude A, the voltage waveform is given as

$$V = A \times \sin(2\pi f t),$$

and its derivative is

$$\frac{dV}{dt} = 2\pi f A \times \cos(2\pi f t).$$

Therefore, the current through a capacitor will be $C(dV/dt)$, which is

$$I_c = C \times 2\pi f A \times \cos(2\pi f t).$$

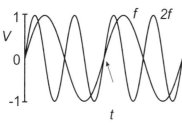

Figure J

Thus, if a 1-V, 1-kHz AC voltage is applied to a 1-μF capacitor, the current amplitude is 10^{-6} F $\times 2\pi \times 1000$/sec $\times 1$ V, or 6.28 mA. If the AC voltage is doubled in frequency to 2 kHz, leaving its amplitude at 1 V, the maximum rate of change of the voltage waveform is likewise doubled (*Figure J, arrow*), and the capacity current is doubled as well. Unlike for a resistor, therefore, the current through the capacitor depends on frequency, with high frequencies producing more capacity current.

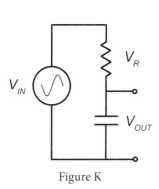

Figure K

2.5 RC filters. When in series, a resistor and capacitor create a voltage divider that can differentially attenuate signals according to their frequency; in other words, the circuit can act as a **filter**. The configuration in *Figure K*, for example, works as a low-pass filter. For a DC input, which has a frequency of zero, the capacitor acts like a circuit element with infinite resistance. Consequently, no current flows through either the capacitor or the resistor, and the voltage drop across the resistor, V_R, is, by Ohm's law, zero. The voltage across the capacitor, $V_{OUT} = V_{IN} - V_R$, will therefore be equal to the applied voltage, V_{IN}. For a low but non-zero AC frequency, the capacitor allows a small alternating current to flow through the circuit. A small voltage drop therefore appears across the resistor and V_{OUT} is slightly attenuated relative to V_{IN}.

At higher and higher frequencies, the capacitor allows more and more current to flow. The voltage drop across the resistor becomes progressively larger, such that the output voltage, V_{OUT} (i.e., the voltage drop across the capacitor), becomes progressively smaller. Thus, the circuit strongly attenuates high-frequency signals but allows low-frequency signals to pass from V_{IN} to V_{OUT}, forming a **low-pass filter**. A **high-pass filter** would result by exchanging the positions of the resistor and capacitor, such that V_{OUT} is the voltage across the resistor.

Note that the phase of the output signal also changes with frequency: at low frequencies, where most of the voltage falls across the resistor, the output voltage is nearly in phase with the applied voltage; at high frequencies, where most of the voltage falls across the capacitor, the phase is shifted by nearly 90 degrees because of the dependence of the capacity current on the derivative of the voltage.

Filters are often characterized by the frequency at which the output signal is attenuated by 3 decibels (dB), approximately a factor of 2. The filter constant, or '3-dB point,' is the inverse of the product of the resistance and capacitance, $1/2\pi RC$. Note that RC is equal to τ, the decay time constant of the current that would flow through the circuit in response to a step change in voltage (see section 2.3).

2.6 LC *circuits*. Inductors, also called 'chokes,' are usually made of small coils of wire. In one sense, they are the complement of capacitors: whereas in a capacitor current is proportional to the derivative of voltage, in an inductor voltage is the derivative of current. With a DC voltage applied to the inductor, a steady (unvarying) current will flow because the metal wire of the coil has almost zero resistance. With an AC voltage, however, the magnetic field in the coil will change as the current changes. A *changing* magnetic field always generates an electric field, which, in this case, induces a voltage across the coil. The quantity **inductance**, **L**, is measured in units of henrys, H, which are $V \times sec/amps$, and an inductor is represented in circuit diagrams as a series of loops or a scallop (*Figure L*). The

Figure L

more loops in the coil of an inductor, the higher the voltage for a given change in current, and the greater the inductance. For a fixed inductance, the greater the rate of change of current, the higher the voltage that is produced. Because the voltage across an inductor is proportional to the change in current, dI/dt, the voltage change lags the current change by 90 degrees. This idea can be grasped intuitively by recognizing that for a sinusoidally alternating current, the peak voltage occurs when dI/dt is maximal (at the zero crossing) such that the voltage is the derivative of the sinusoid.

An everyday example of a voltage induced in a coil by changing a current can be seen in the spark that can result from unplugging a large motor, such as a fan or vacuum cleaner. Reducing the current in the motor's coil instantaneously to zero produces a voltage in the coil, which is briefly large enough (>3000 V) to drive charge across the high-resistance air gap between the socket and plug as it moves away.

Before H&H's work, the idea of a membrane inductance was seriously considered. The action potential and its afterhyperpolarization resemble a highly damped oscillation (see Paper 5, Figure 22), and in electronic circuits, oscillations can be produced by the combination of an inductor and capacitor (*Figure M*). A discharging capacitor in an *LC* circuit generates current flow that builds up a magnetic field in the coil. When the capacitor is fully discharged, the voltage is zero, and the current, like the magnetic field in the coil, is at its maximum positive value (*Figure N*, arrow). At this point, dI/dt is zero, and the magnetic field begins to collapse. As it collapses, it induces a voltage of the opposite polarity. This voltage generates a current flow in the opposite direction, which partially recharges the capacitor. This cycle of charging and discharging the capacitor gives rise to oscillations of voltage with a gradual decay in amplitude as energy is dissipated by the slight resistance of the wires.

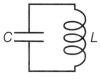

Figure M

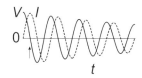

Figure N

It is now clear that axonal membranes do not contain coils or other elements that act as inductors. As H&H demonstrated, the membrane potential oscillations that raised the possibility of a membrane inductance can be explained simply by the time dependence and voltage dependence of ionic conductances without requiring the hypothesis of a dedicated inductive element, and no such element has been found to exist within the membrane.

Appendix 3

ELECTRONIC CIRCUITS

3.1 *Amplifier output impedance and input impedance*. The notion of **gain** (**G**) in an amplifier is familiar, in that the signal (voltage) at the output terminal of the amplifier is proportional to the signal at the
input, but many times larger. Of equal importance to the performance of the amplifier, especially for biological signals, is the **output impedance** of the electrode and axon and their relationship to the **input
impedance** of the amplifier. To understand output impedance, consider a standard alkaline battery. Its

voltage, nominally about 1.6 V, is measured while no
current is being drawn from the battery. If the battery were an 'ideal' voltage source, it would have zero
resistance associated with it, and could be dia-
grammed as the simple battery in *Figure A, left*. If the
two terminals of an ideal battery were connected together (shorted), the current, *I*, equal to *V/R*, would
be infinite. In reality, however, the physical materials
that compose the battery have a small but nonzero

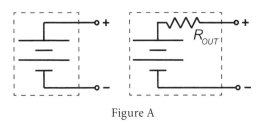

Figure A

resistance; hence, the battery is not ideal, but includes its own internal resistance. When the battery begins to supply current, therefore, the voltage at the terminals drops roughly in proportion to the current,
and the battery acts like the circuit in *Figure A, right*, which includes the internal resistance, R_{OUT}. This
internal resistance, called the **output resistance**, limits the current that the battery can supply.

The term '**output impedance**' is used more commonly than 'output resistance,' because many
electronic devices, unlike batteries, also have capacitive elements that contribute to the effective output resistance (see Appendix 2.1). Output impedance, R_{OUT}, can be measured as the voltage of the
source when no current is flowing divided by the current that flows when the terminals are shorted together. In an alkaline battery, the internal resistance is typically a few tenths of an ohm, and the voltage
at the terminals will therefore drop by a few tenths of a volt when the battery supplies 1 A of current. More
generally, *all* voltage sources—power supplies, the output of an amplifier, waveform generators—have
nonzero output impedances. The axonal membrane, too, is a voltage source with an output impedance,
the sources of which include the seawater, connective tissue, axoplasm, ion channels, and, during electrophysiological recording, the electrode itself. To measure voltage, the voltage source—here, the axon-plus-

electrode combination—must be connected to a voltage-measuring device, such
as an oscilloscope, multimeter, or amplifier. An ideal measuring device would
draw no current. In practicality, however, all measuring devices require energy
to detect and report the voltage, and therefore they draw a nonzero current. In
most voltage-measuring devices, the current that is drawn is proportional to
the voltage being measured. Thus, just as a voltage source has an output impedance, a voltage-measuring device has an **input impedance** (or input resistance) that is effectively in parallel with its input terminals. Hence, the
device can be modeled as an ideal detector (the circle) that draws no current
(i.e., with infinite impedance), in parallel with a resistor that represents the
input impedance (*Figure B*).

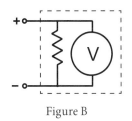

Figure B

This input impedance can cause problems when its magnitude is comparable to the output impedance of the voltage source being measured. In *Figure C*, a voltage source with voltage V_S and
output impedance R_{OUT}, is connected to a measuring device with input impedance R_{IN}. Note that

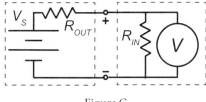

Figure C

R_{OUT} and R_{IN} create a voltage divider such that V, the voltage at the detector, will be

$$V = \frac{V_S \times R_{IN}}{R_{OUT} + R_{IN}}.$$

The voltage divider can be made more explicit by rearranging the elements in the diagram as shown in *Figure D*.

If R_{OUT} and R_{IN} are equal, the detected voltage, V, will be exactly ½ of the voltage of interest, V_S. To measure V_S accurately, therefore, R_{IN} must be far greater than R_{OUT} so that the ratio $R_{IN}/(R_{OUT} + R_{IN})$ approaches 1. For example, if $R_{IN} = 99\ R_{OUT}$, then the detected voltage will be accurate to 1% of V_S.

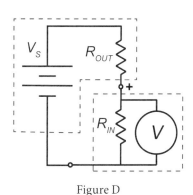

Figure D

Modern patch-clamp electrodes have resistances on the order of 10 MΩ, which dominate the output impedance of the neuron-electrode combination. Accurate recordings of membrane potential therefore demand an amplifier with an input impedance about 100 times greater, around 1 GΩ. Even the much larger squid axon and low-impedance silver-wire electrodes developed by H&H required an amplifier with an input impedance of 100 kΩ or more. Note, however, that the amplifying stages of H&H's voltage-clamp amplifier, formed by the voltage-amplifying vacuum tubes T_4/T_5 and T_6/T_7 (see Paper 1, Figure 6), are embedded in networks of resistors in the range of only 10 to 20 kΩ. If the recording electrodes were connected directly to the inputs of this high-gain portion of the circuit, the axonal membrane potential would be significantly attenuated by the resulting voltage divider. To solve this problem, H&H interpose **cathode followers** (also called 'voltage followers'), which are high-impedance, low-gain (0.9), vacuum-tube amplifiers between the electrodes and amplifying stages of the amplifier. These are the vacuum tubes T_1, T_2, and T_3 (see Paper 1, Figure 6).

Vacuum tubes by their nature, have very high input impedances and are therefore well suited to intracellular measurements. They are, however, expensive, bulky, fragile, variable in their properties, and require power supplies of 200 V or more, as well as a separate, low-voltage power supply to heat the internal filament. Amplifiers made in the era of solid-state electronics use field-effect transistors (FET) or operational amplifiers (see section 3.2) with FET built into their inputs. These are inexpensive, small, sturdy, and fast, and can have input resistances of 10^{12} Ω or greater, high enough to accommodate the highest-resistance electrodes.

3.2 The two-electrode voltage clamp. H&H used a two-electrode voltage-clamp configuration (*Figure E, top*). Its operation can be understood by considering the central component of most modern voltage-clamp circuits, the **operational amplifier**, or **op-amp**. Op-amps have two input terminals, the noninverting input (V_+) and the inverting input (V_-), and they amplify the difference between the voltages at the two input terminals—hence they are also called **differential amplifiers**. Solid-state op-amps available today have dozens of transistors and resistors on a single silicon chip arranged into several stages of amplification, such that the total gain (G) between input and output is in the range of 10^5 or greater. In other words, the output voltage is equal to the difference between the voltage at the two input terminals, multiplied by approximately 10^5:

$$V_{OUT} = (V_+ - V_-) \times G$$

In practice, the range of voltages that V_{OUT} can reach is limited by the power supplied to the op-amp (called $+V_{CC}$ and $-V_{CC}$), which in modern amplifiers is commonly +12 V and –12 V. As a consequence, V_+ and V_- can differ by no more than 120 µV before V_{OUT} saturates or 'hits the rails.' To avoid saturation, most op-amp-based circuits employ negative feedback. In the voltage-clamp circuit, negative feedback is achieved by connecting the output of the op-amp directly to the inverting input of the amplifier. Therefore,

$$V_{OUT} = V_-$$

For $G = 10^5$, combining the two equations gives

$$V_{OUT} = V_- \times \left(\frac{G}{1+G} \right) \approx 0.99999 \times V_+$$

which is, to a close approximation, equal to V_+. Thus, as a result of connecting the output to V_-, the op-amp supplies whatever current is necessary to make the voltage at the two terminals nearly the same.

In H&H's voltage clamp, the voltage-sensing electrode is connected to V_-, and the command potential $V_{COMMAND}$ is connected to V_+, while the output of the amplifier is fed back to the current-passing electrode. If, for example, the membrane resistance is 1000 Ω and the command potential is set to -60 mV, the amplifier will have to supply a steady $-60\ \mu$A of current to hold the inside of the axon at -60 mV relative to the outside. If $V_{COMMAND}$ is changed to -20 mV, the difference between V_+ and V_- will become $+40$ mV, and with a gain of 10^5, the amplifier will rapidly start to raise V_{OUT} toward 40,000 V. It will never reach that value, however: as soon as V_{OUT} begins to rise, the difference at the in-

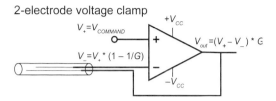

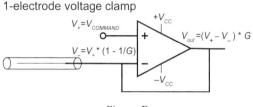

Figure E

puts ($V_+ - V_-$) falls, and the amplifier quickly settles at a new equilibrium with $V_{OUT} = -20$ mV, at which point it will be supplying $-20\ \mu$A. Note that, by $Q = CV$, the total charge necessary to change the membrane potential by 40 mV is $(0.040\ V) \times C$, where C is the capacitance of the membrane. This charge transfer occurs during the brief capacitive transient present in most voltage-clamp records, which is completed well before the membrane starts to respond to the voltage change with changes in ionic conductance.

Once the new membrane potential is reached, the amplifier must supply the necessary current to maintain V_- at V_+. This current is given by $I = V/R$. In the example given above, after a change from -60 mV to -20 mV, the membrane resistance, which is initially 1000 Ω, will begin to drop as voltage-gated conductances activate (ion channels open). Because of the negative feedback arrangement of the circuit, however, the amplifier will draw off exactly the amount of current necessary to hold the membrane potential, V_-, at -20 mV; this current will be equal to the ionic current flowing across the membrane. The experimentalist therefore can record the current supplied by the amplifier to obtain a measure of the ionic current and to derive the resistance (conductance) of the membrane.

3.3 The single-electrode voltage clamp.

In H&H's voltage clamp, the voltage of the current-passing electrode and the voltage-sensing electrode are nearly identical because they are close together inside the axon and the resistance of the axoplasm between them is negligible. The single-electrode voltage clamp, as the name implies, combines these two electrodes into a single electrode connected both to V_- and V_{OUT} (*Figure E, bottom*). A single-electrode configuration was used by Cole and colleagues in their early versions of the voltage clamp. A single wire is more easily threaded down the giant axon than is the large and delicate two-electrode assembly.

As described in Paper 1, note 15, the currents required to clamp a squid axon are so large that a single electrode would undergo significant polarization (from the release and deposition of chloride ions), leading to errors in voltage clamp. Because H&H anticipated this difficulty, they chose to use the two-electrode voltage-clamp configuration in their experiments (see Historical Background). The two-electrode configuration also allowed H&H to have a longer current-passing electrode than voltage-sensing electrode, a crucial design element in the guard-chamber arrangement of Paper 1, Figure 1.

Because of their small size, mammalian neurons rarely have ionic currents that exceed tens of nanoamps, and can be voltage-clamped with feedback currents that are a million-fold smaller than those required to clamp the squid giant axon. The silver-silver chloride wire inserted into a micropipette can easily pass such small currents without undergoing measurable polarization. Modern patch-clamp amplifiers therefore use the single-electrode configuration, making it possible to voltage-clamp small neurons in which the insertion of two electrodes is impractical.

3.4 *The Wheatstone bridge.* Before H&H's experiments, experimentalists, including Cole and Curtis, were interested in measuring the impedance of excitable cells, both at rest and during 'activity,' with a goal of determining whether the nature of the membrane impedance was best approximated as a resistance, a capacitance, an inductance, or some combination of these elements. To make such measurements, the axon (or other excitable membrane) can be introduced as an unknown impedance into a known electronic circuit. The output of the circuit can then be recorded in response to different stimuli, and the value of the impedance can be deduced.

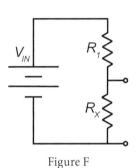

If an unknown impedance functions purely as a resistor (R_X), variations in its resistance can be measured by placing it in series with a resistor of value R_1 to make a voltage divider (*Figure F*). V_{OUT} can then be measured when a known steady voltage, V_{IN}, is applied to the divider, and R_X can be calculated from

Figure F

$$V_{OUT} = \frac{V_{IN} \times R_X}{R_1 + R_X}.$$

If R_X varies, corresponding changes in V_{OUT} would occur.

Measurements of small changes in V_{OUT} relative to a large steady voltage, however, can be difficult to make with accuracy, a problem that is resolved by the circuit known as the Wheatstone bridge (*Figure G*). The Wheatstone bridge involves four resistances: R_1, R_2, R_3, and R_X. The values of R_1 and R_3 are known; R_2 is adjustable; and R_X is unknown. R_1 and R_2 are in series; R_3 and R_X are in series; the two pairs are in parallel with each other. A voltage, V_{IN}, is applied across the pairs of resistances, and the voltage, V_{OUT}, is measured across the point between R_1 and R_3 and the point between R_2 and R_X. The value of R_2 is then adjusted until V_{OUT} is 0. When V_{OUT} is 0, the bridge is said to be 'balanced.' At that point, $R_2/R_1 = R_X/R_3$. The unknown resistance, R_X, can therefore be calculated as $R_2 R_3/R_1$.

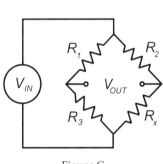

Figure G

Any change in R_X—for example, one that might result from an action potential in an axon—takes the bridge out of balance, resulting in a nonzero V_{OUT}. The change, ΔR_X, can then be calculated from the ΔV_{OUT}. Measuring V_{OUT} relative to the bridge-balanced value of 0 can be accomplished with far greater accuracy than measuring ΔV_{OUT} relative to a large steady V_{OUT}, as described above for the voltage divider.

Thus, a steady voltage applied to the bridge in the circuit of *Figure G* permits a high-precision measurement of the resistance, R_X. An impedance change, however, need not be solely resistive. Changes in capacitance or inductance can be measured if an alternating signal (see Appendix 2.4) is applied to the bridge (*Figure H*). With the bridge balanced, the output signal, V_{OUT}, starts at an amplitude of 0.

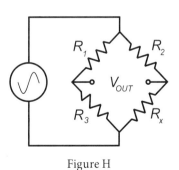

Figure H

Any change in R_X, C_X, or L_X takes the bridge out of balance, generating an output that oscillates at the frequency of the input signal; larger amplitude oscillations indicate a greater change in impedance. Such changes are shown in H&H's reproduction (see Paper 5, Figure 16) of the impedance measurements by Cole and Curtis (1939). As impedance changes, V_{OUT} changes not only in its amplitude but also in its phase relative to V_{IN}, depending on whether R_X, C_X, or L_X is changing. If the unknown impedance is purely a resistance, no phase change will be present, but if it includes capacitance or inductance, a phase shift will occur. Together, the changes in amplitude and phase can be used to deduce which of the three components of the unknown impedance have changed.

Appendix 4

DERIVATIONS OF RATES FOR THE H&H EQUATIONS

A major component of Paper 5 is the conceptualization of 'gates'—that is, the particles that can be either in an open or in a closed state—and the derivation of mathematical expressions that describe how much (magnitude, expressed as a proportion of the maximum) and how fast (kinetics, expressed as a rate) these gates open and/or close at any potential, thereby regulating conductance. The rules that govern the transitions of the gates between open and closed states are the same that govern many chemical processes. The ideas and the mathematical steps in the derivation are included in the annotations, but the progression from Equation 6 through Equation 11 of Paper 5 is presented here for convenient reference.

In Paper 5, H&H indicate that the conductance at any voltage is some proportion of the maximal conductance:

$$g_K = \overline{g}_K n^4 \qquad\qquad \text{(Paper 5, Equation 6)}$$

They express the proportion by introducing the variable n, the 'dimensionless variable which can vary between 0 and 1.' The simplest image that conveys the meaning of the variable n is that of a gate that can open and close; the value n is the proportion of gates that is open, and the gates are clustered such that four such gates are required to open for the potassium conductance to be active. A more precise, modern image comes from molecular studies of the potassium channel, which have identified the potassium channel as a tetramer. Each subunit contains a voltage-sensing domain, whose movement leads to conformational changes that ultimately occlude or open the channel pore; with four such domains, the probability that the channel is open is n^4. The voltage sensors thereby gate the conductance. Note that 'gate' is used here as a verb. Unfortunately for terminology, the word 'gate,' as a noun, is also applied to the structure that physically blocks current flow, but it is molecularly different from the voltage sensors. The physical gate is on the intracellular end of the pore but does not sense voltage itself. It is pulled out of the permeation pathway when the voltage-sensing domain moves.

In the present context, for simplicity, each voltage-sensing domain can be thought of as a gating particle or gate. It has become customary to refer to the particles as n-gates, although H&H do not use this terminology. Thus:

n = the proportion of open gates; equivalently, the probability that a single gate is open

$1 - n$ = the proportion of closed gates; equivalently, the probability that a single gate is closed

The value of n varies with voltage and with time. Hence, if the voltage changes, the probability will change, reaching a new equilibrium value (magnitude) with some rate (kinetics). At any particular voltage, therefore, the goal is to describe how the probability of open gates changes with time, or, mathematically, to describe $n(t)$. Because the probability is a function of time, it should strictly always be written as $n(t)$, but it is referred to in most places below simply as n for simplicity.

The rate of change can be defined by a pair of rate constants that describe the transition between the closed gates, $1-n$, and the open gates, n:

$$1 - n \underset{\beta_n}{\overset{\alpha_n}{\rightleftharpoons}} n$$

Here, α_n is the forward rate constant, which acts on closed gates, $(1-n)$, opening them; thus, $\alpha_n(1-n)$ gives a measure of the rate of gate opening. β_n, the backward rate constant, acts on open gates, n, closing them; thus, $\beta_n(n)$ gives a measure of the rate of gate closing. For convenience in the section below, the subscripts on the rate constants are dropped and they are referred to simply as α and β.

Upon stepping to any new potential, the rate of change of the proportion of open gates, **dn/dt**, can be expressed as the difference between the closed gates that are opening, $\alpha(1-n)$, which adds to the proportion of open gates, and the open gates that are closing, $\beta(n)$, which subtracts from the proportion of open gates. Expressed mathematically,

$$\frac{dn}{dt} = \alpha(1-n) - \beta(n) \qquad \text{(Paper 5, Equation 7)}$$

Therefore, if the rate constants at each voltage were known, it would be possible to obtain a complete description of how fast and to what extent the proportion of open gates changes. The question becomes whether there are experimentally measurable quantities that will permit H&H to find the values of the rate constants. With this goal, the terms of this equation can be rearranged:

$$\frac{dn}{dt} = \alpha - \alpha(n) - \beta(n)$$

$$\frac{dn}{dt} = \alpha - n(\alpha + \beta)$$

$$\frac{dn}{dt} + n(\alpha + \beta) = \alpha \qquad \text{(Equation A)}$$

Equation A is a simple first-order differential equation of the form $dy/dx + ay = C$, where a and C are constants, and dy/dx is the first derivative of the function y. Solving for y in such an equation will always give an exponential function (see Appendix 1.3). Equation A can be solved in the following way, which employs the standard approach for solving such a differential equation:

1. Use the value of the constant a in Equation A to find the 'integrating factor,' $e^{\int a\, dt}$. Since the constant a is $(\alpha+\beta)$, it is necessary to evaluate the integral $\int(\alpha+\beta)\, dt$. The integral of any constant with respect to time is the constant multiplied by time, here, $t(\alpha+\beta)$. Therefore, the integrating factor becomes e raised to this power:

$$e^{\int a\, dt} = e^{t(\alpha+\beta)}.$$

2. Multiply Equation A through by the integrating factor:

$$e^{t(\alpha+\beta)}\frac{dn}{dt} + e^{t(\alpha+\beta)}n(\alpha+\beta) = \alpha e^{t(\alpha+\beta)}.$$

3. Recall that the derivative of the product of two functions, $d(yz)/dt$, is $z(dy/dt) + y(dz/dt)$.

    In the equation above:

    Let $y = n$. Hence, $dy/dt = dn/dt$.

    Let $z = e^{t(\alpha+\beta)}$. Hence, $dz/dt = (\alpha+\beta)e^{t(\alpha+\beta)}$.

    Therefore, the equation in step 2 has the form $z(dy/dt) + y(dz/dt) = \alpha e^{t(\alpha+\beta)}$ and can thus be rewritten as:

$$\frac{d(ne^{t(\alpha+\beta)})}{dt} = \alpha e^{t(\alpha+\beta)}.$$

4. Integrating both sides gives:

$$ne^{t(\alpha+\beta)} = \frac{\alpha}{\alpha+\beta}e^{t(\alpha+\beta)} + C,$$

    where C is a constant.

5. Dividing through to isolate n gives an expression for $n(t)$ in terms of this constant:

$$n(t) = \frac{\alpha}{\alpha + \beta} + Ce^{-t(\alpha+\beta)}.$$

At any voltage, the first term in this equation is a constant, which is determined by the values of the rate constants at that voltage. The second term changes over time, according to an exponential decay of amplitude C, which represents the *change* in the proportion of open gates.

6. At this point, it is possible to ask the question, *what is the proportion of open gates at equilibrium* (i.e., n_∞)? At long times, as t approaches ∞, $Ce^{-t(\alpha+\beta)}$ will tend toward zero, as the exponent becomes infinitely negative. Therefore

$$n_\infty = \frac{\alpha}{\alpha + \beta}. \qquad \text{(Paper 5, Equation 9)}$$

7. It is also possible to solve for the constant C by considering the question, *how does the proportion of gates change with time* (i.e., how does n change from $t=0$ to $t=\infty$)? At $t=0$, one can solve for n_0:

$$n_0 = \frac{\alpha}{\alpha + \beta} + Ce^0, \text{ which is}$$

$$n_0 = \frac{\alpha}{\alpha + \beta} + C.$$

Substituting n_∞ from above,

$$n_0 = n_\infty + C, \text{ or}$$

$$C = n_0 - n_\infty.$$

The constant C therefore is the difference between the starting probability and the ending probability, which is the magnitude of the exponential relaxation (as mentioned above).

8. The full equation for $n(t)$ from step 5 can now be written out, substituting in n_∞ and specifying the constant C:

$$n(t) = n_\infty + (n_0 - n_\infty)e^{-t(\alpha+\beta)}.$$

9. This is an exponential decay of the form $y(t) = y_\infty + Ae^{-t/\tau}$. Therefore, the time constant can be expressed as:

$$\tau = \frac{1}{\alpha + \beta}. \qquad \text{(Paper 5, Equation 10)}$$

The amplitude, A, of the decay is the difference between the initial probability n_0 and the final probability n_∞. The equilibrium value is n_∞. Thus, the equation for $n(t)$ can be expressed as

$$n(t) = n_\infty - (n_\infty - n_0)e^{-t/\tau}. \qquad \text{(Paper 5, Equation 8)}$$

In summary, the proportion of open gates begins at the value n_0 and relaxes with a time constant τ to a value n_∞. Each n value can be expressed as the fourth root of the corresponding conductance value, making the equation

$$g_K = \left\{ (g_{K\infty})^{1/4} - \left[(g_{K\infty})^{1/4} - (g_{K0})^{1/4} \right] \exp\left(\frac{-t}{\tau} \right) \right\} \qquad \text{(Paper 5, Equation 11)}$$

Experimentally, with a step to any voltage, it is possible to measure the initial conductance, g_{K0}, at the onset of the step; the time course of the change in conductance during the step; and the final or steady-state conductance, $g_{K\infty}$, during that step. Since the n values can be extracted by taking the fourth root of the conductance values, one can obtain values of n_0, n_∞, and τ at any potential, and from there calculate α and β.

From $\tau = 1/(\alpha + \beta)$ and $n_\infty = \alpha/(\alpha + \beta)$ above, it follows that

$$\alpha = \frac{n_\infty}{\tau}, \text{ and}$$

$$\beta = \frac{1 - n_\infty}{\tau} \qquad \text{(Paper 5, page 509).}$$

Thus, from experimental measurements, it is possible to find the rate constants that define the kinetics of gate opening and closing and final magnitude of the open probability at any potential, thereby fully describing conductance as a function of time and voltage.

Appendix 5

NUMERICAL METHODS FOR THE H&H EQUATIONS

5.1 Integrating the Hodgkin and Huxley equations. H&H summarize the equations derived from their experiments in Paper 5 at the beginning of Part III. Equation 26 is a differential equation in that it explicitly relates V to its first derivative, dV/dt (see Appendix 1.3). In addition, the scale factors (n, m, and h) that relate V to its derivative are themselves defined by differential equations relating each to *its* derivative (Paper 5, Equations 7, 15, and 16). Finally, the scale factors that relate n, m, and h to their derivatives are in turn dependent on V (Paper 5, Equations 12, 13, 20, 21, 23, and 24).

H&H propose that this system of differential equations *completely* describes the behavior of the membrane on the time scale investigated. To test this hypothesis explicitly, they must solve the equations to derive V as a function of time and show that the resulting function closely matches the voltage of real membranes during voltage-clamp steps, stationary and propagated action potentials, relative and absolute refractory period, and so on.

Some systems of differential equations can be solved explicitly by manipulating the equations and their variables. The differential equation, $dy(t)/dt = -y/\tau$, for example, can be solved to arrive at a function of y that depends only on t: $y(t) = e^{-t/\tau}$. Unfortunately, the H&H equations are too complex to solve explicitly, meaning that one cannot derive a general mathematical expression for V as a function of t that does not still involve V or its derivatives or integrals. The only way to demonstrate that the equations predict the behavior of the membrane, therefore, is to approximate the solution to the equations by making calculations with real numeric values under each different condition of interest. Such approaches fall under the heading of 'numerical methods,' 'numerical analysis,' or 'numerical integration.'

There are many different methods of numerical integration (such as the Hartree method used by H&H). These differ in detail, but all use the same procedure, which is to calculate numerically the values of the derivatives of each parameter, p, at time t_0, and then to use the derivatives to approximate the new value of each parameter at time $t_1 = t_0 + \delta t$, that is, $p(t_1) = p(t_0) + dp/dt \times \delta t$. In the case of the H&H equation, the procedure must be applied to three parameters, n, m, and h. The new values of n, m, and h determine the new value of V, which in turn determines the new values of the derivatives of n, m, and h. This procedure is repeated again and again for numerous steps of δt, for a long enough time to observe the evolving behavior of V. The key here is that δt must be kept small enough so that the derivatives of the parameters do not change significantly over the course of a step. Otherwise each newly calculated value of p will contain small errors, which accumulate and compound each other over the many steps of integration. The smaller δt, the more accurate the integration, but the more laborious the calculation.

The steps required for integrating the H&H equations are illustrated in *Figure A*. Here, Euler's method of integration is illustrated because it is simpler in detail than the Hartree method applied by H&H, but the principles and the results (with an appropriately chosen δt) are the same.

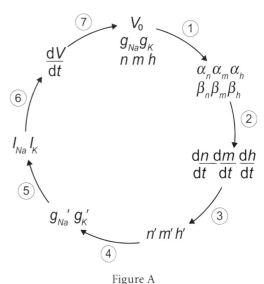

Figure A

For the stationary (membrane) action potential, it is first necessary to assume starting values for V, n, m, and h, in other words $V(t_0)$, $n(t_0)$, $m(t_0)$, and $h(t_0)$, which also give $g_{Na}(t_0)$ and $g_K(t_0)$.

1. From $V(t_0)$, calculate all the α's and β's, the rate constants that govern transition of the n, m, and h gating particles from one state to the other, using Equations 12, 13, 20, 21, 23, and 24.

2. From the α's and β's, calculate the time derivatives of n, m, and h (dn/dt, dm/dt, dh/dt) from Equations 7, 15, and 16.

3. From dn/dt, dm/dt, dh/dt, calculate the new values of n, m, and h at time $t_1 = t_0 + \delta t$:

$$n(t_1) = n(t_0) + \frac{dn}{dt} \times \delta t,$$

$$m(t_1) = m(t_0) + \frac{dm}{dt} \times \delta t, \text{ and}$$

$$h(t_1) = n(t_0) + \frac{dh}{dt} \times \delta t.$$

4. From the new n, m, and h, calculate $g_{Na}(t_1)$ and $g_K(t_1)$ from Equations 6 and 14.

5. From g_{Na} and g_K, calculate the ionic currents $g_{Na}(t_1) \times (V - E_K)$ and $g_{Na}(t_1) \times (V - E_K)$. Note that here $g(t_1)$ and $V(t_0)$ are used in the calculation, but by making δt small enough, the difference between $V(t_0)$ and $V(t_1)$ is small enough to neglect.

6. From I_{Na}, I_K, and I_l, calculate dV/dt from Equation 26, on the assumption that total membrane current during the propagated action potential is zero.

7. Derive the membrane potential, $V(t_1)$, at the new time as $V(t_0) + dV/dt \times \delta t$.

These steps are then repeated, replacing the starting values of V, n, m, and h with the results from the previous iteration.

Implementing this method on a computer and observing action potentials is relatively straightforward. In fact, H&H's initial plan was to use one of the earliest digital computers at Cambridge for the integration, but they were unable to do so. As Hodgkin recorded in his autobiography:

> Finally there was the difficulty of computing the action potentials from the equations we had developed. We had settled all the equations and constants by March 1951 and hoped to get these solved on the Cambridge University computer. However, before anything could be done we heard that the computer would be off the air for six months or so, while it underwent a major modification. Andrew Huxley got us out of that difficulty by solving the differential equations numerically using a hand-operated Brunsviga calculating machine. The propagated action potential took about three weeks to compute and must have been an enormous labour for Andrew. But it was exciting to see it come out with the right shape and velocity and we began to feel that we had not wasted the many months we had spent in analysing records. (C&D p. 291)

Figure B

The mechanical calculator Huxley used was a pre-war era Brunsviga Model 20 (*Figure B*), which is an adding machine with a few features that streamline the calculations. To multiply 1234 by 5678, for example, one must follow these steps:

1. Enter 1234 on the sliding levers (on the machine's curved face), one digit at a time.

2. Position the carriage (at the front of the machine) in the ones position.

3. Turn the crank eight (at the far right) times.

4. Slide the carriage to the tens position.

5. Turn the crank seven times.

6. Slide the carriage to the hundreds position.

7. Turn the crank six times.

8. Slide the carriage to the thousands position.

9. Turn the crank five times.

Thus, 30 individual operations are required to multiply two four-digit numbers. The result, which appears on the registers at the top of the calculator must then be transferred into the sliders to proceed to the next multiplication. Fortunately, the Brunsviga 20 has a feature whereby pulling a combination of the clearing levers on each side of the machine transfers the contents of the results register into the lever settings. Nevertheless, given that each step of the integration requires multiplying many numbers together, not to mention raising m and n to the third and fourth powers, a single action potential, which was integrated in hundreds of small time increments, required tens of thousands of turns of the crank. One further presumes that Huxley must have written down the result of each calculation for error checking and correction.

The Euler method of integration described above requires that δt be small enough so that the first derivatives of V, n, m, and h do not change significantly over the course of a single time step. These small increments accommodate the approximation of step 5 above, keeping $V(t_1)$ reasonably close to $V(t_0) + \mathrm{d}V/\mathrm{d}t \times \delta t$ (and similarly for n, m and h). With laborious hand-calculations, any method that reduces the number of time steps required to generate a reasonably accurate action potential is useful. The Hartree method, outlined in the nine steps in the 'Numerical Methods' section of Paper 5, offers just such a method. The size of the time step can be increased, while still maintaining accuracy of the calculations at the new time point, by incorporating the *second* derivatives of the various parameters. For example, if both $\mathrm{d}V/\mathrm{d}t$ and $\mathrm{d}^2V/\mathrm{d}t^2$ are known, $V(t_1)$ can be estimated not only from $\mathrm{d}V/\mathrm{d}t$ but also from how much $\mathrm{d}V/\mathrm{d}t$ *changes* over the course of δt. These are the so-called trapezoidal integration methods, such as the commonly used Runga-Kutta method or the more obscure Hartree method. This method is the origin of the equation in H&H's step 4. Because the time step, δt, is now longer than in the Euler method, $\mathrm{d}n/\mathrm{d}t$, $\mathrm{d}m/\mathrm{d}t$, and $\mathrm{d}h/\mathrm{d}t$ may change appreciably over the course of the time step, because V—and therefore the α's and β's—change during that time. To deal with this problem, after starting with the Euler method in H&H's step 2, the iterative process in their steps 3, 4, 5, and 6 is required for a successive refinement of the predicted n, m, h, and V at t_1 as follows:

Step 2: The Euler method is used to arrive at an initial estimate of n (and m and h) at t_1, which is in turn used to estimate $V(t_1)$. The difficulty is that H&H do not know $\mathrm{d}n/\mathrm{d}t$ for the *present* time step. As an approximation, they use $\mathrm{d}n/\mathrm{d}t$ from the *previous* time step (i.e., the 'backward differences'). These backward differences are close to the derivatives for the present step, but are not identical, which is the source of a small error that would accumulate with each successive time step if left uncorrected.

Step 3: To improve on this first estimate, H&H use the new $V(t_1)$ to calculate a better value for $\mathrm{d}n/\mathrm{d}t$ at t_1, which provides an estimate of how quickly $\mathrm{d}n/\mathrm{d}t$ is changing (i.e., the second derivative, $\mathrm{d}^2n/\mathrm{d}t^2$).

Step 4: The estimate of the second derivative allows a more refined estimate of n at t_1, by the equation given.

Step 5: If this more refined estimate of n differs significantly from the original estimate from step 2, then the calculation must be further refined. Steps 2–4 are repeated, starting from the newly calculated value of $\mathrm{d}n/\mathrm{d}t$ (and m and h and V) from step 4. The process is repeated until successive calculations are nearly identical, which indicates that the calculations have converged on the correct value.

This iterative process requires additional calculations per time step, but the reduction in the number of required time steps compensates for the additional effort.

References

BOOKS

Ashcroft, F. M. (1999) *Ion Channels and Disease: Channelopathies.* Academic Press: San Diego.

Baylor, S. M. (2020) *Computational Cell Physiology: With Examples in Python (Revised and Expanded).* Kindle Direct Publishing.

Bezanilla, F. (2020) *The Nerve Impulse.* Online. http://nerve.bsd.uchicago.edu/

Carnevale, N. T. and Hines, M. L. (2006) The NEURON Book. Cambridge University Press: Cambridge, UK.

Clarke, E. and Jacyna L. S. (1987) *Nineteenth Century Origins of Neuroscientific Concepts* University of California Press: Berkeley, Los Angeles, London. (NONC)

Cole, K. S. (1968) *Membranes, Ions, and Impulse.* University of California Press: Berkeley, Los Angeles. (MII)

Hille, B. (2001) *Ion Channels of Excitable Membranes.* Sinauer Associates: Sunderland, MA.

Hodgkin, A. L. (1992) *Chance and Design: Reminiscences of Science in Peace and War.* Cambridge University Press: Cambridge, UK. (C&D)

Huxley, A. F. (2004) *Andrew F Huxley.* In: *The History of Neuroscience in Autobiography. Vol. 4.* Ed. Larry R. Squire. Academic Press (Elsevier): San Diego. Pp. 282–319. (HNA)

Moore, J. W. and Stuart, A. E. (2011) *Neurons in Action 2: Tutorials and Simulations using NEURON 2nd Edition.* Sinauer Associates: Sunderland, MA.

SCIENTIFIC ARTICLES AND MONOGRAPHS

Armstrong, C. and Bezanilla, F. (1973) Currents related to movement of the gating particles of the sodium channels. *Nature.* 242(5398): 459–461.

Armstrong, C. M., Bezanilla F., and Rojas E. (1973) Destruction of sodium conductance inactivation in squid axons perfused with pronase. *Journal of General Physiology.* 62(4):375–391.

Armstrong, C. M. and Bezanilla, F. (1977) Inactivation of the sodium channel. II. Gating current experiments. *Journal of General Physiology.* 70(5):567–590.

Bernstein, J. (1868) Ueber den zeitlichen Verlauf der negativen Schwankung des Nervenstroms. *Archiv für die gesamte Physiologie des Menschen und der Tiere.* 1:179–207.

Bernstein, J. (1902) Untersuchungen zur Thermodynamik der bioelektrischen Ströme. *Pflügers Archiv für die gesamte Physiologie des Menschen und der Tiere.* 92:521–562.

Bezanilla, F., Perozo E., and Stefani, E. (1994) Gating of Shaker K+ channels: II. The components of gating currents and a model of channel activation. *Biophysical Journal.* 66(4):1011–1021.

Bezanilla, F. and Armstrong, C. M. (1977) Inactivation of the sodium channel. I. Sodium current experiments. *Journal of General Physiology.* 70(5): 549–566.

Blinks, L. R. (1928) High and low frequency measurements with *Laminaria. Science.* 68(1758):235.

Blinks, L. R. (1930) The direct current resistance of *Nitella. Journal of General Physiology.* 13(4):495–508.

Carter, B. C. and Bean, B. P. (2009) Sodium entry during action potentials of mammalian neurons: incomplete inactivation and reduced metabolic efficiency in fast-spiking neurons. *Neuron.* 64(6):898–909.

Chanda, B. and Bezanilla, F. (2002) Tracking voltage-dependent conformational changes in skeletal muscle sodium channel during inactivation. *Journal of General Physiology.* 120(5):629–645.

Chandler, W. K. and Meves, H. (1970a) Evidence for two types of sodium conductance in axons perfused with sodium fluoride solution. *Journal of Physiology* 211(3):653–678.

Chandler, W. K. and Meves H. (1970b) Slow changes in membrane permeability and long-lasting action potentials in axons perfused with fluoride solutions. *Journal of Physiology.* 211(3):707–728.

Cole, K. S. (1934) Alternating current conductance and direct current excitation of nerve. *Science.* 79(2042):164–165.

Cole, K. S. (1941) Rectification and inductance in the squid giant axon. *Journal of General Physiology.* 25(1):29–51.

Cole, K. S. and Baker, R. F. (1941) Transverse impedance of the squid giant axon during current flow. *Journal of General Physiology.* 24(4):535–549.

Cole, K. S. and Curtis, H. J. (1938) Electric impedance of *Nitella* during activity. *Journal of General Physiology.* 22(1):37–64.

Cole, K. S. and Curtis, H. J. (1939) Electric impedance of the squid giant axon during activity. *Journal of General Physiology.* 22(5):649–670.

Cole, K. S. and Hodgkin, A. L. (1939) Membrane and protoplasm resistance in the squid giant axon. *Journal of General Physiology.* 22(5):671–687.

Cooley, J. W. and Dodge, F. A. (1966) Digital computer solutions for excitation and propagation of the nerve impulse. *Biophysical Journal.* 6(5):583–599.

Curtis, H. J. and Cole, K. S. (1937) Transverse electric impedance of *Nitella. Journal of General Physiology.* 21(2):189–201.

Curtis, H. J. and Cole, K. S. (1938) Transverse electric impedance of the squid giant axon. *Journal of General Physiology.* 21(6):757–765.

Curtis, H. J. and Cole, K. S. (1940) Membrane action potentials from the squid giant axon. *Journal of Cellular Comparative Physiology.* 15:145–157.

Curtis, H. J. and Cole, K. S. (1942) Membrane resting and action potentials from the squid giant axon. *Journal of Cellular Comparative Physiology.* 19:135–144.

Fatt, P. (2014) An interview with Paul Fatt conducted by Jonathan Ashmore on 7 September 2013. *The Oral Histories Project for The Society's History & Archives Committee.* The Physiological Society: London, UK.

Frankenhaeuser, B. and Hodgkin, A. L. (1956) The after-effects of nerve impulses in the giant nerve fibres of *Loligo. Journal of Physiology.* 131:341–376.

Fricke, H. (1925) The electric capacity of suspensions with special reference to blood. *Journal of General Physiology.* 9(2):137–152.

Gaffey, C. T. and Mullins, L. J. (1958) Ion fluxes during the action potential in *Chara. Journal of Physiology.* 144(3):505–524.

Galvani, L. (1791) De viribus electricitatis in motu musculari commentarius. (Commentary on the effect of electricity on muscular motion, translated by Robert Montraville Green, 1953). Elizabeth Licht: Cambridge MA.

Goldman, D. E. (1943) Potential, impedance, and rectification in membranes. *Journal of General Physiology.* 27(1):37–60.

Höber, R. (1910) Eine Methode, die elektrische Leitfähigkeit im Innern von Zellen zu messen. *Pflügers Archiv für die gesamte Physiologie des Menschen und der Tiere.* 133:237–259.

Höber, R. (1912) Ein zweites Verfahren, die elektrische Leitfähigkeit im Innern von Zellen zu messen. *Pflügers Archiv für die gesamte Physiologie des Menschen und der Tiere.* 148:189–221.

Hodgkin, A. L. (1937a) Evidence for electrical transmission in nerve: Part I. *Journal of Physiology.* 90(2):183–210.

Hodgkin, A. L. (1937b) Evidence for electrical transmission in nerve: Part II. *Journal of Physiology.* 90(2):211–232.

Hodgkin, A. L. (1938) The subthreshold potentials in a crustacean nerve fibre. *Proceedings of the Royal Society of London. Series B, Biological Sciences.* 126(842):87–121.

Hodgkin, A. L. (1939) The relation between conduction velocity and the electrical resistance outside a nerve fibre. *Journal of Physiology.* 94(4):560–570.

Hodgkin, A. L. (1947) The effect of potassium on the surface membrane of an isolated axon. *Journal of Physiology.* 106(3):319–340.

Hodgkin, A. L. (1951) The ionic basis of electrical activity in nerve and muscle. *Biological Reviews* 26:339–409.

Hodgkin, A. L. and Huxley, A. F. (1939) Action potentials recorded from inside a nerve fibre. *Nature.* 144:710–711.

Hodgkin, A. L. and Huxley, A. F. (1945) Resting and action potentials in single nerve fibres. *Journal of Physiology.* 104(2):176–195.

Hodgkin, A. L. and Huxley, A. F. (1947) Potassium leakage from an active nerve fibre. *Journal of Physiology.* 106(3):341–367.

Hodgkin, A. L. and Huxley, A. F. (1952a) Currents carried by sodium and potassium ions through the membrane of the giant axon of *Loligo. Journal of Physiology.* 116:449–472.

Hodgkin, A. L. and Huxley, A. F. (1952b) The components of membrane conductance in the giant axon of *Loligo. Journal of Physiology.* 116:473–496.

Hodgkin, A. L. and Huxley, A. F. (1952c) The dual effect of membrane potential on sodium conductance in the giant axon of *Loligo. Journal of Physiology.* 116:497–506.

Hodgkin, A. L. and Huxley, A. F. (1952d) A quantitative description of membrane current and its application to conduction and excitation in nerve. *Journal of Physiology.* 117:500–544.

Hodgkin, A. L., Huxley, A. F., and Katz, B. (1952) Measurement of current-voltage relations in the membrane of the giant axon of *Loligo. Journal of Physiology.* 116:424–448.

Hodgkin, A. L. and Katz, B. (1949) The effect of sodium ions on the electrical activity of giant axon of the squid. *Journal of Physiology.* 108(1):37–77.

Hodgkin, A. L. and Rushton, W.A.H. (1946) The electrical constants of a crustacean nerve fibre. *Proceedings of the Royal Society of London. Series B, Biological Sciences.* 134(873):444–479.

Huxley, A. F. (2000) Sir Alan Lloyd Hodgkin, O.M., K.B.E. *Biographical Memoirs of Fellows of the Royal Society.* 46:219–241.

Katz, B. (1937) Experimental evidence for a non-conducted response of nerve to subthreshold stimulation. *Proceedings of the Royal Society of London. Series B, Biological Sciences.* 124(835):244–276.

Katz, B. (1947) Subthreshold potentials in medullated nerve. *Journal of Physiology.* 106(1):66–79.

Keynes, R. D. and Lewis, P. R. (1951a) The resting exchange of radioactive potassium in crab nerve. *Journal of Physiology.* 113(1):73–98.

Keynes, R. D. and Lewis, P. R. (1951b) The sodium and potassium content of cephalopod nerve fibers. *Journal of Physiology.* 114(1–2):151–182.

Lorente de Nó, R. (1950) The ineffectiveness of the connective tissue sheath of nerve as a diffusion barrier. *Journal of Cellular and Comparative Physiology.* 35(2):195–240.

Lucas, K. (1909a) The 'all or none' contraction of the amphibian skeletal muscle fibre. *Journal of Physiology.* 38(2–3):113–133.

Lucas, K. (1909b) On the refractory period of muscle and nerve. *Journal of Physiology.* 39(5):331–340.

Lucas, K. (1910) An analysis of changes and differences in the excitatory process of nerves and muscles based on the physical theory of excitation. *Journal of Physiology.* 40(3):225–249.

MacKinnon, R. (1991) Determination of the subunit stoichiometry of a voltage-activated potassium channel. *Nature.* 350(6315):232–235.

Marmont, G. (1949) Studies on the axon membrane; a new method. *Journal of Cellular and Comparative Physiology.* 34(3):351–382.

Mauro, A. (1969) The role of the Voltaic pile in the Galvani-Volta concerning animal *vs.* metallic electricity. *Journal of the History of Medicine and Allied Sciences.* 24(2):140–150.

Narahashi, T., Moore, J. W., and Scott, W. R. (1964) Tetrodotoxin blockage of sodium conductance increase in lobster giant axons. *Journal of General Physiology.* 47:965–974.

Osterhout, W.J.V (1922) Injury, recovery, and death, in relation to conductivity and permeability. *Monographs on Experimental Biology.* J.B. Lippincott Co. Philadelphia & London.

Overton, E. (1902) Beiträge zur allgemeinen Muskel- und Nervenphysiologie. *Archiv für die gesamte Physiologie des Menschen und der Tiere.* 92:346–386.

Philippson, M. (1921) Les lois de la résistance électrique des tissus vivants. *Bulletin de l'Academie Royale des Sciences, des Lettres et des Beaux-Arts de Belgique, Classe des Sciences.* 7:387–403.

Schuetze, S. M. (1983) The discovery of the action potential. *Trends in Neurosciences.* 6:164–168.

Young, J. Z. (1936) The giant nerve fibres and epistellar body of cephalopods. *Journal of Cell Science.* 78:367–386.

Yamamoto, D., Yeh, J. Z., and Narahashi, T. (1985) Interactions of permeant cations with sodium channels of squid axon membranes. *Biophysical Journal.* 48(3):361–368.

Zagotta, W. N., Hoshi, T., and Aldrich, R. W. (1994) Shaker potassium channel gating. III: Evaluation of kinetic models for activation. *Journal of General Physiology.* 103 (2): 321–362.

Index

Page numbers in italics indicate tables and figures.